COMPUTER PROCESSING OF HANDWRITING

COMPUTER PROCESSING OF HANDWRITING

Editors

R Plamondon

Département de Génie Electrique
et de Génie Informatique
Ecole Polytechnique de Montréal
Canada

C G Leedham

Department of Electronic Systems
Engineering
University of Essex
United Kingdom

World Scientific

Singapore • New Jersey • London • Hong Kong

Published by

World Scientific Publishing Co Pte Ltd
5 Toh Tuck Link, Singapore 596224
USA office: 27 Warren Street, Suite 401-402, Hackensack, NJ 07601
UK office: 57 Shelton Street, Covent Garden, London WC2H 9HE

British Library Cataloguing-in-Publication Data
A catalogue record for this book is available from the British Library.

COMPUTER PROCESSING OF HANDWRITING

ISBN-13 978-981-02-0408-2
ISBN-10 981-02-0408-6
ISBN-13 978-981-02-0520-1 (pbk)
ISBN-10 981-02-0520-1 (pbk)

PREFACE

The Fourth International Graphonomics Society Conference was held at the University of Trondheim, Norway between 24th-26th July 1989. It was organised and chaired by Professor Nils Søvik of the Department of Education, University of Trondheim, Norway and Dr. Alan Wing of the Medical Research Council - Applied Psychology Unit, Cambridge, England.

The Fourth IGS Conference, like the previous ones, provided a forum to facilitate communication between the different groups of investigators in the field of handwriting research. In total, 94 papers, selected by an international panel of reviewers, were presented by authors from 16 countries under the following headings:

- Individual Variation in Handwriting
- The Lexicon in Handwriting Recognition
- Movement Kinematics in Graphic Skills
- Recognition of Strokes in Handwriting
- Handwriting Assessment
- Handwriting and Personality
- Acquisition of Graphic Skills
- Movement Control
- Pattern Recognition (with calligraphy)
- Development Dysfunction and Handwriting
- Recognition: Special Applications
- Dysfunction and Ageing
- Signature Recognition
- Handwriting Difficulties and Remediation
- Handwriting Dynamics
- Handwritten Letter Formation
- Drawing
- Models of the Handwriting Process

Three major publications are planned as a result of this conference. One is the present book which contains a group of selected papers dealing with *Computer Processing of Handwriting*. The other two major publications are:

— a book of 16 selected papers concerned with *The Development of Graphic Skills*, edited by J. Wann, A. Wing and N. Søvik, published by Academic Press.

— a selection of 8 articles on motor aspects of handwriting, to appear as a special issue of the journal *Human Movement Science* under the editorship of G. Van Galen, A. Thomassen and A. Wing.

This book on *Computer Processing of Handwriting* contains 17 chapters divided into three parts:

Part 1 — Automatic Signature Verification
Part 2 — On-line Symbols and Script Recognition
Part 3 — Context and Knowledge.

It is worth mentioning that each chapter in this book was critically reviewed by at least two experts in the field of handwriting research. This was done to ensure that each paper satisfied the standards required for publication in an international journal. The editors extend their sincerest gratitude to all the reviewers for their valuable contributions.

The preparation of this present volume would not have been possible without the co-operation and assistance of the many people involved in the organisation of the Trondheim conference, particularly the two co-chairmen and Professor Arnold J. W. M. Thomassen, President of the IGS. We would also like to extend our thanks to Miss Hélène Dallaire, secretary at the École Polytechnique, Département of Génie Électrique, Montréal, for her efficient handling of all the "typing" tasks related to the communication with the authors and reviewers during the evaluation process and to Miss Susan Barrington, secretary in the Department of Electronic Systems Engineering, University of Essex for her assistance in the formatting and production of the camera-ready copy of this book.

Part of the preparation of this book was done whilst Réjean Plamondon was a Fellow of the Netherlands Institute of Advanced Study.

The camera-ready copy of this book was produced by Graham Leedham and Susan Barrington using Microsoft WORD 4 on Apple Macintosh computers and LaserWriter printers.

Réjean Plamondon
Graham Leedham
July 1990

CONTENTS

PART 1 — Identity Verification from Automatic Signature Processing

Overview 1

Designing an Automatic Signature Verifier: Problem Definition and System Description 3
Réjean Plamondon and Guy Lorette

Dynamic Approaches to Handwritten Signature Verification 21
Guy Lorette and Réjean Plamondon

Automatic Processing of Signature Images: Static Techniques and Methods 49
Réjean Plamondon, Guy Lorette and Robert Sabourin

Identity Verification From Automatic Processing of Signatures: Bibliography 65
Réjean Plamondon and Guy Lorette

PART 2 — On-Line Symbols and Script Recognition

Overview 87

Measurement of Dynamic Digitizer Performance 89
M. L. Meeks and T. T. Kuklinski

The PAD (Pen And Display) - A Demonstrator for The Electronic Paper Project 111
Colin A. Higgins and R. James Duckworth

A Scale-Space Filtering Approach to Stroke Segmentation of Cursive Script 133
M. Kadirkamanathan and P. J. W. Rayner

An On-line Handwriting-Recognition System Based on Unreliable Modules 167
Hans-Leo Teulings, Lambert R.B. Schomaker, Jan Gerritsen, Hans Drexler and Marc Albers

A Prediction-Verification Strategy for Automatic Recognition of Cursive Handwriting 187
H.Oulhadj, E. Petit, J. Lemoine, M. Gaudaire and G. Lorette

On Determining Distinctive Features of Handprinted Characters using Statistical Techniques 207
Shozo Kondo

A Gesture Based Text and Diagram Editor 221
Lisa K. Welbourn and R. J. Whitrow

Automatic Recognition and Transcription of Pitman's Handwritten Shorthand 235
Graham Leedham

PART 3 — Context and Knowledge

Overview 271

The Use of Orthographic Information for Script Recognition 273
C. J. Wells, L. J. Evett, P. E. Whitby and R. J. Whitrow

A Tree-Based Dictionary Search Technique and Comparison with *N*-Gram Letter Graph Reduction 291
David M Ford and Colin A Higgins

A Task Oriented Approach to Reading and to Handwritten Text Recognition 313
Christophe Parisse, Victor Rosenthal, Abdelmalek Imadache, Evelyne Andreewsky and Francoise Cochu

Automatic Perception of the Structure of Handwritten Mathematical Expressions 337
Claudie Faure and Zi Xiong Wang

A Computer Generation Model of Brush-Used Handwritten Chinese Characters and its Application in Education 363
J. Zeng, X. Zhang, T. Inoue, H. Sanada and Y. Tezuka

Author Index 401

Subject Index 407

PART 1

IDENTITY VERIFICATION FROM AUTOMATIC SIGNATURE PROCESSING

Overview

The purpose of the three chapters of this section is to review the efforts devoted to the solution of identity verification problems via computer analysis of signatures. There has been great interest in these problems recently and, although basic problems like optimization of data acquisition, feature selection, discriminating functions and comparison algorithms are far from being entirely solved, several research and commercial prototypes have demonstrated the feasibility of identification systems, based on the use of new hardware and software facilities that have become available.

The topics included in this review are grouped as follows. The first chapter, by Plamondon and Lorette, deals with the general design of an automatic signature verification system. Some basic handwriting models are described to point out the origin of the assumptions upon which many verification and identification systems are based. The intrinsic difficulties in handwriting generation processes are also highlighted. Presenting the basic parts of a signature verification system leads to a discussion of the underlying theoretical and technical difficulties encountered in the design of these systems.

Achievements in automatic signature verification methods and techniques using computers are then overviewed: the dynamic or on-line signature verification approach in the second chapter, by Lorette and Plamondon, and the static or on-line approach in the third chapter, by Plamondon, Lorette and Sabourin. In both cases, the focus is on data acquisition, preprocessing techniques, feature extraction methods, comparison processes and system performance evaluation.

Throughout these three papers special attention is also given to appropriate definitions and terminology. Finally, new emphasis and research directions are suggested. Other aspects and applications of handwriting and signing processes, such as automatic handwriting recognition, forensic sciences, experimental psychology medicine, etc., are not considered in this section.

To avoid repetition, the complete bibliography on the subject is presented at the end of this section as a separate chapter. The reader should note that this review is limited to techniques based on the automatic use of computer and specialized signature input devices.

Computer Processing of Handwriting
Eds. R. Plamondon & C. G. Leedham

DESIGNING AN AUTOMATIC SIGNATURE VERIFIER: PROBLEM DEFINITION AND SYSTEM DESCRIPTION

RÉJEAN PLAMONDON
Laboratoire Scribens, Département de Génie Electrique, Ecole Polytechnique de Montréal, C.P. 6079, Succ. "A" Montréal QC, CANADA H3C 3A7

GUY LORETTE
IRISA, Université de Rennes I, Campus de Beaulieu, 35042 Rennes Cédex, FRANCE

The handwritten signature is among the most widely used of the personal attributes for identity verification. With the growing need for fast and accurate processing of individual identification, the design of an automatic signature verification system constitutes a real challenge. This paper describes the various difficulties to be faced in designing such a system. In the handwriting generation processes, some of these problems are discussed in terms of intrinsic difficulties. The basic parts of an automatic verification system are also described, to extend the discussion to technical as well as theoretical difficulties.

1. Introduction

As has often been pointed out, Sykes (1978), (1979), in our modern society there is a pressing need for reliable, fast and accurate processing of individual identification in many application fields, i.e. banking transactions, computers, databases or sensitive information access, access to high security buildings or restricted areas. Depending on the target, different criteria apply to the choice of a particular identification system.

Firstly, verification of a customer's identity is necessary to secure credit cards and checks against fraudulent use. This is generally performed through signature or personal identification number (PIN) verification. Often, signature verification is not routinely carried out by cashiers, but only done in the case of large transactions.

The development of security devices is justified by the extensiveness of fraud. It was reported, Beatson (1985), that an average of 10,000 cards were lost or stolen each day in the U.S. and fraud amounted to $716 million per annum. Fraud at point-of-sale terminals (POS) is frequently encountered since some customers are in the habit of writing their PIN on their credit cards, in order to remember it, Little (1974); Warfel (1975).

Computer fraud has been estimated at up to $750 million US per annum world wide, Beatson (1985), and this is still only the visible part of the iceberg since it is an accepted fact that a lot of computer crimes are not reported, Parker (1976). For checks, in particular, it is hoped that random forgeries (i.e. when someone signs with his own signature instead of forging someone else's) will be detected and rejected.

In the near future, when the automation of banking operations becomes generalized, automatic identifier systems will be necessary, Warfel (1975); Miller (1984). However, because of the need to reduce the system costs or to increase the facility of use for customers, there is no guarantee that the most reliable and most effective devices will be selected for banking applications.

Secondly, for electronic mail, computer log-on or data base access via remote terminals or networks, the most common identification means encountered is the use of passwords and PIN, Wood (1977), (1978); Parks et al. (1981). Security can be reinforced by a change of password from time to time or even at every log-on. But, on the one hand, users frequently use a name, number or mnemonic which is easy to memorize (name, address, phone number, etc.) but also easy to discover, and on the other hand, if a password is provided randomly by the machine, the user has a tendency to write it down somewhere for easy retrieval, which is a security risk. In this kind of application, both the convenience and maximum security play a leading role in the design of an identification or verification system.

Finally, for the protection of very sensitive information or for high-security buildings and restricted areas, the major priority is the highest security level to prevent unauthorized access. To increase security, several devices can be combined into multilevel systems, independently of the cost or the convenience of use, Haberman and Fejfar (1976); Woodard and Maier (1979).

In this chapter, we provide the background necessary to understand what an automatic signature verification system is, and the difficulties encountered in designing such a system. First, we review the basics of the handwriting process itself. Then, we look at different identifiers to evaluate the advantages and disadvantages of using signatures. Finally, we describe the various modules of a signature verification system, and we conclude by discussing the problems related to this type of application.

2. Description of the Handwriting Process

The complexity of the processes involved in handwriting generation has challenged many groups of researchers over the past decades: experimental psychologists, physicists, engineers, neurologists, computer scientists, graphologists, grapho-analysts and cognitive scientists. On the one hand, handwriting has been analyzed as a specific human skill to study motor behavior, control and learning, to detect neurological diseases, to develop diagnostics or tests in education and even to design anthropomorphic robots, Thomassen et al. (1983); Martlew (1983); Kao et al. (1986); Morasso and Tagliasco (1986). On the other hand, handwriting has also been studied as an output communication channel to develop new man-computer interfaces, character recognizers, signature verifiers, writer identifiers and handwriting generators, Suen et al. (1980); Klement et al. (1980); Plamondon et al. (1987a), (1989); Tappert et al. (1988); Nouboud and Plamondon (1990); Plamondon and Yergeau (1990).

Several conceptual, physical and empirical models have been developed as a result of these studies. Although the model may differ greatly in the way they describe the phenomena, depending on the context and the purpose for which they have been developed, most of them derive from a commonly accepted view of the biophysical processes involved in fast handwriting generation, as depicted in Figure 1.

This simplified scheme reflects some of the basic hypotheses assumed in the study of handwriting. As with any highly skilled motor process, fast handwriting is considered as a ballistic movement. That is, a motion controlled without instantaneous position feedback from a motor program, Lashley (1917); MacDonald (1964); Denier van der Gon and Thuring (1965); Yasuhara (1975). According to this model, some central nervous system mechanisms within the brain fire with a predetermined intensity and duration, the nerve network that activates the proper muscles in a predetermined order. The motion of the pen on the paper, resulting from muscle contraction/relaxation, leaves a partial trace of the trajectory of the pen tip.

Although it is not clear where the boundary of fast handwriting stands with respect to slower writing processes where position and visual feedback would probably apply, Smith and Schappe (1970), this representation has resulted in segmentation of the analysis process into several steps, from movement planning to muscle actuation. Depending on whether the emphasis is on the behavior of the central nervous system or on that of the hand in the resulting models, two complementary approaches have been used to study handwriting: top-down and bottom-up analysis.

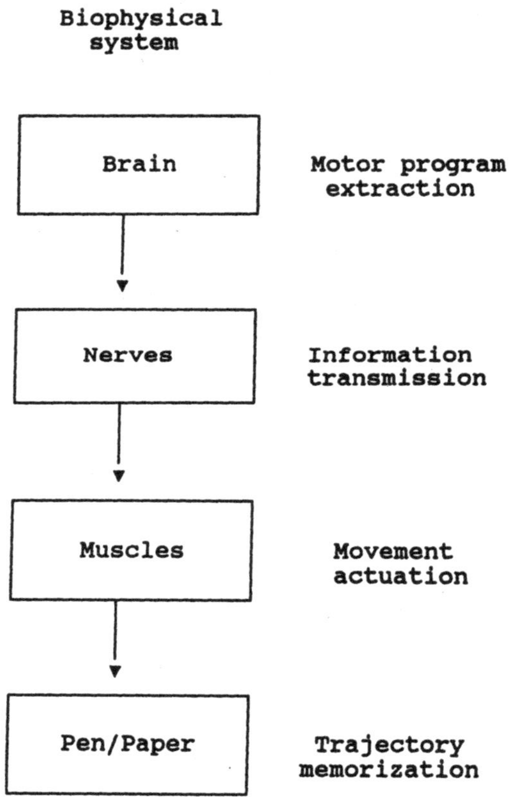

Figure 1. Biophysical processes involved in signature generation.

The top-down approach has been developed and used mainly, by researchers, interested in the study of the motor program itself: fundamental unit of coding, response timing, data retrieval, learning, handwriting generation from high-level information, etc. (for examples see: Eden (1968); Wing (1978); Van Galen (1980), (1984); Van Galen and Teulings (1983); Maarse and Thomassen (1983); Teulings et al. (1986)). In this context, research is concerned with the high-level representation of information, Morasso et al. (1983); Schomaker et al. (1987). The biophysical processes are grouped in a blackbox dedicated to movement generation. This blackbox has to cope with the activation of the proper nerves and muscles to generate movements corresponding to the grapheme/allograph representation of a program.

The bottom-up approach has been used mainly by physicists and engineers interested in the study and application of the biophysical processes. Several models of hand motion under the activation of muscles have been proposed. For a recent review, see Plamondon and Maarse (1987, 1989). In their simplest forms, these models encompass the study of a point mass under the action of muscular forces. Some refinements have incorporated, at least theoretically, the viscous force, the friction force, the elastic force and also, to a certain extent, nerve/muscle interface phenomena. More recent models, Plamondon (1987), (1989a), (1989b) characterize the whole biomechanical processes in terms of transfer functions. This approach to some extent bridges the gap between the top-down and bottom-up approaches.

Figure 2 shows a generalization of these models, depicting the two typical approaches previously described. The middle schematic (2b) is used in top-down studies of motor behaviour during handwriting. According to this conceptual approach, Van Galen and Teulings (1983), handwriting involves several levels of control that can be grouped in three steps resulting in additive effects on reaction-time experiments: first, an abstract motor program is retreived from long-term motor memory, then parameters like actual size, accuracy and speed are fed into this program to make it more concrete, and, finally, this information (program and parameters) is translated for the recruitment and activation of the proper muscles.

The bottom schematic (2c) is a general representation of the bottom-up approach. Several variants of this scheme have been reported in the literature so far, see for examples: Yasuhara (1975); Hollerbach (1981); Dooijes (1983); Plamondon and Lamarche (1986); Edelman and Flash (1987), etc.. According to this view, research efforts are concentrated on the mathematical descrip-

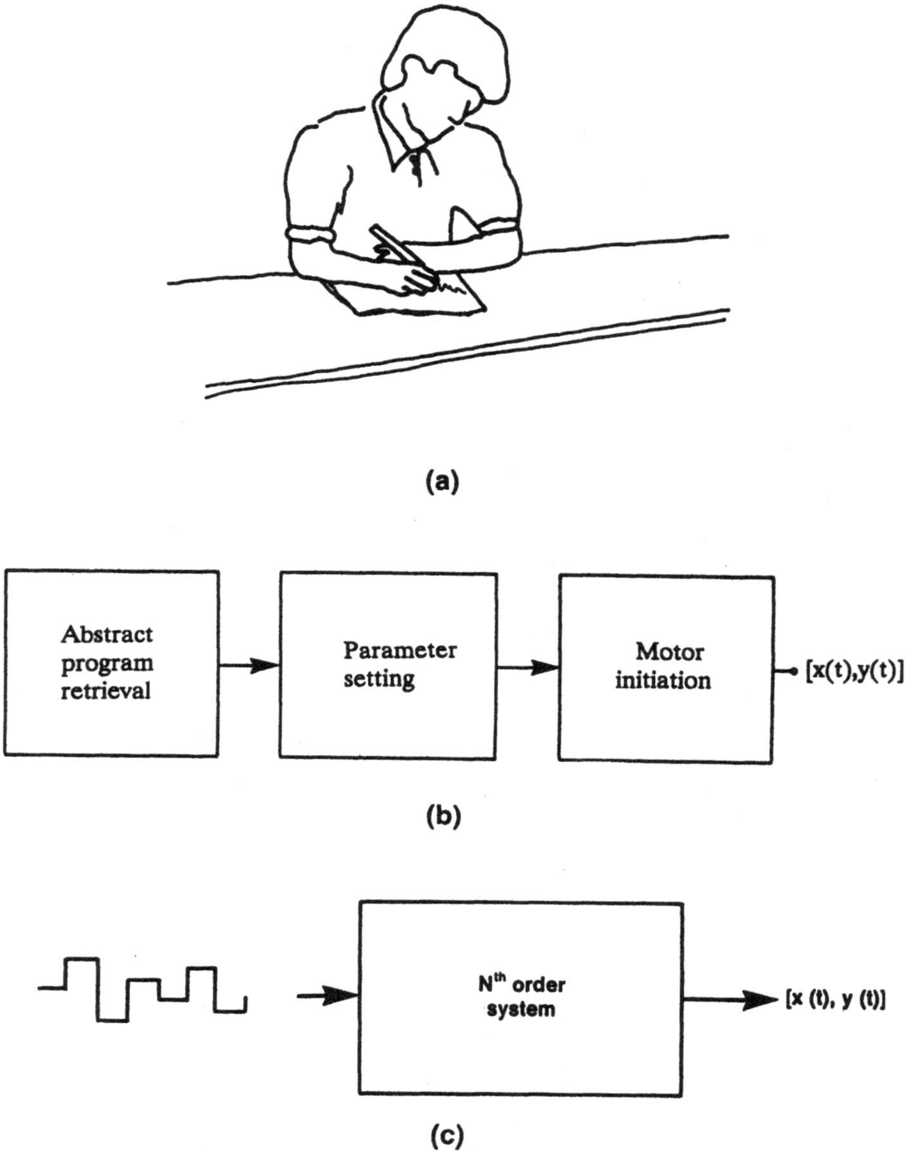

Figure 2. Generalized handwriting models.

tion of the nerve-muscle-pen-paper system represented by movement generators: choice of differential equations, parameter extraction, movement analysis and synthesis, etc..This system is assumed to be controlled by a motor program which can be described as a series of stimuli. These stimuli can be represented as a sequence of input step functions, Plamondon and Maarse (1987), (1989), and result in the production of strokes when fed into the biomechanical model.

From these models, handwriting may be viewed as the output of a space-time-variant system equivalent to the writer, where the input is the learned motor program. How this program is constructed, used and influenced by other psychophysiological mechanisms is still an open question, but the design of a signature verification system is based on the fact that people do not write according to a standard penmanship and that deviation from the norm is individual-dependent, for examples see: Wolf (1948); Mendel (1961); Duvernoy (1973); Saudek (1978); Stockolm (1979).

Whatever the model used, it is readily admitted that signatures reflect not only the semantic information related to the name of the signer but also some intrinsic and unique features related to the specificity of the biomechanical system used to produce the message, the physiological, for examples see: Legge et al. (1964); Margolin and Binder (1984) and even the psychological state of the writer, for examples see: Zweighenhaft and Marlowe (1973); Wallner (1975); Simpson et al. (1980). What parts of the process are writer-dependent and writer-independent is still an open question. What is the best way to extract this specific information and how is it reflected in some local or global characteristics of an image or a signal are also questions that remain unanswered.

It is generally admitted that both static, i.e. geometric, and dynamic information may be necessary for identification, and dynamic information is more difficult to imitate than static information, Locard (1936); Harrison (1981); Salvage (1984). Indeed, a memorized pentip trace on a piece of paper can be generated from numerous motor programs, as long as timing information is not taken into account. On the contrary, one should be able to mimic the whole process itself, as it evolves with time.

Handwriting modeling is of great practical interest in the extraction of discriminant and invariant information from a handwritten specimen. For example, top-down models can be used to segment a piece of text on the basis of the definition of a fundamental unit of a motor program, Eden (1968); Blesser et al. (1976); Morasso and Mussa Ivaldi (1982); Morasso et al. (1983); Plamondon and Yergeau (1990). In the bottom-up approach, knowing the transfer function of the system, one can try to recover and use the stimuli from the observed outputs, Yasuhara and Oka (1977); Yasuhara (1983); Lamarche and Plamondon (1984); Lamarche (1985); Plamondon (1989b). For linear systems this can be done by deconvolution.

In a recent study, Leclerc (1989); Leclerc and Plamondon (1990), a signature generation model has been validated with the help of analysis/synthesis experiments. Parameters representing peripheral and central processes involved in the production of a signature have been automatically extracted and used to represent a signature.

The top-down and bottom-up modeling approaches have also been combined and used to evaluate the intrinsic difficulty in imitating a signature, Brault (1988); Brault and Plamondon (1989a). A parametric coefficient, based on data from experimental psychology and handwriting generation models, has been proposed to estimate this difficulty and to identify beforehand some problematic signers.

3. The Signature as an Identifier

Personal identification means are now well known, (Riganati (1975); Bean and Prell (1977a), (1977b); National Bureau of Standards (1977), (1980); Warfel (1979); Woodard and Maier (1979); Woodard et al. (1981); Shaw (1980); Picodata (1984)), but so far none of them can be considered as ideal. The identifiers proposed to date may be classified into two differents classes:

Indirect techniques in which individuals are identified by something that they have (a name), that they possess (an access card, a key, a token) or that they know (a PIN, a password), Wood (1977), (1978); Morris and Thompson (1979). In these cases, what is effectively verified is only a piece of knowledge or a possession and not the person himself. This is the weakness of these identifiers since objects can be stolen or lost and knowledge can be forgotten or discovered by other users. A partial comparison, Beatson (1985), indicates that the use of a one-time password in conjunction with a PIN could theoretically provide the highest level of security with this type of identifier.

Direct or positive identification techniques, Forsen et al. (1977) in which individual identification can be carried out by means of some biometric or physiological characteristic of the person (finger lengths, finger prints,

Benson (1977), finger lengths, palm prints or hand geometry, Modern Data (1973), face geometry, Harmon et al. (1978), (1981), or retina vascularisation, Davis and Price (1984)) or by characteristics related to something that he is able to do: signature, Fejfar (1978); Botnick (1983), voice print, Haberman and Fejfar (1976); Foodman (1977).

The use of other characteristics (sebaceous secretions, enzyme electrophoresis, blood analysis, encephalography, etc.) have also been proposed, Shaw (1980), but the measurement and analysis of these characteristics is often too complex for use on a daily basis.

Some systems have been tested and their relative performances have been measured, Informatique et Sécurité (1986). These experiments have shown that the retina vascularisation recognition method seems to be the most powerful but it is far from being the most convenient, the least expensive, the fastest, and the one requiring the minimum amount of stored data. From these points of view, the signature verification system seem to be more suitable.

Parizeau and Plamondon (1987, 1989a) have studied what kind of handwritten specimens could be used as the identifier: the signature itself, the initials or a practiced handwritten password. This study shows that, at least for a verification system based on the comparison of position signals via three well-known comparison algorithms, the signature outperforms handwritten passwords and initials.

Raphaël and Young (1974) have proposed nine criteria for an identifier to succeed in automatic personal identification:

1. Ubiquity (possessed by an entire population)
2. Availability (readily obtained)
3. Indispensability (the mean of identification must be retained)
4. Acceptability (by the majority of the population)
5. Uniqueness (unique to each person)
6. Permanence (no change during a lifetime)
7. Reliability (errors must be detected)
8. Security level (to avoid use by others)
9. Brevity (identification should be short).

The handwritten signature satisfies many of these requirements. Obviously a signature is the most widespread technique for identity verification in literate populations. A signature is virtually always available and cannot be forgotten. Moreover, a signature is legally accepted as a proof of identity and signature verification has been socially very well accepted for many years. This procedure involves no possibility of danger or criminal connotations unlike retina vascularisation investigation or fingerprints verification. Signatures are in current use, and signature verification systems should be more convenient to users.

If the aim is not merely the verification of a signature but rather an identification of the writer, static, i.e. geometric, features can be used to predict, in a first step, the identity of the person corresponding to the image and, in a second step, to confirm or invalidate this identity with dynamic data, Bonnefoy et al. (1981), (1984); Lorette (1983); Bonnefoy (1986). However, as will be shown later, most of the work done so far in the domain of signature dynamics is related to signature verification rather than identification. The claimed identity of the individual is established through an indirect identifier (e.g. PIN, magnetic card, password) before verification takes place.

Although signatures are not completely stable over time, Evett and Totty (1985), this drawback can be overcome by gradually updating the reference signatures, Liu (1978); Liu et al. (1979); Crane and Ostrem (1983); Worthington et al. (1985). If signature verification can be implemented for a large population using a one-to-one comparison, then it should be practicable to apply identification, to a limited number of users. For signature verification, the information can be reduced and fast verification is possible with an acceptable security level and at a reasonable cost. For this reason, signature verification systems can be expected to be put into general use. Several experimental and commercial prototypes already, in fact, exist and have been partly validated by tests upon sets of hundreds to thousands of users.

4. System Schematics

To facilitate understanding of the various signature verification processes proposed in the literature, a general data flow diagram of such a system is depicted in Figure 3, Plamondon (1984); Plamondon and Lorette (1989a).

Signatures can be acquired in two ways:

"ON-LINE" i.e. the acquisition occurs during the signing process itself

"OFF-LINE" i.e. acquisition is carried out from the signature paper after writing.

Usually, signature verification is referred to as "DYNAMIC" and "STATIC" according to the two forms of data and process that are involved. In the first case, one or several signals, varying with time, give a representation of the written signature. In the second case, a signature on paper appears as a 2-D image which can be picked up by optical means. There are, therefore, two classes of specialized acquisition devices: one for signal acquisition, the other for image acquisition. Dynamic systems use a digitizer or an instrumented pen to generate signals; static systems produce an image of a signature with the help of a camera or a scanner. The raw data obtained are then preprocessed to remove spurious information, to filter the data and to validate the acquisition. The next step is referred to as the feature extraction process. At this stage, specific and discriminant characteristics are extracted from a signature signal or image, resulting in the generation of a feature set. As we will see later on, the relative complexity of the feature extraction process differs greatly from one design to another. In some systems, this step is almost non-existent since the whole signals (in a compressed or uncompressed format) are directly used as the feature set. In other systems, complex parameters are extracted from the preprocessed data. The resulting parameter set may take any one of various forms, reflecting some local or global characteristics of the signature.

Whatever its content, the feature set is used for two purposes, depending on the state of the system: an enrolment state (dotted lines in Figure 3) or a comparison state (solid line). Indeed, prior to any comparison, a signature reference set must be generated for each user of the system. This set is generally computed from a few signature specimens and it is then linked to the reference data base with the help of an ID number fed into the system according to the proper identification input protocol of the enrolment process.

For verification systems, in current use, the reference data base is already available. For an on-line prototype, a user signs his name after entering his indirect identifier into the system. For an off-line system, this identifier might be automatically read from the paper form on which the signature has been written. This ID information is used to extract the proper reference set from the data base. The feature set just collected is then compared to this reference set, with the help of the appropriate comparison algorithm. Finally, a decision process compares the resulting distance measurement to a decision threshold and the signature is accepted or rejected. In both cases, the proper action is executed by the output system according to the application environment.

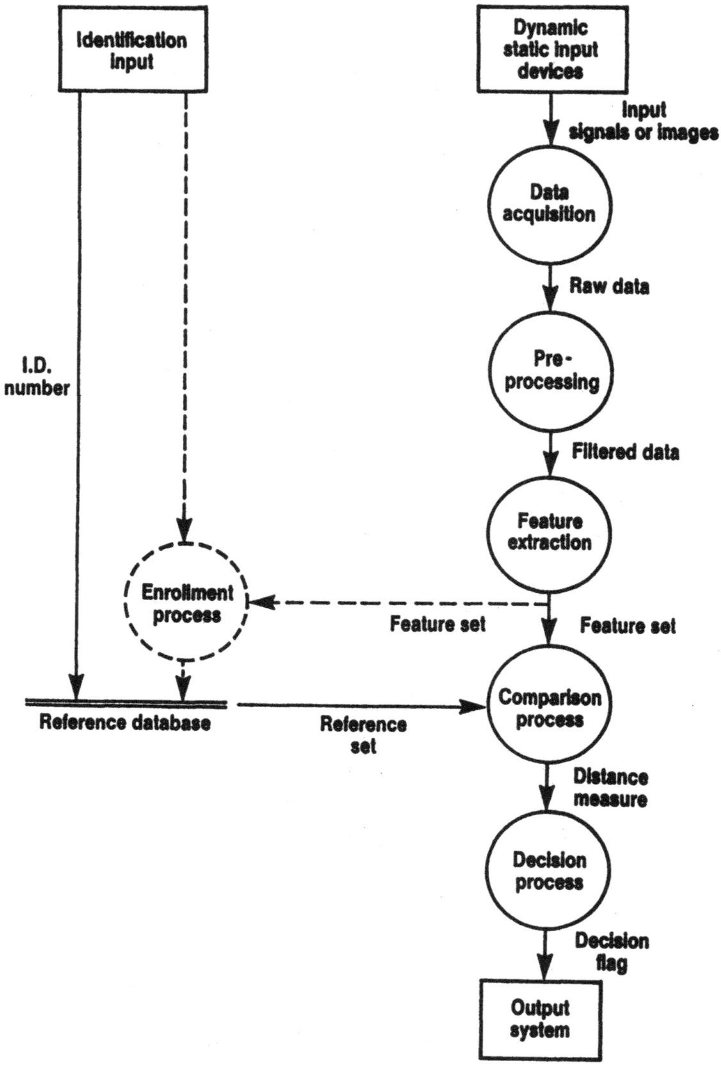

Figure 3. Data flow diagram of a signature verification system.

This general protocol hides a lot of practical details, e.g. the design and updating of the data base, the trade-offs between general or personalized decision thresholds and the selection of the feature set, etc. Moreover, many modifications could be made in such a protocol, depending on environmental and experimental constraints. In an ideal identification system, no indirect identifier would be used and the feature set would be compared to all the references of the data base. In a credit card verification system, the card could be used not only as an indirect identifier but also as the memory support for the reference set of the card holder, Herbst and Liu (1980); Reitter et al. (1988). This option is particularly promising with the advent of smart cards and point-of-sale terminals, Weinstein (1984). In a similar fashion, the comparison process could be performed automatically by the machine, or interactively, under the control of a human operator. For some applications, more than one trial may be allowed prior to any final decision about the signature and the signer.

Theoretically, two decisions can be made: *accept* or *reject*.

Accepted signature specimens may be:

AUTHENTIC if the specimen has been performed by the author of the reference and if its feature set has a good similarity with the reference set.

IMITATED if the specimen has been performed by someone who is not the author of the reference but its feature set has a good similarity with the reference set.

Rejected signature specimens may be:

DEGENERATE if the specimen has been performed by the author of the reference and if its feature set is different from the reference set. The term "DISGUISED" is often used when this degeneracy is voluntarily performed, Herkt (1986).

FALSE if the specimen has been performed by someone who is not the author of the reference and its feature set is different from the reference set.

Depending on the testing conditions and the environment, some authors include different types of forgeries to validate their systems:

simple forgery: where the forger makes no attempt to simulate or trace a genuine signature.

substitution or random forgery: where the forger uses his own signature instead of the signature to be tested. Some authors, Chorley et al. (1975); Beatson (1985) also call this a "zero effort forgery".

skilled or freehand forgery: where the forger imites as closely as possible the static and dynamic information of a signature. When specifying this type of forgery, information about training and testing should be given (with or without visual feedback, training period, etc.).

5. Discussion Problems Related to Signature Verification

Signing is not a process that is conducive to the production of perfectly precise, repeatable and identical characteristic data from successive trials. The only certainty in that when two signatures are identical one of them is a forgery, i.e. probably a copy, Locard (1936); Conway (1959); Harrison (1981). In fact, a great deal of variability can be observed in signatures Plamondon et al. (1987b), according to:

COUNTRY

Signature verification is appropriate for literate populations only. Differences can be noticed between countries, e.g. in Japan the signature is not in current use, (Picodata (1984)), in North America written names or personal symbols are common,while in Europe a personal symbol is often used as a signature. Figure 4 shows a few types of signature specimens.

AGE

For youngsters, important changes in signature occur since signatures are not well established, and can even completely change from time to time. After that, signatures are at their most stable, but again become less consistent in age.

TIME

For adults, signatures may be considered to be stable in the short term, this is not true over long periods, and it is more realistic to consider the signature as the result of an evolution process.

HABITS

Depending on their work, some people sign documents every day while others do not. This may have an influence on the consistency of the signature specimens.

PSYCHOLOGICAL OR MENTAL STATE

These factors may modify the appearance of signatures and increase their variability. Conversely, these factors may be used for graphological investigations. Also it has been noticed that a person may have developed several different versions of his own signature and that the range of the characteristics of a signature differs from one individual to another, Chorley et al. (1975); Lorette (1983), (1984).

PHYSICAL CONDITIONS

Temporary or permanent illness or physical deficiency, change in physical condition due to fatigue, cold, drugs, alcohol and medicine are also known to affect the writing process.

PRACTICAL CONDITIONS

Factors such as the writing position, the weight or size of the pen, its wiring, the nature of the writing surface, etc. can notably affect signatures. These factors must be taken into account in data base comparison.

Two types of signature variability must to be clearly distinguished, National Bureau of Standards (1980):

1. *intraclass* or *intrapersonal* variability, i.e. the variation observed within a class of signature specimens of one individual.

2. *interclass* or *interpersonal* variability, i.e. differences that exist between the genuine signatures of different writers.

In theory, intraclass scatter must be as low as possible and interclass scatter large enough to be used for class separation. In practice, classes are not obviously well separated. Features must be selected so as to reduce the class overlap. Thus, the acceptance and rejection thresholds must be determined according to the signer (depending on his intrapersonal variability), the type of application and the security level that is required.

In spite of this, a certain stability has been observed in signatures, Lorette (1983), (1984). Some characteristics can even be recovered in two different signature specimens from the same person. This situation is encountered in some countries for women within a few years of marriage. Where they previously signed with their maiden name, they now sign with their married name, however the new signature still reflects some characteristics of their previous signature Lorette et al. (1979); Brocklehurst (1984), (1985).

For several reasons, signature verification cannot be considered as a trivial pattern recognition problem. This is due, first of all, to the great variability of the signing process as mentioned above. Ideally, signature verification can be presented as a two-class partitioning problem. Let Ω be the set of all the signatures relative to a signer. A binary decision would allow the partition of Ω into: ω_1 the class which consists of genuine signatures and ω_2 the class of forgeries.

Consistent with that representation, the performance of a verification system is generally evaluated according to the error representation of a two-class pattern recognition problem, that is, with the type I (FRR: False Rejection Rate) and type II (FAR: False Acceptance Rate) percentage error rates.

Since good imitiation samples (static and dynamic) are not usually available,or not in sufficient number, ω_2 is never wholly known and statistical techniques, like maximum likehood classification, are not appropriate. In most cases, random forgeries (i.e. signatures of others individuals stored in the data base) are used instead of skilled forgeries for probabilistic law inference and system performance evaluation. This has the disadvantage of artificially reducing the type I and type II error rates.

One has also to cope with another difficulty. For practical reasons, the learning and training sets of signatures are often small and the generalization of the recognition process from the initial set is not an obvious one since statistical models cannot be assumed to be valid, Nagel (1976); Nagel and Rosenfeld (1977); Crane and Ostrem (1983); Brocklehurst (1984), (1985). This is overcome by allowing a certain amount of tolerance in the decision process by relaxing the threshold acceptance values. This results in a compromise between type I and type II errors.

Finally, the generation of one or a few reference feature sets from a group of signatures collected during a person's enrolment is a problem in itself, Liu (1978). Using mean values is not necessarily the best solution. Similarly, the update process of the reference data base must be studied to permit normal evolution of the genuine reference signatures and to reject attempts by any forger to progressively substitute his own signature for someone else's, Anthony et al. (1978).

Since any evaluation system involves people, these experiments are not easy to manage. A lot of time is spent conducting significant tests, and the motivation of the users involved is an important factor since, as has been

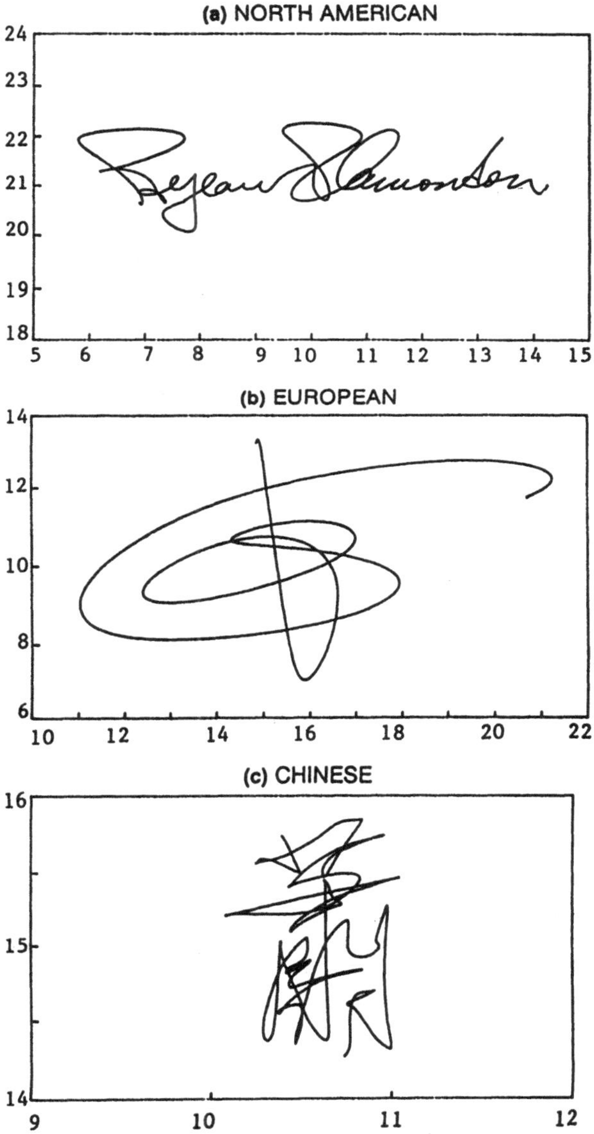

Figure 4. Typical signature specimens.

pointed out, mental state can affect the process. Also, it is not obvious how a representative sample of all potential users (men and women of different ages and activities, right- and left-handers) as found in a whole population, can be obtained.

6. Conclusion

Automatic signature verification is a difficult problem. Taking into account the intrinsic variability of a signature according to country, age, time, habits, psychological or mental state, physical conditions and practical signing conditions constitutes the major challenge for any system designer, quite apart from the fact that an automatic signature verification system also has to be capable of dealing with the various types of forgeries. What is the best signature acquisition device for a typical application? What type of signal or image processing algorithm should be used within a specific representation space? How should a signature be compared and a decision made on its authenticity, while taking into account the intrinsic evolution of a signature? What classification protocol performs best for a two-class pattern recognition problem with one of the classes unknown? These are some of the questions that require thorough study .

Further advances in this field will probably result from a better understanding of the fundamental psychophysical processes and the application of this knowledge to the specific tasks to be automated. This will require joint effort by multidisciplinary teams.

Acknowledgements

This work was supported by grant OGP0000915 to R. Plamondon from NSERC and by grants from the French National Education and Research Ministry tó G. Lorette. Parts of this work were done when Professor Plamondon was a visiting scientist at IRISA and a Fellow of the Netherlands Institute for Advanced Study and when Professor Lorette was a visiting scientist at the Laboratoire Scribens, Ecole Polytechnique de Montréal.

References

A complete bibliography appears at the end of this section.

Computer Processing of Handwriting
Eds. R. Plamondon & C. G. Leedham

DYNAMIC APPROACHES TO HANDWRITTEN SIGNATURE VERIFICATION

GUY LORETTE
IRISA, Université de Rennes I, Campus de Beaulieu, 35042 Rennes Cédex, FRANCE

RÉJEAN PLAMONDON
Laboratoire Scribens, Département de Génie Electrique, Ecole Polytechnique de Montréal, C.P. 6079, Succ. "A" Montréal QC, CANADA H3C 3A7

Dynamic signature verification consists of on-line signature acquisition and verification i.e., verification of the author's identity occurs during the signing process itself. The present paper reviews developments in dynamic signature verification methods and techniques. The various approaches are classified in two separated categories dealing with functions and parameters respectively. In both cases the data acquisition, preprocessing, feature extraction and comparison steps are analyzed and the problems of each are pointed out. The performance of existing systems is presented and some remaining problems are highlighted.

1. Introduction

The dynamic techniques apply to on-line data acquisition and processing during the signing process. These techniques involve two steps: reference signature enrolment and signature verification *vs* reference, Plamondon and Lorette (1989b).

Since signing is generally considered to be the result of ballistic movements (see preceding chapter), it is assumed that dynamic characteristics are the most specific to each signer and sufficiently stable and repetitive, Herbst and Liu (1977b). In other words it is expected that intrapersonal variations of these characteristics are noticeably fewer than interpersonal ones, Plamondon and Lorette (1989a). Consequently, signature verification can be carried out by comparing dynamic characteristics. As shown in the preceding chapter, dynamic signature verification may be viewed as a particular handwriting recognition system. It may be described as a five stage process consisting of: data acquisition, preprocessing, feature extraction, comparison to a reference signature and performance evaluation.

Moreover, dynamic signature verification methods may be classified into two principal groups. In the first group are methods dealing with functions as features. In this case, the complete signals (position, pressure, speed, and acceleration versus time), are considered as, or represented by, mathematical time functions whose values directly constitute the feature set. In the second group, the methods refer to parameters as features. These parameters are computed from the measured signals. Both global (e.g. total duration, number of components, overall dimensions, etc.) and local (e.g. maximum values, peak curvatures, starting direction angle, etc.) information is either explicitly or implicitly taken into account, together or separately, Bonnefoy et al. (1981). It is obvious that this function *vs* parameter distinction leads to straightforward feature extraction and comparison procedures specific to each of these classes, and some particular problems are related to each type of method.

2. Data Acquisition

In previous studies on signature verification, a considerable amount of acquisition equipment has been proposed, Plamondon and Lorette (1989a). ON-LINE or DYNAMIC systems are based on extraction and processing during signing of one or several signals varying with time (position, velocity, acceleration, force, pressure, pen tilt angle). Several types of devices such as digitizers, specific data input devices and instrumented pens have been used for signature data acquisition.

2.1 Digitizers, Interactive Flat-Panel Displays and Specific Input Devices

Standard digitizing tablets have been used for handwritten signature verification. Most often these devices provide the digitized position of the pentip as X,Y coordinate pairs at a constant rate or at a constant spatial increment. The absolute coordinates (X,Y) of any point are known with an accuracy varying from 0.005 to 0.025 inches. The most frequently used sampling rate for digitizers is about 100 Hz, but frequencies between 60 Hz and 200 Hz have been used. This is in accordance with the Shannon theorem, since it is generally accepted that the upper bound of frequency in a signature spectrum is below 20 Hz, Maamari (1986); Bills et al. (1987). Oversampling will give a large amount of unecessary data which will complicate and slow down the identification process, while undersampling will debase the signals and drastically lower the performance of the system, Teulings and Maarse (1984); Kim and Tappert (1984).

In some cases, auxiliary signals are also available from these input devices such as a pen proximity signal to monitor when the pentip is in the vicinity of the writing surface, a pen-down signal when it is in contact with the paper and even information on the tilt angle of the pen. This information may be used to determine the parts of the signature that are traced on the document and those, if any, in the air and the user's habits when holding the pen during signing.

Tablets have been designed for specific purposes, to deliver a pressure signal, for example, Chorley et al. (1975); Maarse et al. (1988); Leedham and Downton (1986), or speed components in two orthogonal directions, Stuckertt (1979); Bechet (1984). Recently de Bruyne (1988a) has used a low-cost piezoelectric film underneath a writing surface to detect both pressure and velocity information. The same author (1988b) has also proposed a system using the principle of frustrated total internal reflection to monitor the pentip displacement with a camera.

Electronic set-ups have also been designed to be connected to a commercial digitizer output to extract position, speed and acceleration information on a parallel port and to provide flexibility of the sampling frequency, Maamari and Plamondon (1986). In another prototype, a finger is used, instead of a pen, to interrupt an infra-red beam on a touch-sensitive screen, Ibrahim and Levrat (1979). Handwritten specimens can also be drawn on a C.R.T. screen with a light-pen, Nevill (1984). In both of these latter systems, position information is computed from the time elapsed since the previous synchronization signal of the screen sweep.

Recently, the use of interactive flat-panel displays, has also been proposed for handwriting acquisition, Komada et al. (1985); Nagayama et al. (1985); Schlosser (1985); Tappert et al. (1986). These devices combine a flat-panel display over an opaque digitizer or a transparent digitizer over a flat-panel display. The overall system is intended to behave like an electronic deskpad. The trace of the pentip during writing is monitored by the digitizer and displayed on the panel by activation of the appropriate pixels, thus generating an "electronic ink". Interactive flat-panels have still not been used for signature acquisition but probably will be in the near future.

Figure 1 shows the various output signals extracted with a digitizer from the specimen presented in (1a). The X,Y coordinates, sampled at 120 points/second, are plotted in different formats: the pen-down trace (continuous curve) and the pen-up trace (dotted curve) within 0.25 inches of the digitizer surface (1b), the complete trace (dots) in a X,Y cartesian system (1c) the X(t) signal (1d). The velocity $V_X(t)$ and acceleration $A_X(t)$ components, as derived from an analog differentiation circuit, Maamari and Plamondon (1986), are also shown in (1e), (1f), respectively. The pen-up/pen-down signal is depicted in (1g).

2.2 Instrumented Pens

To measure specific physical characteristics of handwritten signatures specialized styluses have been developed. These provide force, acceleration or pressure information during signing. Forces are measured with strain-gauges, Crane et al. (1977); Ostrem and Crane (1981), Crane and Ostrem (1983), or derived from pressure and position, Yasuhara and Oka (1977), while acceleration is detected by small accelerometers, Herbst and Liu (1975); Brault (1983), and pressures are measured by piezoelectric or semi-conductor sensors, Sternberg (1975); Haberman and Fejfar (1976); Eernisse et al. (1977); Sternberg (1975); Leedham et al. (1984). Optimal sensor layouts have been studied for a few prototypes, Lew (1977), (1978a), (1978b), (1979), (1980a), (1980b); Brault et al. (1983); Plamondon et al. (1983).

These components and even some electronic circuits, are embedded in the stylus. This often has the drawback of increasing the size, the form or the weight of the pen. The information provided by these systems reflects, at least partially, the global movement of the pen and not only that of the pentip. This is especially true for instrumented pens, where transducers are fixed at some specific position within the whole pen, Edwards and Paterson (1981); Brault (1983); Baron (1986); Telmosse (1986).

Instrumented pens provide dynamic information as measured in a moving reference system, as opposed to the digitizers which produce position versus time signals as measured in a fixed coordinate system. Although information in a moving reference system can theoretically be recovered, (provided the three Euler angles are known), in practice, it is not always feasible or meaningful, Lew (1978c), (1980b); Telmosse (1986); Telmosse and Plamondon (1986).

(a)

(b)

(c)

Figure 1. Signature signals as extracted with a digitizer.
(a): handwritten word specimen (b): pen-down (continuous) and pen-up (dotted) components (c): sampled version (120 points/second)

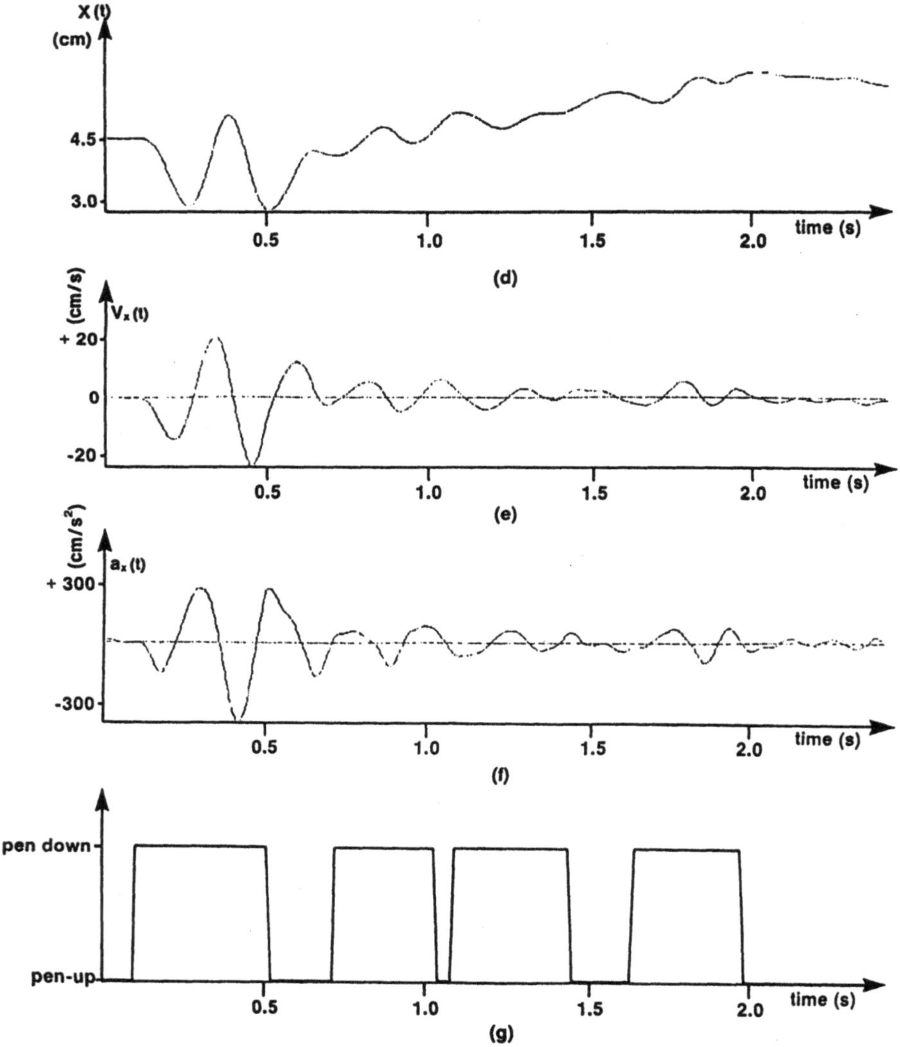

Figure 1. Signature signals as extracted with a digitizer. (continued)
(d): X(t) signal (e): $V_X(t)$ signal (f): $A_X(t)$ signal (g): pen-up/pen-down signal.

For instrumented pens, a high sampling frequency f = 640Hz, Worthington et al. (1985), has been used in order to make the preprocessing or filtering easier, but after decimation the frequency is reduced to f = 80 Hz, since it has been suggested that under 60 Hz the sampling frequency is not sufficient to give a correct representation of a signature.

Figure 2 (b,c) shows the various output signals available from an accelerometer pen, Brault and Plamondon (1986), when writing the specimen in Figure 2a. Two graphics represent examples of the output of thin accelerometers incorporated in the pen casing (in 2b the accelerometers are in the pentip, in 2c they are at the other end of the pen). Figure 2d shows the pen-up/pen-down signal.

Table 1 presents a summary of the properties of the various instrumented pens and signal conditioning units used for signature verification. The second column of this table lists the specifications of the acquisition system as a whole: operating principle, sampling frequency, bandwidth, resolution, number of channels, dimensions, weight, etc. The last column highlights some specific characteristics of a particular implement. For example, depending on the number, the position and the orientation of acceleration transducers on the one hand, and on the type of signal processing techniques used on the other, accelerometer pens may be classified as asymmetric, Herbst and Liu (1977a), (1977b), (1977c), or symmetric, Plamondon et al. (1983); Worthington et al. (1985).

3. Preprocessing

This term includes all the algorithms that are carried out on raw data to produce relevant information. Preprocessing is generally not described in detail in application papers. It can be executed by software, or by hardware, Maamari and Plamondon (1986), to reduce spurious noise on signals, and by thresholding, to detect gaps in pen-down signal. It consists of signal amplifying, filtering, conditioning and digitizing, Lorette (1983); Plamondon and Brault (1983); Worthington et al. (1985); Achemlal et al. (1986); Bonnefoy (1986), signal truncation and normalization, new signal computation, Yasuhara and Oka (1977); Sato and Kogure (1982); Zimmerman and Varady (1985), direction encoding, Farag and Chien (1972a), start-up and end-of-signature detection, pen-up and pen-down detection, signature segmentation into components, segments or strings Lamarche and Plamondon (1984); Worthington et al. (1985); Feig (1985); Chainer et al.(1985); Bonnefoy (1986).

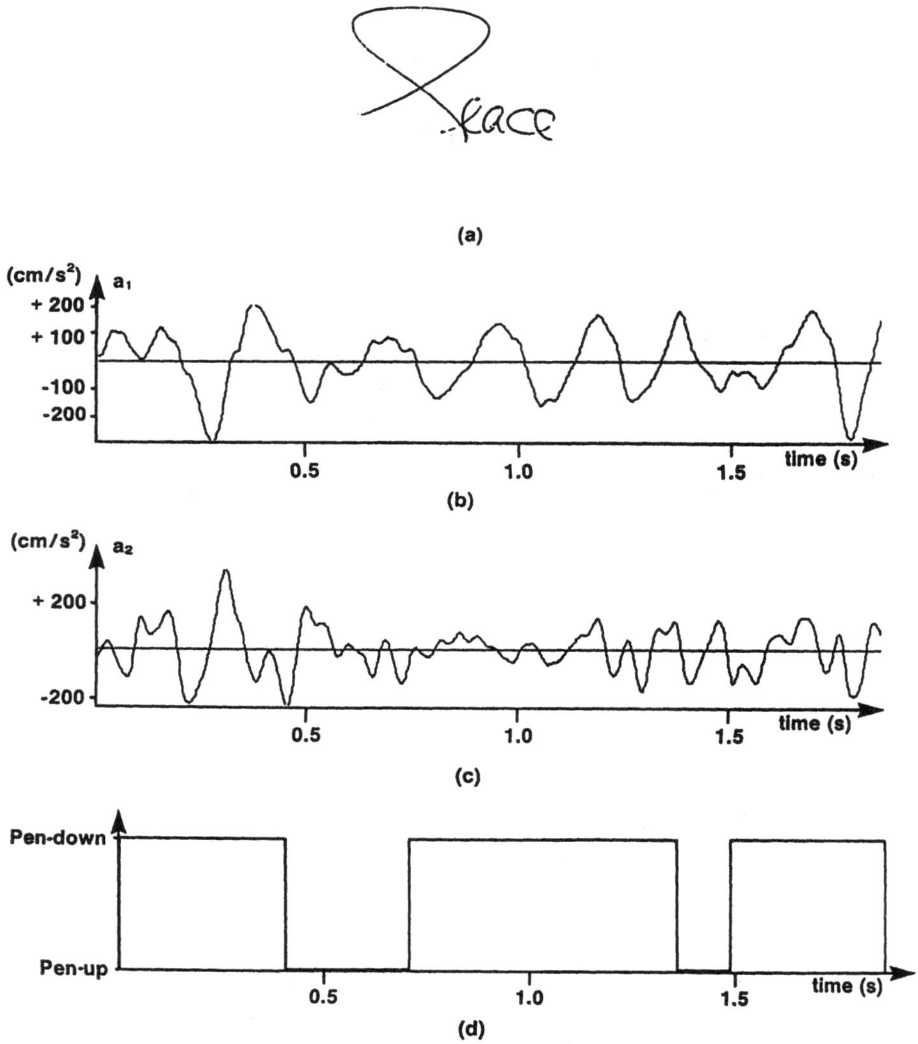

Figure 2. Signature signal as extracted with an accelerometer pen.
(a) : handwritten word specimen (b,c) : pentip acceleration signals (d) : pen-up/pen-down signal.

INSTRUMENTED PENS		
REFERENCES	SPECIFICATIONS	NOTES
BRENNER DE BRUYNE (1970)	- ultrasonic pen - resolution: 0.2mm - 200 Hz switch for pen-up/pen-down	orthogonal pair of plane microphones and a pen capable of generating a short sonic pulse used for XY pentip localization
CRANE WOLF OSTREM (1977)	- three strain-gauges - 50 to 100 Hz	3 noise-free force signals: F_X, F_Y, F_Z
EERNISSE LAND SNELLING (1977)	- piezoelectric pressure transducers	high cut-off frequency: 30 Hz detection of flexing in the X and Y directions
HERBST LIU (1977a) (1977b) (1977c)	- two orthogonal semi-conductor strain-gauge accelerometers - pen contact switch - 200 Hz - pressure-sensitive	asymmetric design (preferred orientation) on/off force 30 g one filter with roll-off 6 dB per octave at 30 Hz linear phase recursive digital filter at 40 Hz 8 bit data ouput 3 signals - a version with variable capacitance sensor has also been proposed by Gjone and Grossman (1981) and one with four accelerometers by Liu and Herbst (1977)

Table 1. Specifications of instrumented pen.

INSTRUMENTED PENS		
REFERENCES	SPECIFICATIONS	NOTES
LEEDHAM DOWNTON BROOKS NEWELL (1984)	- displacement transducer used to measure pressure	instrumented clutch pencil connected to a digitizer
MAARSE SCHOMAKER TEULINGS (1986)	- piezoelectric pressure transducer	signal conditioning unit interfaced to a digitizer
MAUCERI (1965)	- one accelerometer mounted on a commercial ballpoint pen paper contact	cylindrical accelerometer (1 inch long. 1/2 inch diameter) ridge on the pen closure of a conductive loop using Teledatas conductive paper
PLAMONDON BRAULT ROBILLARD (1983)	- four accelerometers - mechanical switch for pen lift up signal - 160 Hz	high cut-off frequency: 20 Hz low cut-off frequency: 0 Hz 5 signals symmetric design (no preferred orientation) 8 bit data output
STERNBERG (1975)	- one pressure transducer at the top of the pen	detection of axial pressure

Table 1. Specifications of instrumented pen. (continued)

INSTRUMENTED PENS		
REFERENCES	SPECIFICATIONS	NOTES
WORTHINGTON CHAINER WILLIFORD GUNDERSON (1985)	- piezoelectric biaxial accelerometer	symmetric design (no preferred orientation)
	- piezoelectric force sensor	2 pole Butterworth low-pass filter at 30 Hz
	- measure of the pressure derivative	8 bit data output
	- 640 Hz reduced to 80 Hz	3 signals high pass frequency: 2.5 KHz

Table 1. Specifications of instrumented pen. (continued)

This latter problem is very important in the sense that a lot of function comparison techniques rely directly on the segmentation of a signature. These segments are often defined as portions of signals concurrent with pen-down movements and are deduced from pressure signals, Liu and Herbst (1977); Herbst and Liu (1977b), (1977c); Chainer et al. (1985); Feig (1985); Worthington et al. (1985). Herbst and Liu (1977a) have also processed these pen-up/pen-down signals to remove random or artistic variations and to limit the maximum duration of a segment. Lamarche and Plamondon (1984) have proposed segmenting signatures using a handwriting model. In this approach, segments are defined as portions of signature movements that result from the stimuli of a motor program. Recently, Brault and Plamondon (1987a,b) have developed a method for detecting singularities in the handwritten curve. This method is based on both geometric and sequential information and results in a possibility function which may be used for segmentation.

4. Feature Extraction

Feature extraction methods stress the distinction between the two approaches previously mentioned: function and parameter representations. In most cases, signature representation and information compression do not allow recovery or reconstruction of the original signature from acquired data.

4.1 Functions

The function representation is comprised of, or computed from, signals either measured directly with suitable sensors or derived from other acquired signals, Brault and Plamondon (1986), according to some implicit assumptions and approximations. For example, the horizontal and vertical accelerations can either be measured directly with accelerometers or calculated from the horizontal and vertical displacements by two successive numerical derivations. It is assumed that numerical derivations do not introduce errors or uncertainties in the final results. Thus from a mathematical point of view, signatures are represented by parametric or complex functions. The principal signals considered for the feature function are as follows:

Position: In some investigations (see Table 2), the set of X,Y coordinate pairs of successive points obtained from sampling at regular time intervals have been taken as signature features, Farag and Chien (1972a); Ibrahim and Levrat (1979); Watson (1979); Lorette (1984); Beatson (1985); de Bruyne (1985). This is a natural way of proceeding since these data are easily obtained from standard devices such as commercial graphic tablets. For the geometric aspects of a signature, an intrinsic function using cumulative phase versus curvilinear abscissa has also been proposed, Lorette (1983).

The pentip position when the tip is in the air and in the vicinity of the writing surface can also be recorded and used. Strangio (1976), Herbst and Liu (1977b,c) have noticed that these parts of traces are certainly harder to imitate since they have no trace, but they are also much more variable, less precise and less repeatable. De Bruyne (1977), (1978), de Bruyne and Messmer (1980) have derived timing information from the position signal: time sequences of contact duration, time sequences bounded by reversal of X or Y displacements. In their approach, Sato and Kogure (1982), use coordinates in a complex function, z(t)= X(t) + i Y(t), which is then normalized to be independent of location in the plane, of rotation, and of size and constant movement from left to right.

Pressure: The first test based on a strain-gauge pressure pen was conducted by Haberman and Fejfar (1976), using the Veripen system (see Table 2). Pressure has also been used in conjunction with other signals such as acceleration, Liu et al.(1979); Worthington et al. (1985) or position, for new function computation Yasuhara and Oka (1977); Sato and Kogure (1982), to improve the verification process. Pressure signals can be affected by pen inclination and by finger pressure on the pen or on the tablet, depending on the way they are measured. The time derivative of the pressure has also recently been used by Worthington et al. (1985) with acceleration signals.

Forces: It is assumed that forces result from predetermined stimuli transmitted from the brain to the muscular system and that they are related to writing size, Denier van der Gon and Thuring (1965); Herbst and Liu (1977b), (1977c). In their experiment, Yasuhara and Oka (1977) have proposed force for signature verification (see Table 2). It was not measured with strain-gauge sensors but was derived from position and pressure via a mathematical model of the mechanical handwriting process. This assumes that some physical parameters, i.e. equivalent mass of the hand-pen couple, viscosity and friction coefficients, are known and constant throughout a signature.

Velocity: The velocity signal is also of interest since it is one of the most well-accepted invariants in movement generation, Morasso (1981); Abend et al. (1982). A comparative simulation of fourteen handwriting generation models has recently shown that velocity-controlled models yield the best reconstruction output in an analysis-by-synthesis experiment, Plamondon and Maarse (1989). Local speed is either deduced from positions variations, Lorette (1984); Strangio (1976); Maamari and Plamondon (1986), or directly measured with special devices, Stuckert (1979); Bechet (1984); Thompson-Titn (1984) (see Table 2). In these systems, each sampled point is represented in the plane of two velocity component vectors.

Acceleration: Several techniques are based upon the fact that signing is essentially a ballistic phenomenon. According to this model, position second-time derivatives are assumed to be very significant functions Lew (1980a). The direct use of acceleration signals was first proposed by Mauceri (1965) and then by Herbst and Liu (1975), and Evangelisti et al. (1976) (see Table 2).

Their methods were later further enhanced with the use of writing pressure, Liu et al. (1979). Some authors have also studied and optimized an accelerometric pen, Lew (1979), (1980b); Plamondon et al. (1983), or developed a hardware analog device, Maamari and Plamondon (1986) for acceleration measurements intended for signature verification. Zimmerman and Werner (1978) have used an FIR filter of order 25 with coefficients weighted by a Blackman window to extract this information from position data.

Up to now, no general study has been carried out to compare the advantages and disadvantages of the performance obtained from analog or numerical derivations of position signals versus direct measurement with instrumented pens. However, practical difficulties in matching results obtained in a mobile reference system related to a pen in the former case, and those obtained in a fixed reference system related to the writing surface in the latter, have been studied, Lew (1978c), (1980a); Brault (1983); Telmosse (1986); Telmosse and Plamondon (1986).

Plamondon and Parizeau (1988) have recently studied the performance of position velocity and acceleration signals in signature verification. They have shown that the most discriminating functions are signals that reflect the vertical activity of the signing process, and that the best representation space for a 2D signature verification system is probably the velocity domain.

4.2 Parameters

Geometric parameters, timing parameters and the more specific dynamic parameters related to some distinctive properties of a signal are involved in signature verification, Howbrook (1978); Ostrem and Crane (1981); Bonnefoy and Lorette 1984; de Bruyne (1985); Achemlal et al. (1986) (see Table 2). For example, Crane and Ostrem (1983) have derived forty-four original parameters from the three signals generated by a force-sensitive pen (scaled mean, numbers of zero crossing, time up, etc.). De Bruyne (1985) has extended the use of geometric parameters in static signature verification. Some authors have also applied specific transform techniques to signals and use as features a set of coefficients derived from these mathematical transformations. For example, a rough approximation of the analog pressure signal obtained by one-bit quantization, i.e. on/off paper contact, has been proposed by Zimmerman and Varady (1985) as a first stage for a signature represented in the frequency domain via a Walsh transform, and they considered the first forty low frequency coefficients of the power spectra as features. The

AUTHORS	SIGNAL AND FEATURE DESCRIPTION	TRAINING AND/OR TEST DATA BASE SPECIMENSxWRITERS	COMPARISON METHOD	ERROR RATES	COMMENTS
ACHEMLAL MOURIER BONNEFOY (1986)	X,Y 10 personalized parameters selected among 40	600 genuines training set: (10S x 60W) 600 forgeries 1 imitator	weighted city-block distance	Type I :11% Type II: 8%	use of real forgeries 3 trials
BEATSON (1985)	X, Y movements of the pen in the air included, spatial and dynamic features	1000 genuines 100 forgeries		Type I : 1% Type II: 2%	no information on training set commercialized by Signify Corp. as SIGNIFY system
BECHET (1984)	V_X, V_Y normalization	3000 genuines training set: (4 or 5Sx40W) 1500 forgeries 12 imitators	Euclidean distance and score	Type I : 5% Type II: 5%	over 3 months
BONNEFOY JOUNET LORETTE GAUDAIRE (1981)	X,Y parameters selected from a training set of 18	342 genuines (15S x 14W): training (12Sx11W):test no forgeries	sequential recognition decision-tree	Type I:0 to 6% Type II:----	only one trial over one year
BRAULT PLAMONDON (1984a) (1984b)	average acceleration, sum of acceleration, number of samples	243 genuines (5S x 50W)	histogram classifier likeliness global and local likeliness coefficients	Type I :1.2% Type II: 1%	among 243 signatures random forgeries
CHORLEY OLDING PARKS POBGEE WATSON (1975)	X,Y 10 parameters selected among 100	2000 genuines (5S x 70W): training 549 forgeries 40 imitators	windows in 10 dimensions space	Type I :4% Type II:0-.05%	after 4 weeks reference: 5 consistent signatures 3 trials random and real forgeries commercialized by Transaction Security as VERISIGN and SECURISIGN systems

Table 2. Performance of on-line signature verification systems.

AUTHORS	SIGNAL AND FEATURE DESCRIPTION	TRAINING AND/OR TEST DATA BASE SPECIMENSxWRITERS	COMPARISON METHOD	ERROR RATES	COMMENTS
COLLANTIER (1984) SALVAGE (1984)	X, Y signature coded into 40 bytes of data	5656 genuines (5S x 460W): training 3509 S: test	correlation 0 to 99	Type I : 3.5% Type II: ----	reference 3 consistent signatures 3 trials over one year credit card in a public store commercialized by Quest Micropad, as Q-SIGN system
CRANE OSTREM (1983)	FX, FY, P 25 parameters selected among 44 reference vector	5220 genuines (2x10S) x 58W: training set 648 forgeries 12 imitators	weighted Euclidean distance (standard deviation distance)	Type I : 1.5% Type II: 1.5% Type I | 2.25% Type II | to 3%	over 4 months - 3 trials random forgeries real forgeries
DE BRUYNE (1985)	X,Y 4 parameters selected among 18	71 genuines from 11W, 52 forgeries	weighted parameters grading maxi likelihood ratio test	Type I : 3% Type II: 2%	reference parameters evaluated from 10 specimens
DYCHE (1969)	X,Y 40 parameters based on time length and on position, speed and acceleration self and cross moments	training: 333 genuines from 1 signer 167 forgeries from 3 amateur forgers tests: signatures from each classes	likelihood ratio nearest neighbor linear boundary	Type I : 0% Type II: 0%	similar results using a subset of 15 parameters
FARAG CHIEN (1972a)	256 (X,Y) coordinates duration 2 s chain coded vector histogram	200 genuines (20S x 10W) 250 forgeries: 10 users imitated 5 imitators	city-block distance Mahalanobis distance	Type I: 14-23% Type II:14-23%	
HABERMAN FEJFAR (1976)	pressure 9 local or global features	209 W (170 M, 39 F) 2645 type I tests 106,505 type II tests		Type I :6.81% Type II:3.19%	higher error rates for females using the VERIPEN system

Table 2. Performance of on-line signature verification systems. (continued)

AUTHORS	SIGNAL AND FEATURE DESCRIPTION	TRAINING AND/OR TEST DATA BASE SPECIMENSxWRITERS	COMPARISON METHOD	ERROR RATES	COMMENTS
HALE PAGANINI (1980)	force amplitude 15 Haar coefficients 18 waveform physical features	typical experiments 500 genuines 97 S 951 verification attempts 181 non observed signer forgery attempts 59 observed signer forgery attempts	weighted distance	Type I :1.5% Type II: 1.2% 2.5%	 non-observing signer observing signer general result form 9976 signatures
IBRAHIM LEVRAT (1979)	X, Y, V, A global parameters peak matching zero crossing	47 genuines (1S x 10W): training 36 forgeries 2 imitators	Euclidean distance	Type I : 19% Type II: 5.5%	signing with one finger instead of a pen
LAM KAMINS (1989)	X,Y 15 Fourier coefficients	8 genuines from 1 signer 152 skilled forgeries from 19 signers	discriminant analysis and Jacknife procedure	Type I : 0% Type II: 2.5%	test on preprocessing also performed
LIU HERBST ANTHONY (1979)	a_X, a_Y, P a_X, a_Y	1)6000 genuines (6S x 201W): training forgeries 40 users imitated 10 S/user 40 imitators 2)1042 genuines (5S x 40W): training 750 forgeries	regional correlation	Type I : 1.7% Type II: 0.4% Type II: 0.022 Type I : 2.4% Type II: 3.2%	3 trials - over 6 months real forgeries with random forgeries 3 trials
LORETTE (1983)	X,Y global & local geometric & dynamic 7 parameters	210 genuines (5S x 14W): training (10S x 14W): test no forgeries	clustering analysis	Type I: 6%	only one trial

Table 2. Performance of on-line signature verification systems. (continued)

AUTHORS	SIGNAL AND FEATURE DESCRIPTION	TRAINING AND/OR TEST DATA BASE SPECIMENSxWRITERS	COMPARISON METHOD	ERROR RATES	COMMENTS
MAUCERI (1965)	acceleration pen paper contact, 10 indices related to frequence spectra	2350 genuines 45S x 40W no forgeries	squared distance	60% recognition rate	partial results from 10 subjects only
	19 local and global features	20 S/W learning 25 S/W test	weighted squared distance	90% recognition rate	
MITAL HIN LENG (1988)	Time interval Number of peaks Distance and area measures on pressure signal	200 genuines from 50 50 signers for reference Number of test unknown up to three trials per test	correlation	Type I : 0% Type II: 0%	test over a six weeks period
MJØLNES SØBERG (1987)	System A see Bechet (1984)	3215 genuines from 69 signers 414 forgeries from 60 persons	see Bechet (1984)	Type I : 10.5% Type II: 2%	test of two commercial systems over six weeks, global results after a maximum of three attempts per test
	System B see Chorley et al (1975)	3706 genuines from 69 signers 424 forgeries from 60 persons	see Chorley et al. (1975)	Type I : 18.4% Type II: 3%	
PARIZEAU PLAMONDON (1989a)	XY	50 genuines signatures initials and passwords from 39 signers	Regional Correlation Tree Matching Dynamic Time Warping	Total error:4% on signature with Y signal	comparative study on type of scripts
PLAMONDON PARIZEAU (1988)	XY V_X V_Y A_X A_Y	50 genuines from 39 signers	Regional Correlation	on Y signals: Type I : 2.1% Type II: 2.8%	comparative study on type of representation space
			Tree Matching	Type I : 1.8% Type II: 1.2%	other results on velocity and acceleration signals
			Dynamic Time Warping	Type I : 1.4% Type II: 1.6%	

Table 2. Performance of on-line signature verification systems. (continued)

AUTHORS	SIGNAL AND FEATURE DESCRIPTION	TRAINING AND/OR TEST DATA BASE SPECIMENSxWRITERS	COMPARISON METHOD	ERROR RATES	COMMENTS
SATO KOGURE (1982)	X, Y, P complex normalized functions shape, motion and pressure vector	110 genuines (10S x 11W) 330 forgeries 1 user 10 S/users 3 imitators	D.P. Matching Mahalanobis pseudo-distance	Type I : 1.8% Type II ≈ 0%	no information on training set
STERNBERG (1975)	pressure signal reduced to a few "measures" via mathematical transformation	1000 genuines (10S x 100W) 50 imitators each one forging 8 subjects sign 3 times random forgeries		Type I : 0.7% Type II: 1.8% Type II: 1.8%	over 2 months with deliberate forger commercialized by VERIPEN as signer system
WORTHINGTON CHAINER WILLIFORD GUNDERSON (1985)	X, Y, P 5 similarity measures	5000 genuines (6Sx108W): training 4700S: test 2133 forgeries	similarity measure	Type I : 1.77% Type II: 0.28% 2.33%	tests over 9 months
YASUHARA OKA (1977)	X, Y, P calculated force function	100 genuines (3S x 10W): training (7S x 10W): no training	D.P. matching Euclidean distance	Type I : 3% Type II: < 1%	key words used instead of signature
ZIMMERMAN VARADY (1985)	one bit pressure P walsh functions power spectrum 40 low sequency harmonics	90 genuines (5S x 9W): training	Fisher discriminant functions linear classifier	Type I :30-50% Type II: 4-12%	over several weeks
ZIMMERMAN WERNER (1978)	X,Y acceleration derived from FIR filter	not specified	straight-forward digital correlation	less than 10% miss-recognition	preliminary tests

Table 2. Performance of on-line signature verification systems. (continued)

same group, Bills (1986); Bills et al. (1987) has analyzed signatures and script using the Fourier transform, autoregressive and time-varying spectral analysis. Similarly, Hale and Paganini (1980) have applied the Haar transform to force patterns, and use, among other parameters, the first fifteen coefficients as a feature vector. Lam and Kamins (1989) use fifteen harmonics derived from Fast Fourier transforms of a normalized signature.

Generally speaking, two types of parameters may be distinguished:

local parameters: maximal or minimal values of signal features extracted from specific segments like number of peaks, local curvature, starting direction, etc.

global parameters: total time, number of segments or pen lifts, means and standard deviations, number of zero crossings, number of maxima and minima for each segment, area, proportions, etc.

In several experiments, an a priori list of parameters is proposed. This set can have up to one hundred or more parameters, Chorley et al. (1975); Bonnefoy (1986) (see Table 2). From this initial set, a subset of selected parameters is considered for signature verification. Theoretically, parameter selection must be carried out according to a maximal discriminant power criterion between genuine and skilled forged signatures, Crane and Ostrem (1983). In practice, random forgeries are often used since they are easier to obtain, but this must surely result in an overestimation of system performance.

In some works, parameters are not considered as vector components, but other representation techniques are used to analyze and compare the data. For example, Brault and Plamondon (1984a,b) have proposed using histograms for signature parameter representation. In the work of Requier and Lorette (1979), a graph description is used. In their study, Bonnefoy and Lorette (1984) use a decision-tree representation.

For most applications, the parameters used are general, i.e. the same for all writers. Crane and Ostrem (1983), and more recently, Achemlal et al. (1987) have proposed the use of personalized parameters for each user. But even in this type of method, the three basic problems of parameter selection are encountered.

The first problem is the choice of a criterion for selecting pertinent and efficient parameters: type I (FRR) versus type II (FAR) error equality, the most stable parameters, the most discriminant parameters between true and freehand or skilled forgeries (in this context, an automatic feature selection method has also been proposed by Bonnefoy and Lorette (1984)).

The second problem is the determination of an optimal number of parameters. In the case of a small set of users, Bonnefoy (1986) has proposed retaining a minimum number of parameters adapted to best discrimate between enrolled writers. This reduces the dimension of the feature vector representation space without notable performance degradation of the system. This should also simplify and speed up the recognition process. However, this approach still has the disadvantage of being difficult to generalize to a large set of users.

The last problem, but not the least, is the lack of valuable theoretical statistical models of true signature versus forgery distribution. This makes it impossible to use statistical feature selection methods (e.g. Bayes test). To overcome this problem, jacknife, i.e. leave-one-out, procedures, Crane and Ostrem (1983); Achemlal et al. (1986), or sequential procedures, Bonnefoy (1986), have been considered.

4.3 Functions *versus* Parameters

It has been pointed out that, owing to information compression, the use of the parameter method leads to a reduction of the memory space necessary for reference signature storage, usually of the order of tens of bytes instead of hundreds or thousands of bytes per user as in the case of function representation. This has the advantage of being efficient in terms of algorithmic simplicity, computation speed, and hardware and software requirements, Crane and Ostrem (1983); Bonnefoy (1986); Achemlal et al. (1986). But, as will become apparent, signature verification performance using the function method is rather higher. In practice,the choice of a particular method depends on the nature and constraints of a particular application, i.e. number of users, memory space available, type I (FRR) versus type II (FAR) error rate acceptability. Both types of method, have particular problems to be overcome. In future systems, the two types of approach could also be combined.

5. Comparison and Decision Processes

These processes also differ from the previous function/parameter distinction. Generally speaking the comparison technique evaluates the fit between the reference signature and the test specimen, to calculate the similarity of or the distance between two signatures. At the decision stage, similarity values or distances are compared to a threshold fixed a priori to decide if the test signature should be accepted or rejected.

5.1 Function Comparison and Decision Approaches

Obviously, the straightforward method for function comparison is the use of linear correlation, but this is not a valid choice for signatures for a number of reasons.

The first problem is related to overall signal duration. In spite of a certain amount of consistency, Herbst and Liu (1977b), (1977c); Lorette et al. (1979), have pointed out that the total duration varies from one signature to the next, even when they are produced by the same writer. Since signals are sampled at a constant rate this provides a final number of sampled points which is different in each specimen. Consequently, comparison between two signals is more cumbersome. The second problem has to do with the non-linear time axis distortions between two different signature specimens. Random variations exist, and these can create portions of signals, deletions, additions and gaps due to the pauses or hesitations of the writer.

Various approaches have been proposed to solve these two problems and to determine simultaneously the best fit between two corresponding functions. For compressing or expanding the time axis, non-linear correlation through dynamic programming matching (D.P.M) previously used for speech recognition has been extended to signature verification. In the work of Yasuhara and Oka (1977), this comparison technique is proposed to match the force functions and to compute a minimum distance between them. In their approach, Sato and Kogure (1982) used D.P.M to determine the minimum distances between two signatures which are compared in terms of shape, motion and writing pressure information. In Strangio's experiment (1976) a peak-matching technique via a peak-match table using a global mean-square error criterion has been considered. Hanan et al. (1978) have proposed a finite-state machine approach to achieve dynamic time warping. To minimize the response time, they suggested representing a signature signal by a sequence of n-tuples of K-class slope categories and to evaluate the probability of generating the test sequence from the reference sequence. If this probability is greater than a specific threshold, the signature is accepted. Brault and Plamondon (1987a) have proposed a two-step approach to dynamic programming using an algorithm with a variable window.

Herbst and Liu (1977b), (1982) have proposed a regional correlation algorithm which has been improved by Lew (1983), Chainer et al. (1985) and Grossman et al. (1981). The algorithm uses a method of correlation in the complex plane resulting in weighted coefficient vectors. The acceleration and pressure signals of a reference and a test signature are segmented. The corresponding segments or subsegments are shifted with or without overlapping to minimize a root-mean-square criterion of fitting, Herbst and Liu (1977b), (1977c), or a similarity coefficient, Worthington et al. (1985). Their algorithm also solves the instrumented pen rotation problem since the form of correlation that is used is rotationally invariant. In Worthington et al. (1985), a symmetrically designed pen with no privileged orientation is used and signatures are compared via five dimension vectors of similarity coefficients. These coefficients reflect segment alignment, pressure and acceleration correlation and coherence. Mital et al. (1988) used a simplified form of correlation on the pressure signal.

Parizeau (1987) and Parizeau and Plamondon (1989b) have recently compared the advantages and disadvantages of three popular comparison algorithms (regional correlation, dynamic time warping and tree matching) in a signature verification system based on pentip position representation. This study shows that differences between algorithms are signal- and script-dependent.

In the system developed by Bechet (1984) a normalization technique introduces variable time-shifts and compression or expansion coefficients to optimize the global discrepancy between a pair of functions. In this method, after representing each sampled point in the velocity plane, the final proximity is evaluated upon the percentage of the number of points lying in circles of predetermined radius.

Finally, there are also problems with geometric transformations such as translation, rotation, scale or dimension changes. These difficulties are solved either by normalization, Sato and Kogure (1982), or by the use of invariant geometric properties, Lorette (1983).

5.2 Parameter Comparison Decision Approach

The parameter comparison is very straightforward. Since signatures are often described by vectors in a feature space, their proximity is evaluated using a distance measure, Chorley et al. (1975); Ibrahim and Levrat (1979); Crane and Ostrem (1983); Achemlal et al. (1986).

De Bruyne (1985) has used two techniques for signature comparison, the first one being the maximum likelihood ratio, the second consisting of weighted parameter grades or global likeness coefficients. In his work, Lorette (1983, 1984) has considered a normalized principal component analysis and an ascendant hierarchical classification method, a step-wise discriminant analysis or a dynamic cluster method. In Bonnefoy's experiments (1986), identification is carried out through the use of an n-ary decision tree deduced from the specimen signature training set.

Brault and Plamondon (1984a) have defined a set of likeness coefficients to compare histograms produced from signatures. This approach also compensates partially for the change in initial conditions from one signature to another from the same writer. Zimmerman and Varady (1985) use a Fisher linear classifier of the forty coefficients extracted from the Walsh power spectra as their comparison method. Requier (1979) has developed a method using graph and subgraph isomorphism for signature verification. Others techniques for comparison have also been tried. Farag and Chien (1972a,b) have proposed comparing the chain-coded representation of the pentip positions in a two-step process consisting of global detection via histograms, and the use of the simple city-block distance between chains. A signature is evaluated on a fixed scale with a score varying from 0 to 99.

5.3 Threshold Problems

In most experiments, the decision threshold is unique and general in the sense that it has one single value for all users. Loosening or tightening this threshold for each signer, has also been proposed to take into account the specificity of intrapersonal variations, Chorley (1975). Liu (1978) has proposed a method for building a signature reference data base with personalized thresholds for each subject. Yasuhara and Oka (1977) have used personalized thresholds to make sure that their algorithm could accept all signature specimens from the same signer. In all of these cases, the security level will differ for each signer. Similarly, Farag and Chien (1972a) have studied a sequential verification system for which a two-threshold scheme has been considered.

Obviously, the threshold choice depends on the application being considered,and according to the false acceptance or false rejection risk. Thresholds leading to equal type I (FRR) and type II (FAR) errors are often proposed, Ostrem and Crane (1981). The thresholding problem is very important because ultimately it largely determines the performance of the system.

6. Performance Evaluation

Actually, owing to the lack of natural, ergonomic and convenient devices, the results of signature verification experiments differ considerably from laboratory to field test. In the laboratory, the process can be easily controlled, while a field test involves a stand-alone process with almost no control on the signers. Moreover, tests may differ from one application to another since signature verification systems cover a large spectrum of devices, from small systems for local applications with a limited number of users to very large systems controlling the identities of many people from remote terminals via networks.

The performance of signature verification systems may be evaluated from two measurements; i.e. type I (FRR) and type II (FAR) errors. As has been mentioned previously, these ratios vary with the acceptance/rejection threshold. The ideal situation occurs when the "best features" are selected to completely separate true signatures from forgeries. But, as forgeries are not necessarily available and as an adequate solution has not yet been found to the problem of feature selection, the feature choice is not optimal in practical situations and often no threshold leads to zero error rates. The difficulty of evaluating the performances of different signature verification systems through type I and type II errors rates, since thresholds are different from one experiment to the next and field tests are applied to different signature data sets.

Performance of signature verification systems in existence so far are given in Table 2. It will be noted that results vary from a few percent to less than one percent for type I/type II errors. However, for several reasons final results cannot all be directly compared.

Except for the comparative studies by Parizeau (1987), Parizeau and Plamondon (1987), (1989a,b) and Plamondon and Parizeau (1988) there is no data common to the experiments. This is due, in particular, to the lack of public signature data bases. To our knowledge, Mjølnes and Søberg (1987) are the only authors to have reported on the comparative performance of two available dynamic signature verification devices. This test involved sixty-nine subjects signing several times a day, over a period of six weeks. These independent field tests revealed error rates higher than those performed in laboratory environments, Chorley et al (1975); Bechet (1984). Neither of the systems was found to be satisfactory at this stage of development.

In general, the experimental protocols vary considerably from one experiment to another. Some authors perform verification tests, and others, identification tests. The tests are not carried out in similar conditions: some are laboratory tests, others field-tests or even industrial or commercial tests. These results also show very large differences in the size of training and test subsets and users sets, i.e. from a few signers to several thousand. In some experiments, the reference consists of only one reference signature, while in others from two to six signature specimens are kept.

Moreover, recognition trials vary from one to three per test. Since results could depend on conditions like writer position, user motivation, and the fact that pertinent information needed to analyze data base acquisition and test procedures is not always available. Some experiments do not involve the use of any imitations,and the types of imitations, when they are used, are not the same (i.e. simple forgeries, skilled forgeries, etc.). Another difficulty arose with forgeries: finding skilled forgers is not an easy task. They should imitate not only the final static drawing of a signature but also its dynamic characteristics, Bonnefoy et al. (1981), which is very much harder to do.

7. Conclusion

One typical result has been pointed out in several studies, Chorley et al.(1975); Crane and Ostrem (1983); Collantier (1984); Worthington et al. (1985); Achemlal et al. (1986). This relies upon the fact that performance degradation in dynamic signature verification systems is only due to a few users with very bad signature consistency. Some proposals have been made to deal with this problem, i.e. users whose reference signatures are too variable are not enrolled in the system or the threshold is largely relaxed for them. However, this latter suggestion has the disadvantage of increasing system vulnerability, Chorley et al. (1975). Some coefficients have been proposed to evaluate forgery vulnerability, Chorley et al.(1985) (improvement with training and admission rate). More recently, a new approach has been proposed by Brault and Plamondon (1989a), to identify the problematic signer before enrolment and also to facilitate performance results comparison by data base characterization. These authors use a coefficient to estimate the difficulty experienced by a forger in reproducing a signature visually and dynamically.

The rule in the domain of dynamic signature systems so far is the lack of normalization and, obviously, this constitutes one of their greatest disadvantages. However, compared with other identity verification systems, Fejfar and Myers (1977); Shaw (1980); Davis and Price (1984); Informatique et Sécurité (1986), dynamic signature verification systems show better reliability and performance.

Acknowledgements

This work was supported by grant nos. A-0915 and CRP2097 to R. Plamondon from NSERC and FCAR, federal and provincial programs, and by grants from the French National Education and Research Ministry to G. Lorette. Part of this work was done when Professor Lorette was a visiting scientist at Laboratoire Scribens, Ecole Polytechnique de Montréal and when Professor Plamondon was a visiting scientist at the Institut de Recherche en Informatique et Systèmes Aléatoires (IRISA), at the Laboratoire de Génie Électrique de Créteil (LGEC) and a Fellow of the Netherlands Institute for Advanced Study. The authors wish to thank Mrs. G. Monté, Mr. R. Baron, Mr. R. Richard and Mr. R. Sabourin for their assistance with Figures 1 and 2. Special thanks go to Mrs. M. Chaussy, Mrs. A. M. Clement and to Mrs. H. Dallaire for typing the manuscripts and to Mrs. C. Hunt for her text revision.

References

A complete bibliography appears at the end of this section.

Computer Processing of Handwriting
Eds. R. Plamondon & C. G. Leedham

AUTOMATIC PROCESSING OF SIGNATURE IMAGES: STATIC TECHNIQUES AND METHODS

RÉJEAN PLAMONDON
Laboratoire Scribens, Département de Génie Electrique, Ecole Polytechnique de Montréal, C.P. 6079, Succ. "A" Montréal QC, CANADA H3C 3A7

GUY LORETTE
IRISA, Université de Rennes I, Campus de Beaulieu, 35042 Rennes Cédex, FRANCE

ROBERT SABOURIN
Laboratoire Scribens, Département de Génie de la Production Automatisée, Ecole de Technologie Supérieure, 4750, Henri-Julien, Montréal QC, CANADA H2T 2C8

This paper surveys the various techniques and methods used for the automatic verification of signature images. The problems discussed are grouped under three main headings. Preprocessing problems deal with the preparation of the relevant information, that is, extracting the signature from the background. The feature extraction methods are presented under two complementary subheadings: global and local approaches. The comparison processes, based on different similarity and distance measures are then described, and the performance of the systems is reported. Particular emphasis is given, at the end, to the description of an image understanding system dedicated to signature verification.

1. Introduction

Problems of off-line signature verification are generally considered to be more difficult than on-line ones. A signature image can be easily copied optically or mechanically. Moreover, timing and dynamic information is highly degraded in a static specimen. Part of this information can be recovered by expert document analysts, using special skills and techniques, Osborn (1929); Locard (1936); Harrison (1981); Hilton (1982); Hagan (1984), but most of their methods cannot be implemented easily in a computer environment. A few papers have been reported in the literature to cope with this problem. On the one hand, Sakai et al. (1974) have proposed an interactive hybrid system for collecting and displaying the static and dynamic features of a test signature and a reference on a CRT screen, using a color code. The system was used to help clerks examine specific features of a signature. The final decision about the specimen's authenticity was left to him. On the other hand, Ammar et al. (1985, 1986) have tried to indirectly extract and use pressure information from a signature images. Similarly, Sabourin and Plamondon (1990a) have recently proposed an image understanding system where pseudo-dynamic features are used. Except for these recent papers, most of the work done in this area so far has been concerned mainly with static feature analysis only.

As for the dynamic approach, the design of an off-line signature verification system requires the solving of five types of problems: data acquisition, preprocessing, feature extraction, comparison process and performance evaluation.

In the first part of this paper, we will review the various data acquisition and preprocessing techniques that have been used to extract and filter the signature image. Then, we will present several feature extraction methods, followed by a comparative performance analysis in the next section. To do this, we will refer to a review of work published on off-line signature verification, as summarized in tabular form in Table 1. Before concluding, we review in more detail the description of an image understanding system as applied to signature verification, to highlight the advantages of this new approach.

2. Data Acquisition

Different types of acquisition devices are used for off-line signature verification systems, Lorette and Plamondon (1989). Obviously, the most commonly used input system is the vidicon tube T.V.camera. The size of the image is normally 512 x 512 pixels wide, and 64 to 256 grey-levels of quantization are currently obtained.

Figure 1 shows the data available from specimens of Mr. Peace's signature, as obtained with a camera. Figure 1a shows the original sample. Figure 1(b) is a plot of the corresponding luminance histogram. Since the specimen was written on a white sheet of paper, most of the signature information is masked by the large high-luminance peak. With the proper enhancement technique, the signature luminance signal can be extracted, as shown in Figure 1(c). Figure 1(d) represents the binary image obtained from the previous graph after proper thresholding. Finally a skeleton of the specimen, as obtained after a typical thinning preprocessing is plotted in Figure 1(e).

Cameras with C.C.D matrix array sensors perform better in terms of stability, sensitivity, and lack of geometric distortions. Electro-optical scanners with linear C.C.D sensors are also used for document digitization. Their standard resolution is in the 100 to 800 dots per inch range (D.P.I.) and they are able to quantize up to 256 grey-levels.

An optical sensor consisting of a linear array of 64 photodiodes has been used in a commercial product for check verification, but such a resolution seems to be too low to detect several aspects of a signature which can be of major significance in forgery detection, Totty and Hardcastle (1986). Another possibility would be to use a graphic-tablet digitizer and to acquire X,Y coordinate pairs without considering time information, i.e. that the points will be ordered from the top-down and from left to right, as with a raster scanning. Thus an off-line data entry system simulation would be done, Marcozzo (1964); Lorette (1983); de Bruyne (1985).

It is noticeable, however, that in the off-line signature verification approach, all of the time sequence and most of the dynamic information are lost. This leads inevitably to a higher degree of complexity in the off-line, i.e. static, signature verification process.

3. Preprocessing

This step deals with the preparation of the relevant information: non-uniformity correction for sensor elements, localization of the signature in the picture, extraction of the signature from the background, slicing, thresholding and filtering problems, segmentation and data reduction.

Since most of the applications covered by these studies are related to the automatic validation and processing of specific paper forms (like checks), the finding of a signature is not a difficult problem and can be solved with the help of a window operator, Nagel and Rosenfeld (1973). However, as suggested by Nagel (1976), this problem might become more serious if the form standard permits a great deal of variations. If no artifice is used (like

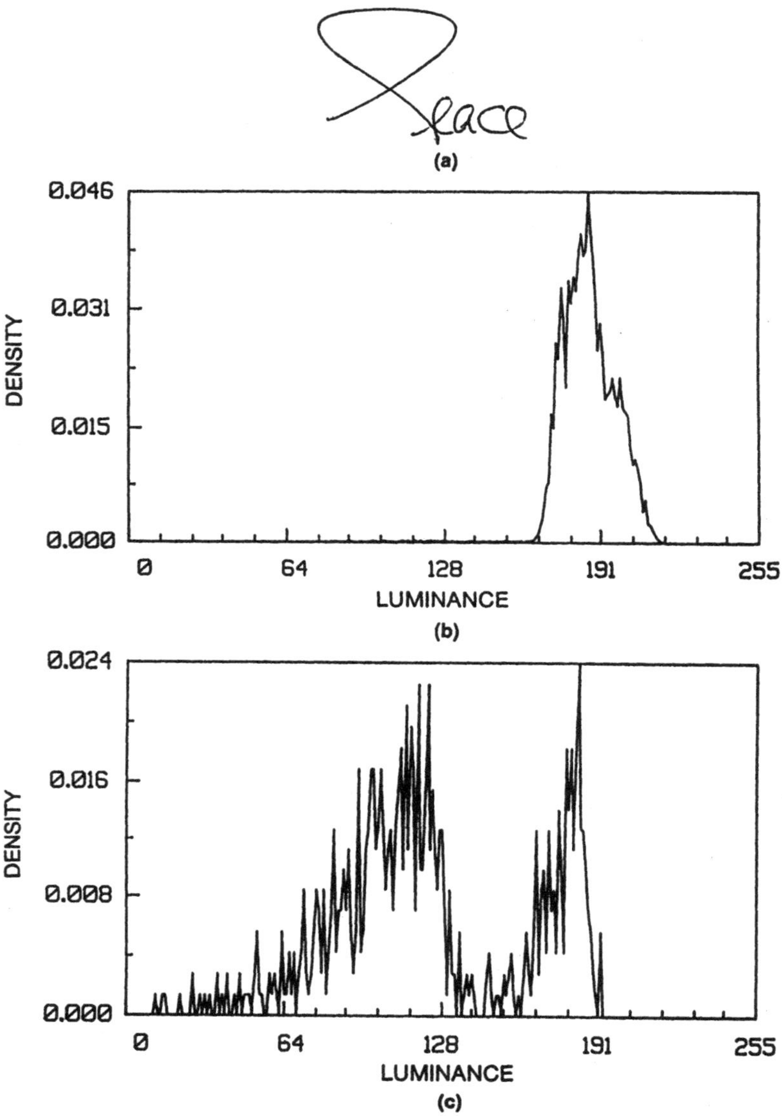

Figure 1. Signature signal as extracted with a camera.

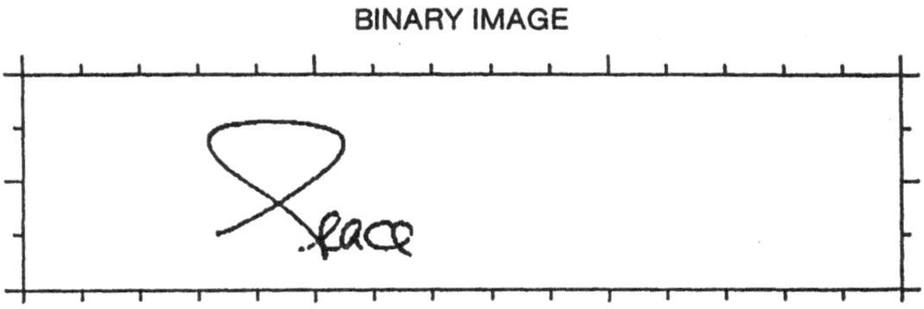

(d)

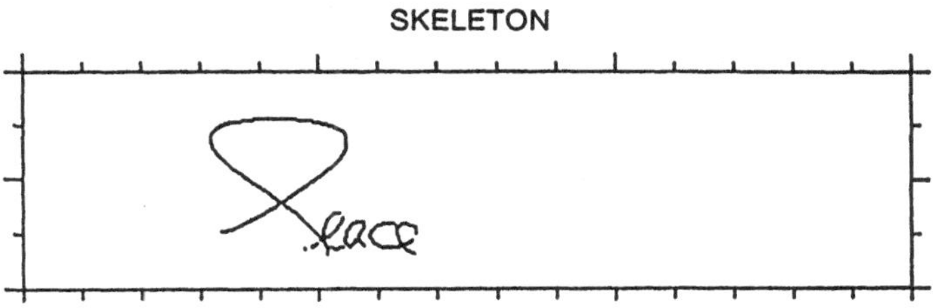

(e)

Figure 1. Signature signal as extracted with a camera. (continued)

signing with special ink), the problem of extracting a signature image from its background is not a trivial one. Thresholding and slicing techniques have been proposed, Nagel (1973); Ammar et al. (1985); Sabourin and Plamondon (1986), but these are generally insufficient in a practical noisy background for which restoration techniques are also necessary, Ammar et al. (1985).

Indeed, the efficiency of slicing or thresholding techniques is directly related to the selection of the cut point from a grey-level histogram of the picture. The problem is then reduced to finding a point on the frequency distribution which separates the peaks of a bi-modal curve where the peak representing the signature is almost lost in noise. The use of the Laplacian or the Sobel operator has been proposed as a border detector. In the first case, the threshold is fixed to select pixels above the 85% value in the cumulative Laplacian distribution, Nagel and Rosenfeld (1973), (1977). In the second, the modulus gradient vector of the grey-level signature is used, with the threshold fixed at the point where the modulus variations in the pixel distribution function become less than 1%, Sabourin and Plamondon (1986). This latter approach offers the advantage of being independent of the thickness of the strokes and of the overall length of the signature.

However, for practical cases where specimens are noisy, these techniques have to be improved. A four-step preprocessing operation has recently been proposed by Ammar et al. (1985) (background equalization and reduction, noise reduction by averaging, automatic thresholding and image extraction). This approach has been found to be successful in removing the overlapping in a signature as well as the signing line, at the cost of eliminating some parts of the signature in about 2% of samples.

Depending on the features needed at the next stage, the preprocessing algorithms can perform more specific actions on the extracted image. Prescreening operators have been used to eliminate points that are judged to contain little information about writer identity, Nemcek and Lin (1974); Phelps (1982). Window operators have been used by Brocklehurst (1984, 1985) to improve the image. A tracking algorithm has been applied by Nagel and Rosenfeld (1973) to identify and label all connected components in the picture. Based on the hypothesis that high pressure regions are reflected by higher grey-levels in an image, a high pressure threshold selection algorithm has been proposed by Ammar et al. (1985) to extract the high pressure regions of a signature. This mechanism is based on the evaluation of a pressure factor, i.e. the ratio of the high pressure area to the signature region. Recently, Sabourin and Plamondon (1988, 1989) have proposed a version of centroidal linkage region-growing-with-merging algorithm to perform signature segmentation. Using the statistics of directional data, textured regions characterized with local uniformity in the orientation of the gradient are extracted.

4. Feature Selection

The division of feature extraction methods into text sensitive and text insensitive features, as has been proposed for text document analysis, Klement et al. (1980), cannot be applied strictly to signatures since the number of characters or symbols within a specimen is limited and often irrelevant (in an alphabetical context). Similarly, since the time information is not available from the image data, the function approach to feature extraction is not really an appropriate distinction for static techniques. Attempts at feature selection are better categorized by their global or local approaches to the problem, although an efficient system would probably incorporate global and local feature extraction, (see, for example, the first column of Table 1). Moreover, none of the feature sets selected so far keeps enough information to permit a reconstruction of the handwritten specimens.

In the global approach, the signature image is taken and processed as a whole, or features are deduced from the global aspect of a signature, or some local parameter values are averaged over the signature. Nemcek and Lin (1974) have used the Hadamaar transform on binary images as a means of data reduction and feature selection, the resulting spectrum constituting the feature set. Sobel's operator has been applied by Sabourin and Plamondon (1986) to grey-level signature images, and the angular behavior of the resulting intensity gradient (assumed to reflect the stroke orientation) has been used as a feature set. A close-fitting polygon has also been computed to evaluate the overall shape of a signature and to serve as the basic features, once normalized, Phelps (1982). A feasibility study on the use of some graphometric techniques, particularly the rythmic line of the signature, has been performed, Sabourin and Plamondon (1987). Nouboud (1989) has also used similar characteristic curves to describe the signature envelope.

Chuang (1977) has divided the signature into upper, middle and lower zones, initial and end strokes (a process familiar to grapho-analysts) and performed sequential analysis of each sub-image to extract features related to each resulting zone and their proportionalities. A single study has been reported so far to extract some global dynamic features from a static image. Indeed, Ammar et al. (1985, 1987) have worked with a set of features mainly related to pressure measurements, as extracted from grey-level images: the vertical position of the baseline of the high-pressure and of the binary image, a pressure factor and threshold, the highest grey-level in the extracted signature image, the dynamic range on the grey scale for a signature, and its area measured as number of pixels. In a recent paper, these authors, Ammar et al. (1988), described signature features and the relations among them using a hierarchical structure.

In the second approach, a set of local attributes is computed. A lot of these features are known to be stable, according to document examiner expertise, Mally (1955). However, most of the studies dealing with local features also

AUTHORS	INPUT IMAGE RESOLUTION FEATURE DESCRIPTION	TRAINING AND/OR TEST DATA BASE (specimensxwriters)	COMPARISON METHOD	ERROR RATES	COMMENTS
AMMAR, YOSHIDA, FUKUMURA (1986)	256 x 1024 grid 256 grey levels 7 global pressure features	200 genuines (20 S x 10 W) 200 forgeries (20S x 10 imitators)	weighted distance	Type I = 6% Type II= 4%	freehand forgeries leave-one-out method with threshold for maximizing performance
BROCKLEHURST (1985)	60 pixels/cm binary 7 local characteristics	2820 genuines (60 S x 47 W) ("equal number of unseen forgeries")		acceptance training threshold at 95% true acceptance Type II: 5%	6 S/W as reference
CHUANG (1977)	100 x 300 grid binary global features	2400 genuines (6 S x 400 W) 1600 forgeries (4 S/user)	weighted distance	training threshold fixed at 95 % genuine acceptance test: Type I, II= 20	3 S/W as reference freehand forgeries
NAGEL, ROSENFELD (1977)	500 pixels/inch 64 grey levels 2 types of local and global features	11 genuines (6S x 1W + 5S x 1W) 14 forgeries (9S x 1W + 5S x 1W)	weighted distance	Type I: 8% Type II: 0% Type I: 12% Type II: 0%	freehand forgeries average result of single and double jacknife method
NEMCEK, LIN (1974)	128 x 256 grid binary 14 features from Hadamar spectra	600 genuines (40 S x 15 W) 120 forgeries (10 user imitated 3 S/user by 4 imitators	maximum likelihood classifier	training threshold at 96% genuine acceptance Type I : 11% Type II: 41%	30 S/W training, 10 S/W test freehand forgeries
NOUBOUD (1989)	384 x 510 grid 64 grey levels signature envelope	1000 genuines (10 S x 100 W) 6 S/W for reference	dynamic programming	Type I : 5% Type II: 8%	with personal thresholds
PHELPS (1982)	resolution not specified close-fitting polygon	3 genuines (3 S x 1 W) 6 forgeries (3 S x 2 W)	similarity of overlapping area		incomplete results

Table 1. Performance of static signature verification systems.

AUTHORS	INPUT IMAGE RESOLUTION FEATURE DESCRIPTION	TRAINING AND/OR TEST DATA BASE (specimensxwriters)	COMPARISON METHOD	ERROR RATES	COMMENTS
SABOURIN PLAMONDON (1986)	128 x 512 grid 256 grey levels angular histogram image gradient	760 genuines (40 S x 19 W)	maximum likelihood classifier and linear discriminant	Type I: 1.5% Type II: 7%	random forgeries 15 S/W for training 25 S/W for test
SABOURIN PLAMONDON (1987)		800 genuines (40 S x 20 W)	nearest neighbor classifier	Type I: 0.57% Type II: 0.03%	5 ref./signer
	rythmic line	63 genuines from 1W, 117 simulated forgeries from 6 amateur forgers	nearest neighbor classifier	Type I: 3.5% Type II: 0%	features extracted interactively at a graphic terminal 6 ref./signer
		17 skilled forgeries from 1 professional forger		Type I: 1.8% Type II: 47.1%	6 ref./signer
SABOURIN PLAMONDON (1990a) (1990b)	128 x 512 grid 256 grey levels set of primitives	224 genuines, pen #1 (28S x 8W) 160 tracing forgeries (10S/user, pen #1) (10S/user, pen#2) 224 photocopies (28S/user)	structural matching static	ϵ_1 = 0.0% ϵ_2 = 1.34% ϵ_2 = 100% ϵ_2 = 98.9% ϵ_2 = 96%	3S/W as reference random forgeries α = 1.0 β = 0.0 comparison threshold selected from the random, forgeries experiment, tracing forgeries from 1 writer
		224 genuines, pen #1 (28S x 8W) 160 tracing (10S/user, pen #1) (10S/user, pen #2) 224 photocopies (28S/user)	structural matching static + pseudo-dynamic	ϵ_1 = 8.93% ϵ_2 = 1.79% ϵ_2 = 6.25% ϵ_2 = 78.8% ϵ_2 = 61.2%	3S/W as reference random forgeries α = 1.0 β = 5.0 comparison threshold selected from the random, forgeries experiment, tracing forgeries from 1 writer

Table 1. Performance of static signature verification systems. (continued)

incorporate some global ones. For example, Brocklehurst (1984, 1985) has proposed the use of the overall length of the signature (global) with a set of local features likes slope measurements, the distance from the left-hand end of the designated space provided for a signature to the point at which the signature begins, concavity measures, etc. Nagel and Rosenfeld (1977) and Rosenfeld (1976) have also worked with characteristics extracted from signatures segmented into vertical and horizontal zones. Assuming that a signature spelling was known, these authors detected tall letters in a specimen. Two global features (the ratio of signature width to short stroke height and to long stroke height) and two types of local features (ratios of the height of a long stroke to the height of the immediately preceding short strokes and slope features of appropriate long letters) were retained for their studies, the number of these local features being dependent on the number of tall letters in a signature.

5. Comparison Processes and Performance Evaluation

In the off-line system, the comparison process is based on the assumption that features sets extracted from genuine signatures are more stable than those of forgeries, that is intrapersonal variations are smaller than interpersonal variations. So, an unknown signature can be classified according to its similarity to the reference set first specified by an indirect identifier.

To evaluate this similarity, three types of approaches have been used. Considering signature verification as a non-standard two-class problem (since the class of forgeries cannot be specified), one group of researchers, Chuang (1977); Nagel and Rosenfeld (1977); Ammar et al. (1985) has used weighted distances. These methods compute a distance measure between the reference set and the test set. The decision is taken according to the value of this distance criterion with respect to a decision threshold, this threshold being pre-established with the help of a training set of genuine signatures. Assuming some a priori probability, Nemcek and Lin (1974), or limiting their studies to random forgeries only, Sabourin and Plamondon (1986), some authors have used maximum likelihood classifiers and linear discriminant functions. This second approach also assumes the statistical interdependence of gaussian distributed features. Nouboud et al. (1988) have used a third approach: dynamic programming.

Table 1 summarizes the results obtained so far by the research groups involved in off-line signature verification. Since the performance evaluation method differs greatly from one approach to the other, these experimental results are again difficult to compare. The table shows, for each author, the type of input data used in the study, a description of the data base used for training or for the test experiments or both, the number and the type of features, with the comparison method, and the error rates as reported according to type I and II definitions. In the last column, comments are reported that might be helpful in analysing the results.

Generally speaking, it is observed that type I and type II errors of the order of a few percent can actually be obtained, for systems working with simple and random forgeries.

Some authors have commented on their system failures, Nemcek and Lin (1974); Brocklehurst (1985). Two major causes are reported: 1) scanning problems or signal quantization errors, resulting in uncompleted specimen acquisition, due to the thinness of the trace or the ink color used, and 2) excessive natural variations in the signatures of a few specific users.

Only one system has been tested with skilled forgeries, and these specimens were not reliably detected, Brocklehurst (1985). According to evaluation by expertdocument analysts, skilled forgery detection would not be efficient with these systems, as long as dynamic (or pseudo-dynamic) and static features are exhaustively extracted from a written specimen, Harrison (1981); Totty (1986); Totty and Hardcastle (1986). This is also corroborated by observations from human perception experiments to the effect that knowledge of the drawing methods can greatly influence the recognizability of distorted characters, Freyd (1983).

Incorporating some knowledge from the expert document analysts sounds an interesting way of arriving at a significant improvement in the performance of automatic handwritten signature verification systems, Plamondon and Lorette (1989a). Locard (1959) has proposed many characteristics belonging to genuine handwritten signatures which are imperceptible to the average forger and very difficult to imitate. These characteristics are related to local variations in aspect ratio, orientation, relative position, shading, etc. taken between pairs of characters. This scheme seems to be more powerful than taking the letter shapes individually, since these characteristics are very easily perceptible to the average forger and therefore easier to imitate, Hilton (1982); Harrison (1981).

6. Image-Understanding System

An image-understanding system based on the extraction of a new signature representation from grey-level images has been proposed recently for automatic signature verification, Sabourin and Plamondon (1989), (1990a), (1990b). By an image-understanding system is meant the separation of evaluation functions for the analysis of handwritten signature into local and global functions, Fischler and Elschlager (1973). Two classes of features, static and pseudo-dynamic, are taken into account here for the representation of the signature. To the former belong the geometric shape and the spatial relations (North, South, East, West, North-East, North-West, South-East and South-West) between some primitives extracted from the signature line. The latter is associated with the grey-level variation and the texture inside the primitive, Sabourin and Plamondon (1990a). Considering simultaneously these two classes of features, and the separation from the local interpretation of the primitives followed by the global interpretation of the signature, an image-understanding system for the interpretation of handwritten signature images has been designed.

In this system, the spatial sampling of handwritten signatures produces an image format of 128 x 512 pixels. The preprocessing stage is responsible for producing a signature image where each pixel from the grey-level image has a label signifying that a specific pixel is a background or a signal picture element. The preprocessing phase is subdivided into two stages, the gradient computation and the background elimination processes. The gradient is evaluated over the entire grey-level image with the Sobel operator. The resulting filtered image is therefore analyzed by the background elimination process. Because the handwritten signature is characterized by a high gradient activity, the density function is computed and a threshold is automatically settled. Finally, those pixels where the gradient activity is high are associated with a signal-pixel, or otherwise, with the background.

The signature image is then analyzed by the primitive extraction process responsible for the production of the primitive sets necessary for the structural analysis of the handwritten signature by the subsequent comparison process. The strategy adopted for extracting primitives takes into account the collinearity of neighboring signal-pixels in the directional plane of the gradient space. This task is realized in two steps.

First, the signal-pixels are merged into atomic regions characterized by the uniformity of a local property: the orientation of their gradient vectors. The resulting atomic regions either belong to the signature line or represent spurious noise elsewhere in the scene. The second step of the primitive extraction process is responsible for generating the primitive set. The high-level merging process starts with the elimination of sparse atomic regions. Afterwards, it continues with a hierarchical merging strategy in growing collinear atomic regions. The growth is limited by the resulting circular variance of the directional data. The latter sub-process is used repeatedly in varying a growing constraint, considering a fixed directional constraint related to the local organization in the directional data located at the common border of neighboring atomic regions. This scheme acts as a zoom procedure focusing the attention of the high-level merging process on the biggest homogeneous regions from the atomic region set. The resulting primitive set is therefore characterized by a collection of arbitrarily-shaped primitives. The degree of merging in the process depends on whether the system is used to enrol a new signer or if it is used to test the signature of a previously enrolled signer. In the former case, the resulting set is a reference primitive set where the growth is limited only by the presence of directional discontinuities along the handwritten signature line. In the second case, the process limits the growth of the test primitive depending on a circular constraint. This second approach generally results in the segmentation of the test signature into smaller and more numerous primitives.

The comparison process is responsible for the structural match between a test primitive set and a reference primitive set resulting in a global similarity measure. The first stage of the comparison process, the local interpretation of primitives, is related to the final merging of test primitives. This sophisticated template-matching mechanism, Sabourin and Plamondon (1990a), (1990b), produces a local similarity measure for all synthesized test primitives located around each test primitive, attempting to label them as reference primitives. The similarity measure takes into account the similarity in shape, position, orientation and aspect-ratio between the merged test primitives and the reference primitives.

The second stage of the comparison process, the global interpretation of the scene, completes the analysis with the evaluation of a solution path produced by the structural matching of the test primitive set to the reference set and followed by the validation of the resulting solution with the comparison of some corresponding pseudo-dynamic features taken along the solution path.

The first stage of the global interpretation process proceeds with the structural match of a reference primitive set and a test primitive set, taking into account the local interpretations of test primitives labelled as reference primitives, and the relational similarity evaluated between pairs of primitives. This scheme results in a static similarity measure and a static solution path.

A pseudo-dynamic similarity measure is then computed using this static solution path. A pseudo-dynamic feature is evaluated from the local analysis of the grey-level texture inside pairs of primitives located on the solution path. The pseudo-dynamic feature adopted here is the proportion of "black labelled pixels" obtained from a discriminant analysis of grey-level histograms computed for each primitive. This feature is called "pseudo-dynamic" because it reflects the effect of the writing process dynamic which produces grey-level variations along the signature line. Results obtained with this system are shown in the last row of Table 1. This is the first study where results dealing with photocopies and tracing forgeries are reported.

7. Conclusion

Some of the prototypes developed to date for signature verification are promising. One can see, from the latter results, for example, the appeal of working with grey-level images, of using a new signature representation that permits the interpretation of the local texture along the signature line in terms of static and pseudo-dynamic features and, finally, of separating the local interpretation of primitives and the global interpretation of the signature.

However, future research efforts would be required in order to use them in practical environment and on a daily basis. Several problems still have to be solved. So far, the most promising systems seem to be those that extract some pressure factor, Ammar et al. (1986) or some pseudo-dynamic characteristics, Sabourin and Plamondon (1990b) from grey-level signature images. These interpretations of grey-level changes constitute an original working hypothesis that needs to be systematically tested. To do so, basic studies should be planned to correlate static and dynamic characteristics and eventually to link them in a signature generation model.

Similarly, more comparative studies are needed to evaluate the relative performance of different algorithms under similar experimental conditions, using a single data base where various types of forgeries are included. Faster algorithms and optimized hardware are also necessary to make these systems really efficient in their particular environments. Moreover, further studies are necessary to adapt these systems to a wide variety of signed documents, and types of pens, ink and paper.

Acknowledgements

This work was supported by grant OGP0000915 to R. Plamondon, from NSERC, by grants from the French National Education and Research Ministry to G. Lorette and by FIR grant to Robert Sabourin from the Ecole de Technologie Supérieure. Parts of this work were done when Professor Plamondon was a visiting scientist at IRISA and a Fellow of the Netherlands Institute for Advanced Study and when Professor Lorette was a visiting scientist at Laboratoire Scribens, Ecole Polytechnique de Montréal.

References

A complete bibliography appears at the end of this section.

Computer Processing of Handwriting
Eds. R. Plamondon & C. G. Leedham

IDENTITY VERIFICATION FROM AUTOMATIC PROCESSING OF SIGNATURES: BIBLIOGRAPHY

RÉJEAN PLAMONDON
Laboratoire Scribens, Département de Génie Electrique, Ecole Polytechnique de Montréal, C.P. 6079, Succ. "A" Montréal QC, CANADA H3C 3A7

GUY LORETTE
IRISA, Université de Rennes I, Campus de Beaulieu, 35042 Rennes Cédex, FRANCE

Abend, W., Bizzi, E., and Morasso, P., "Human arm trajectory formation", *Brain* **105**, 1982, pp. 331-348.

Achemlal, M., Mourier, M., and Lorette, G., "Dynamic signature verification", *Proc. 3rd Int. Symp. on Handwriting and Comput. Applic.*, Montreal 1987, pp. 170-172.

Achemlal, M., Mourier, M., Lorette, G., and Bonnefoy, J.P., "Dynamic signature verification", *Proc. 4th Int. Conf. Exhib. on Computer Security*, Monte-Carlo 1986, pp. 433-441.

Ammar, M., Yoshida, Y., and Fukumura, T., "Description of signature images and its application to their classification", *Proc. 9th Int. Conf. on Patt. Recognition*, Rome 1988, pp. 23-26.

Ammar, M., Yoshida, Y., and Fukumura, T., "Feature extraction and selection for simulated signature verification", *Proc. 3rd Int. Symp. on Handwriting and Comput. Applic.*, Montreal 1987, pp. 167-169.

Ammar, M., Yoshida, Y., and Fukumura, T., "A new effective approach for off-line verification of signatures by using pressure features", *Proc. 8th Int. Conf. on Pattern Recognition*, Paris 1986, pp. 566-569.

Ammar, M., Yoshida, Y., and Fukumura, T., "Automatic off-line verification of signatures based on pressure features", *Proc. tech. group meeting of Patt. Rec. Learn of IECEJ*, Tokyo 1985.

Angell, J.B., Teny, S.C., and Bath, P.W., "Silicon micromechanical devices", *Scientific American*, **248**, no 4, 1983, pp. 44-55.

Anthony, N.J., Herbst, N.M., and Liu, C.N., "Supervised adaptation for signature verification system", *IBM Tech., Disclosure Bull.* (USA), **21**, no 1, 1978, pp. 424-425.

Baron, R., and Plamondon, R., "Acceleration measurement with an instrumented pen for signature verification and handwriting analysis", *IEEE Trans. on Instrumentation and Measurements*, Vol 38, No. 6, 1989, pp. 1132-1138.

Baron, R., and Plamondon, R., "Problématique et solution reliées au conditionnement du signal d'un crayon accélérométrique pour la vérification des signatures", *3e Conf. Can. Informatique Industrielle*, Montréal 1986, pp. 66.1-66.5.

Baron, R., "*Design d'un système expérimental de vérification de signature*", Mémoirede M.Sc.A., Laboratoire SCRIBENS, Ecole Polytechnique de Montréal, 1986.

Bean, C.H., and Prell, J.R., "Personnel access control - criteria and testing", *Final Proceedings of the Second International Conference on Crime Countermeasures -Science and Engineering*, 3A, Oxford 1977a, pp. 03-08.

Bean, C.H., and Prell, J.R., "Personnel access control", *Proc. 1977 Carnahan Conference on Crime Countermeasures*, University of Kentucky ORES, Lexington 1977b, pp. 7-17.

Beatson, R., "Signature dynamics in personal identification", *Proc. 4th World Cong. Computer and Communications Security and Protection*, Paris 1985, pp. 179-196.

Bechet, L., "Reconnaissance dynamique de la signature", *6th Convention of the European Financial Marketing Association*, Montreux 1984, pp. 93-97.

Benson, P., "Test results - advanced development models of BISS identity verification equipment", *Automatic Fingerprint Verification, MTR-3442*, IV, The MITRE Corporation, Bedford 1977.

Bills, G.A., Varady, M.J., and Zimmerman, K.P., "Spectral analysis of script and signatures", *Proc. 3rd Int. Symp. on Handwriting and Comput. Applic.*, Montreal 1987, pp. 157-159.

Bills, G.A., "*Spectral analysis of script*", M.S. Thesis, University of Missouri, Columbia 1986.

Blesser, B.A., Ward, J.R., "Design criteria for an electronic tablet technology with acceptable performance for handwriting capture and analysis", *Proc. 3rd Int. Symp.on Handwriting and Comput. Applic.*, Montreal 1987, pp. 178-180.

Blesser, B.A., Kuklinski, T.T, and Shillman, R.J., "Empirical tests for feature selection based on a psychological theory of character recognition", *Pattern Rec.* **8**, 1976, pp. 77-85.

Bonnefoy J.P., "*Stratégie de recherche de formes caractéristiques dans une image: étude et mise en oeuvre d'un système d'apprentissage et de reconnaissance automatique hiérarchisé application à la reconnaissance de formes de signatures manuscrites*", Thèsede doctorat d'état, Université de Paris XII, 1986, 384 p.

Bonnefoy, J.P., and Lorette, G., "Reconnaissance automatique d'images et designatures manuscrites: un nouveau critère de détermination de variables discriminantes", *Premier Colloque Image*, Biarritz 1984, pp. 421-426.

Bonnefoy, J.P., Jounet, P., Lorette, G., and Gaudaire, M., "Reconnaissance automatique en temps réel de signatures manuscrites: définition et mise en oeuvre d'un critère général", *3e Congrès Reconnaissance des Formes et Intelligence Artificielle*, Nancy 1981, pp. 267-275.

Botnick, D., "Automated signature to bow in '84 as current systems sharpen technology", *Bank Systems and Equipment* **20**, no. 8, 1983, pp. 95-104.

Brault, J.J., and Plamondon, "How to detect problematic signers for automatic signature verification", *Proc. Int. Carnahan Conf. on Security Technology*, Zurich 1989a, pp. 127-132.

Brault, J.J., and Plamondon, R., "Segmentation des signatures manuscrites", *Proc.Vision Interface*, London 1989b. pp. 110-116.

Brault, J.J., "*Proposition et vérification d'un coefficient de difficulté d'imitation dessignatures manuscrites*", Ph.D. Thesis, Ecole Polytechnique de Montréal, 1988.

Brault, J.J., and Plamondon, R., "Global and local time warping function for handwritten curves comparison", *Proc. 3rd Int. Symp. on Handwriting and Comput. Applic.*, Montreal 1987a, pp. 56-58.

Brault, J.J., Plamondon, R., "Handwritten curve partitioning based on geometric and sequential information", *Proc. of 3rd Int. Symp. on Handwriting and Computer Applications*, Montreal 1987b, pp. 50-52.

Brault, J.J., and Plamondon, R., "Coupling visual and accelerometric features to study handwritten signatures", *Proc. Vision Interface '86*, Vancouver 1986, pp. 375-379.

Brault, J.J., and Plamondon, R., "Histogram classifier for characterization of handwritten signature dynamics", *Proc. 7th Intern. Conf. Pattern Recog.* 1, Montreal 1984a, pp. 619-622.

Brault, J.J., and Plamondon, R., "Utilisation de la dynamique de la signature manuscrite dans les systèmes de sécurité, algorithmes préliminaires", *Proc. Can. Conf. on Indust. Comp. Syst.*, Ottawa 1984b, pp. 291-296.

Brault, J.J., Plamondon, R., and Robillard, P.N., "Handwriting modelling and simulation for the development of an instrumented pen", *Proc. of Int. Symp. on Modeling and Simulation*, Bermuda 1983, pp. 3-7.

Brault, J.J., "*Design d'un crayon accélérométrique pour la vérification de signatures*", Mémoire de maîtrise, Ecole Polytechnique de Montréal, Département de Génie Electrique, 1983.

Brenner, A.E., and De Bruyne, P., "A sonic pen: a digital stylus system", *IEEE Trans. on Comp.*, 1970, pp. 546-548.

Brocklehurst, E.R., "Computer methods of signature verification", *J. of Forensic Sciences*, 1985, pp. 445-457.

Brocklehurst, E.R., "Computer methods of signature verification", *Nat. Phys. Lab., Report* no: NPL-DITC-41/84, Teddington 1984, 12 p.

Chainer, T.J., Gundersen, S.C., and Worthington, T.K., "Improved algorithm for the correction of the segmentation of signatures in a signature verification system", *IBM Tech. Discl. Bull.* **27**, no. 8, 1985, pp. 4655-4656.

Chorley, B.J., Olding, S.E., Parks, J.R., Pobgee, P.J., and Watson, R.S., "*Verisign: a system for automatic verification of signatures*", Technical report, 1975, 20 p.

Chuang, P.C., "Machine verification of handwritten signature image", in *Proceedings 1977 International Conference on Crime Countermeasures - Science and Engineering*, J.S. Jackson and R.W. De Vore, (Eds.), University of Kentucky, Lexington 1977, pp. 105-109.

Collantier, P., Groupement Carte Bleue, "*L'expérience de reconnaissance dynamique de signature des établissements Carte Bleue*", Rapport final, 1984, 42 p.

Conway, J.V.P., "*Evidential documents*", C.C. Thomas, Springfield, Illinois, 2nd printing, 80 fig., 1959, 267 p.

Crane, H.D., and Ostrem, J.S., "Automatic signature verification using a three-axis-force-sensitive pen", *IEEE Trans. on Systems Man and Cybernetics*, SMC-13, no3, 1983, pp. 329-337.

Crane, H.D., Wolf, D.E., and Ostrem, J.S., "The SRI pen system for automatic signature verification", *Symp. Proc. NBS Trends and Appl.* **32**, Gaithersburg 1977.

Davis, D.W., and Price, W.L., "*Identity verification in security for computer networks- an introduction to data security in teleprocessing and electronic funds transfer*", John Wiley and Son, ch. 7, 1984, pp. 179-218.

De Bruyne, P., "Dynamic signature verification using your own pen", *Proc. Carnahan Conf. on Crime Countermeasures*, 1988a.

De Bruyne, P., and Meyer U.P., "A signature verification digitizer using frustrated internal reflexion", *Proc. Carnahan Conf. on Crime Countermeasures* , 1988b.

De Bruyne, P., "Signature verification using holistic measures", *Computers &Security* **4**, 1985, pp. 309-315.

De Bruyne, P., and Messmer, P., "Authentication of handwritten signatures with sonar", *Proceedings of the Third International Conference: Security through Science and Engineering*, 169-72, Jackson J.S., Fellbaum, K., (Eds.), Publ. Univ. Kentucky, Lexington 1980, 313 p.

De Bruyne, P., "Developments in signature verification", *Security Management* **22**, no.6, 1978, pp. 57-61.

De Bruyne, P., "A new method of signature verification", *Proc. 1977 Int. Conf. on Crime Countermeasures - Science and Engineering*, J.S. Jackson and R.W. De Vore, (Eds.), University of Kentucky, Lexington 1977, pp. 99-103.

Denier van der Gon, J.J., and Thuring, J., "The guiding of human writing movements", *Kybernetik* **4**, no.2, 1965, pp. 145-148.

Dooijes, E.H., "Analysis of handwriting movements", U. Amsterdam, Netherlands, *Acta Psychologica* **54** 1983, pp. 99-114.

Duvernoy, J., "Extraction d'invariants morphologiques d'une information par corrélation optique et filtrage spatial", *Opt. Commun* **7**, no2, 1973, pp. 142-145.

Dyche, J.W., "Positive personal identification by handwriting", *Proc. of the Carnahan Conference on Electronic Crime Countermeasures*, University of Kentucky ORES, Lexington 1969, pp. 114-126.

Eden, M., "Handwriting generation and recognition", in *Recognizing Patterns* M.I.T. Press, Kolers, P.A. and Eden, M. (Eds.), Cambridge 1968, pp. 138-154.

Edelman, S., and Flash, T., "A model of handwriting", *Biol. Cybernetics* **57**, 1987, pp. 25-36.

Edwards, J.C., and Paterson, R.L., "Position resolver for accelerometer pens", *IBM Tech. Disclosure Bull.* (USA) **24**, 1981, no7B.

Eernisse, E.P., Land, C.E., and Snelling, J.B., "Piezoelectric sensor pen for dynamic signature verification", *Proc. IEEE Int. Electron Devices Meeting*, IEEE, New York 1977.

Evangelisti, C.J., Herbst, N.M., and Morrissey, J.M., "Personal identification system using signature stylus dynamics", *IBM Tech. Disclosure Bull.* **18**, 1976, pp. 2689-2690.

Evett, I.W., and Totty, R.N., "A study of the variation in the dimension of genuine signatures", *Forensic Science Society*, 1985, pp. 207-215.

Farag, R.F., and Chien. Y.T., "On-line signature verification", *Proceedings of the International Conference on On-line Interactive Computing*, Brunel University, London 1972a, p. 403.

Farag, R., and Chien, Y.T., "*Real-time analysis and verification of handwritten signature*", Report CS-72-1, University of Connecticut School of Engineering, 1972b.

Fejfar, A., "*Test results - advanced development models of BISS identity verification equipment*", III, Automatic Handwritting Verification (Technical rept.), Report no: MTR-3442-VOL-3; ESD-TR- 78-150-VOL-3, The Mitre Corp, 1978, 178 p.

Fejfar, A., and Myers, J.W., "The testing of three automatic identity verification techniques for entry control", *Final Proceedings of the Second Conference on Crime Countermeasures - Science Engineering*, 3A 09-3B 05, Oxford 1977.

Feig, E., "Statistical segmentation procedure for signature verification systems", *IBM Tech. Discl. Bull.* **27**, no. 8, 1985, pp. 4626-4628.

Fischler, M.A., and Elschlager, R.A., "The representation and matching of pictorialstructures", *IEEE Trans. on Computers* **C-22**, no.1, 1973, pp. 67-92.

Foodman, M.J., "Test results - advanced development models of BISS identity verification equipment", **II**, *Automatic Speaker Verification*, MTR-34-42, vol. II, The MITRE Corporation, Bedford MA 1977.

Forsen, G.E., Nelson, M.R., and Staron, R.J., Jr., "*Personal attributes authentication techniques*", RADC-TR-77-333, Rome Air Development Center, Griffiss Air Force Base, New York 1977.

Fox, P.F., "*A practical method of personal identification by signature validation*", Report from Quest Automation Group, England.

Freyd, J.J., "Representing the dynamics of a static form", *Memory & Cognition* **11**, 1983, pp. 342-346.

Gjone, R., and Miraglia, J., "Pen for signature verification by acceleration andpressure", IBM Corp. Armonk, *IBM Tech. Disclos. Bull.* **24**, 1981, p. 902.

Grossman, R.S., Hill, D.E., and Martino, M.J., "Personal identification systems performance improvement algorithm", IBM Corp. Armonk, NY, *IBM Tech. Discl. Bull.* (USA) **24**, no2, 1981, pp. 914-917.

Haberman, W., and Fejfar, A., "Automatic identification of personnel through speaker and signature verification: system description and testing", *Proceedings, 1976 Carnahan Conference on Crime Countermeasures*, J.S. Jackson and R.W. DeVore (Eds.), Univ. of Kentucky, 1976, pp. 23-30.

Hagan, W.E., "A treatise on disputed handwriting", *Foundations of Criminal Justice*, AMS Press Inc., New York 1984, 289 p.

Hale, W.J., and Paganini, B.J., "An automatic personal verification system based on signature writing habits", *Proceeding of the 1980 Carnahan Conference on Crime Countermeasures*, University of Kentucky, Lexington 1980, pp. 121-125.

Hanan, M., Herbst, N.M., and Kurtzberg, J.M., "Finite state machine decision procedure model for signature verification", *IBM Tech. Discl. Bull* **20**, 1978, pp. 3355-3360.

Harmon, L.D., Khan, M.K., Lasch, R. and RAMIG, P.F., "Machine identification of human faces", *Pattern Recognition* **13**, 1981, p. 97.

Harmon, L.D., Kuo, S.C., Raning, P.F. and Randkivi, V., "Identification of human face profile by computer", *Pattern Recognition* **10**, 1978, p. 301.

Harrison, W.R, "*Suspects documents, their scientific examination*", Nelson-Hall Inc., Chicago 1981, 583 p.

Herbst, N.M., and Liu, C.N., "Automatic signature verification", in *Computer Analysis and Perception* (Eds.): C.Y. Suen and R. De Mori, CRC Press, Boca Raton, FL 1982, pp. 83-105.

Herbst, N.M., and Liu, C.N., "Card-based personal identification system", *IBM Tech. Disclosure Bull.* (USA) **22**, no 9, 1980.

Herbst, N.M., and Liu, C.N., "Automatic verification of signatures by means of acceleration patterns", *Proceedings, IEEE Computer Society Conference on Pattern Recognition and Image Processing*, 1977a, pp. 332-337.

Herbst, N.M., and Liu, C.N., "Signature verification based on complete accelerometry", *IBM Tech. Disclosure Bull.* (USA) **19**, no 12, 1977b.

Herbst, N.M., and Liu, C.N., "Automatic signature verification based on accelerometry", *IBM J. Res. Develop.* **21**, 1977c, pp. 245-253.

Herbst, N.M., and Liu, C.N., "*Automatic signature verification*", IBM Research, Rep. RC-5810, 1975.

Herkt, A., "Signature disguise or signature forgery?", *J. Forensic Science Soc.* **26**, 1986, pp. 257-266.

Hilton, O., "*Scientific examinations of questioned documents*", Elsevier New-York, Revised edition 1982, 424 p.

Hollerbach, J.M., "An oscillation theory of handwriting", *Biological Cybernetics*, no. 39, 1981, pp. 139-156.

Howbrook, E. "Signature recognition for security systems", *Proceedings of the Conference on Microprocessors in Automation and Communication*, University of Kent, England 1978, pp. 377-384.

Ibrahim, B., and Levrat, B., "Improved security through on-line signature recognition", *Proc. of the Eng. Conf. on App: Inf. Technol. of the IFIP*, London 1979, pp. 491-498.

Informatique et sécurité, "Système d'identification biométrique: l'état de l'art", Analyses et synthèses éditeur, Paris 1986, no.5, pp. 2-4.

Kao, H.S., Van Galen, G.P., and Hoosain, R., "*Graphonomics, contemporary research in handwriting*", North-Holland, Hong Kong 1986, 403 p.

Klement, V., Steinke, K., and Naske, R., "The application of image processing and pattern recognition techniques to the forensic analysis of handwriting", *Proceedings, 1980 International Conference: Security through Science and Engineering*, West Berlin1980, pp. 5-11.

Kim, J., and Tappert, C.C. "Handwriting recognition accuracy versus tablet resolution and sampling rate", *Proc. 7th Int. Conf. on Patt. Rec.*, Montreal 1984, pp. 917-918.

Komada, S., Mori, M. and Watanabe, T. "An electrostatic transparent tablet", 1985 *SID Int. Symp. Digest of Tech. Papers*, 1985, pp. 28-31.

Lam, C.F., and Kamins, D., "Signature recognition through spectral analysis", *Pattern Recognition* **22**, no1, 1989, pp. 39-44.

Lamarche F., "*Caractérisation d'initiales captées par une tablette graphique*", Mémoire M. Ing., Ecole Polytechnique de Montréal, 1985.

Lamarche, F. and Plamondon, R., "Segmentation and feature extraction of handwritten signature patterns", *Proc. 7th Intern. Conf. Pattern. Recogn.* 2, Montréal 1984, pp. 756-759.

Lashley, K.S., "The accuracy of movements in the absence of excitation from the moving organ.", *Am. J. Physiol.* **20**, 1917, p. 169.

Leclerc, F., "*Validation d'un modèle générateur de vitesse à profil gaussien sur dessignatures manuscrites*", Mémoire de M.Sc.A., Ecole Polytechnique de Montréal, 1989.

Leclerc, F., and Plamondon, R., "Validation d'un modèle de la fonction de transfert du bras sur des signatures manuscrites", *Proc. Vision Interface'90*, pp. 105-111, Halifax 1990,

Leedham, C.G., and Downton, A.C., "On-Line Recognition of Pitman's Handwritten Shorthand: an Evaluation of Potential", *Int. J. Man-Machine Studies* **24**, 1986, pp.375-393.

Leedham, C.G., Downton A.C., Brooks C.P. and Newell. F., "On-line acquisition of Pitman's handwritten shorthand as a means of rapid data entry", *Proc. 1st IFIP Conf. on Human Comp. Interaction*, London 1984, pp. 2.86-2.91.

Legge, D., Steinberg, H., and Summerfield, A., "Simple measures of handwriting as indices of drug effects", *Perceptual and Motor Skills* **18**, 1964, p. 549.

Lew, J.S., "An improved regional correlation algorithm for signature verification which permits small speed changes between handwriting segments", *IBM J. Res.and Dev.* (USA) **27**, no. 2, 1983, pp. 181-185.

Lew, J.S., "Signature verification algorithm", *IBM Tech. Disclosure Bull.* (USA) **25**, no7B, 1982, pp. 3844-3847.

Lew, J.S., "Kinematic theory of signature verification measurements", *Math. Biosciences* **48**, 1980a, pp. 25-51.

Lew, J.S., "Optimal accelerometer layouts for data recovery in signature verification", *IBM J. Res. and Dev.* (USA) **24**, no. 4, 1980b, pp. 496-511.

Lew, J.S., "Optimal designs of instrumented pens for signature verification", *IBM Tech. Disclosure Bull.* (USA) **21**, no. 8, 1979, pp. 3415- 3419.

Lew, J.S., "*Accelerometer layouts for signature verification*", Research Report RC-7154, IBM Thomas J. Watson Research Center, Yorktown Heights, NY 1978a.

Lew, J.S., "*Optimal accelerometer layouts for signature verification*" IBM Research Report RC 7374, IBM Watson Research Center, Yorktown Heights, NY, 1978b.

Lew, J.S., "Signature motion recovery with incomplete initial data", *IBM Tech. Disclosure Bull.* **20**, 1978c, pp. 5441-5442.

Lew, J.S., "Design of instrumented pens for signature verification", *IBM Tech. Disclosure Bull.* **20**, 1977, pp. 819-822.

Little, A.D., "*An assessment of less cash/less check technology*", National Science Foundation, 1974.

Liu, C.N., Herbst, N.M., and Anthony, N.J., "Automatic signature verification and field test results", *IEEE Trans. System, Man. Cybern.* **SMC-9**, 1979, pp. 35-38.

Liu, C.N., "Reference design procedure for signature verification", *IBM Tech.Disclosure Bull.* (USA) **21**, no.1, 1978, pp. 426-429.

Liu, C.N., Herbst, N.M., "Automatic signature verification based on accelerometry", *IBM Journal of Research and Development* **21**, 1977, pp. 245-253.

Locard, E., "*Les faux en écriture et leur expertise*", Payot, Paris 1959.

Locard, E., "*Traité de criminalistique - L'expertise de document écrit*", Joannes Desvigue et Cie, Lyon 1936.

Lorette, G., and Plamondon, R., "Electronic handwriting: trade-offs and new perspectives for education", *4th IGS Conference*, Trondheim, Norway 1989, p. 67.

Lorette, G., "On-line handwritten signature recognition based on data analysis and clustering", *Proceedings of 7th I.C.P.R. 2*, Montreal 1984, pp. 1284-1287.

Lorette, G., "*Méthode de traitements automatiques d'images de points et de lignes enreconnaissance de formes*", Thèse de doctorat d'état, Université Paris XII, 1983.

Lorette, G., Requier, J.P. and Gaudaire, M., "Etude expérimentale pour la caractérisation tion de signatures manuscrites", *Actes du 2e congrès AFCT-IRIA -Reconnaissance des Formes et Intelligence Artificielle*, **tome III**, 1979, pp. 290-297.

Maamari, F., "*Extraction en temps réel des caractéristiques dynamiques de l'écriture manuscrite à partir d'une tablette numérique*", Mémoire de M.Sc.A., Ecole Polytechnique de Montréal, 1986.

Maamari, F., and Plamondon, R., "Extraction of the analog pentip position, speed and acceleration signals from a digitizer", *Graphonomics: Contemporary Research in Handwriting*, H.S.R. Kao, G.P. Van Galen, R. Hoosain (Eds.), Elsevier Sci. Pub., North-Holland 1986, pp. 199-211.

Maarse, F.J., Janssen, H.J.J., and Dexel, F. "A special pen for an XY tablet", in *Computer in Psychology: Methods, Instrumentation and Psychodiagnostics*, F.J.Maarse, L.J.M. Mulder, WPB Sjouw and AE Akkerman (Eds.), Swets and Zeithlinger, Lisse, Netherlands 1988, pp. 133-139.

Maarse, F.J., and Thomassen, A.J.W.M., "Produced and perceived writing slant: difference between up and down strokes", *Motor Aspect of Handwritting, Acta Psychologica*, North-Holland, 541, 1983, pp. 131-147.

MacDonald, J.S., "*Experimental studies of handwriting signals*", Ph.D. Thesis, M.I.T.,1964.

Mally, R., "*Der Handschriften-Erkennungsdienst*", BKA-Schriftenreihe **2**, 1955, p.145.

Margolin, D.I., and Binder, L., "Multiple component agrapha in a patient with a typical cerebral dominance: an error analysis", VA Hosp, Neurology Section, Fresno, CA, *Brain & Language* **22**, 1984, pp. 26-40.

Martlew, M., "*The psychology of written language, developmental and educational perspectives*", J. Wiley & Sons, New York 1983, 428 p.

Marzocco, I.F., "Computer recognition of handwritten first names", *WESCON Tech.Papers* **8**, no 8/2, 1964, pp. 1-7.

Mauceri, A.J., "*Feasibility study of personnel identification by signature verification*", Report no SID 65 24 RADC TR 65 33, Space and Information System Division, North American Aviation, Anaheim, Calif. 1965.

Mendel, A.O., "*Personality in handwriting*", Stephen Daye Press, New York 1961.

Miller, D., "Automated banking sharpens need for accurate customer ID", *Bank Syst. and Equip.* (USA) **21**, no7, 1984, pp. 93-98.

Mital, D.P., Hin, C.P., and Leng, W.K., "Computerized signature verification system", *IEEE Control System Mag.* **8**, no3, 1988, pp. 54-57.

Mjølsnes, S.F., and Søberg, G., "A comparative performance experiment of dynamic signature verification", *Proc. 3rd Int. Symp. on Handwriting and Comput. Applic.*, Montreal 1987, pp. 173-175.

Modern Data, "*Security in the Palm of Your Hand*", 1973, p.57.

Morasso, P., and Tagliasco, V., "*Human movement understanding*", North-Holland, Amsterdam 1986, 382 p.

Morasso, P., Mussa Ivaldi F.A., and Ruggiero, C., "How a discontinuous mechanism can produce continuous patterns in trajectory formation and handwritting", *Acta Psychologica* **54**, 1983, pp. 83-98.

Morasso, P., and Mussa Ivaldi, F.A., "Trajectory formation and handwriting: a computational model", *Biol. Cybern* **45**, 1982, pp. 131-142.

Morasso, P., "Spatial control of arm movements", *Exp. Brain Res.* **42**, 1981, pp.223-227.

Morris, R., and Thompson, K., "Password security: a case history", *Communications of the Association for Computing Machinery*, New York 1979, pp. 594-597.

Nagayama, T., Shibuya, J. and Kawakita, T., "Pen-touch-type electromagnetic transparent touch panel", *SID Int. Symp. Digest of Tech. Papers*, 1985, pp. 32-35.

Nagel, R.N., and Rosenfeld, A., "Computer detection of freehand forgeries", *IEEE Trans. Computers* **C-26**, 1977, pp. 895-905.

Nagel, R.N., "*Computer detection of freehand forgeries*", Doctoral Dissertation, Dept. of Computer Science, Univ. of Maryland, 1976.

Nagel, R.N., and Rosenfeld, A., "Steps toward handwritten signature verification", *Proc. 1st. IJCPR*, Washington 1973, pp. 59-66.

Nagel, R.N., "*Computer screening of handwritten signatures: a proposal*", Comput. Sci. Center, Univ. of Maryland, College Park, Tech. Rep. 220, 1973.

National Bureau of Standards, "*Guidelines on user authentication techniques for computer network control*", Federal Information Processing Standards Publication 83, 1980, pp. 3-38.

National Bureau of Standards, U.S. Department of Commerce, "*Guidelines on Evaluation of Techniques for Automated Personal Identification*", Federal Information Processing Standards Publication 48, 1977.

Nemcek, W.F., and Lin., W.C., "Experimental investigation of automatic signature verification", *IEEE Transactions on Systems, Man, and Cybernetics* **SMC-4**, 1974, pp. 121-126.

Nevill, D.G., "*Signature processing*", Computer Education, 1984, pp. 29-31.

Nouboud, F., "Contribution à l'étude et la mise au point d'un système d'authentification de signatures manuscrites", Thèse de Doctorat, Université de Caen, 1989.

Nouboud, F., and Plamondon, R., "On-line recognition of handprinted characters: survey and beta tests, (in press), *Pattern Recognition*, 1990.

Nouboud, F., Cuozzo, F., Collot, R., and Achemlal, M., "Authentification de signatures manuscrites par programmation dynamique", *Congrès PIXIM*, 1988, pp. 345-360.

Osborn, A.S., "*Questioned documents*", Nelson Hall, 2nd Edition, Chicago 1929, 1043 p.

Ostrem, J.S., and Crane, H.D., "*Automatic handwriting verification (AHV)*", Final technical rept. Oct.79-Mar. 81, Sponsor: Rome Air

Development Center, Griffiss AFB, NY., Report no. RADC-TR-81-328, 1981, 81 p.

Parizeau, M., and Plamondon, R., "*What types of scripts can be used for personal identity verification?*", Computer Recognition and Human Production of Handwriting, R. Plamondon, C.Y. Suen, M.L. Simner (Eds.), World Scientific Publish., Singapore 1989a, pp. 77-90.

Parizeau, M., and Plamondon, R., "A comparative analysis of regional correlation, dynamic time warping and tree matching algorithms for signature verification", (in press) *IEEE Trans. on Pattern Analysis and Machine Intelligence*, 1990.

Parizeau, M., "*Etude comparative d'algorithmes pour mesurer la similitude de signauxdans un contexte de vérification de l'identité par la signature manuscrite*", Mémoire de M.Sc.A., Ecole Polytechnique de Montréal, 1987.

Parizeau, M., and Plamondon, R., "Relative performances of signature handwritten password and initials for personal identification", *Proc. 3rd Int. Symp. on Handwriting and Computer Applications*, Montreal 1987, pp. 164-166.

Parker, D.B., "*Crime by computer*", New York: Scribner's, 1976.

Parks, J.R., Carr, D.R., and Fox, P.F., "Positive identification of terminal users", *IEE Colloquium on Techniques and Implications of Digital Privacy and Authentication Systems*, 8/1-3, London, 1981.

Petersen, K.E., "*Silicon as a mechanical material*", Proc. of the IEEE 70, no.5, 1982, pp. 420-457.

Phelps, R.I., "A holistic approach to signature verification", *Proceedings of the 6th International Conference on Pattern Recognition* 2, 1982, p. 1187.

Philips, M.J., and Ward, J.R., "Digitizer technology: performance characteristics and the effects on the user interface", *IEEE Computer Graphics and Applications*, Vol. 7, No. 4, April 1987, pp. 31-44.

Picodata, "*Reconnaissance dynamique de signature, Phases: Etude de l'offre*", Rapport final, 1984, 45 pages.

Plamondon, R., "Handwriting control: a functional model", in *Models of Brain Function*, R. Cotterill (Ed.), Academic Press, London 1989a, pp. 563-574.

Plamondon, R., "A handwriting model based on differential geometry", in *Computer Recognition and Human Production of Handwriting*, R. Plamondon, C.Y. Suen, M. L. Simner (Eds.), World Scientific Publishing, Singapore 1989b, pp. 179-192.

Plamondon, R., and Lorette, G., "Automatic signature verification and writer identification: the state of the art", *Patt. Rec.* **22**, no. 2, 1989a, pp. 107-131.

Plamondon, R., and Lorette, G., "On-line signature verification: how many countries are in the race?", *Proc. Carnahan Conf. on Security Technology*, Zurich 1989b, pp. 183-191.

Plamondon, R., and Maarse, F.J., "An evaluation of motor models of handwriting", *IEEE Trans. on Syst. Man Cybernetics* **19**, no.5, 1989, pp. 1060-1072.

Plamondon, R., Suen, C.Y., and Simner, M.L., "*Computer recognition and humanproduction of handwriting*", World Scientific Publishing, Singapore 1989, 395 p.

Plamondon, R., and Yergeau, P., "A system for the analysis and synthesis of handwriting", *Proc. Int. Workshop on Frontiers in Handwriting Recognition*, 1990, pp. 167-179.

Plamondon, R., and Parizeau, M., "Signature verification from position, velocity and acceleration signals: a comparative study", *Proc. 9th Int. Conf. on Patt. Recog.*, 1988, pp. 26-265.

Plamondon, R., "What does differential geometry tell us about handwriting generation", *Proc. 3rd Int. Symp. on Handwriting and Computer Applications*, Montréal 1987, pp. 11-13.

Plamondon, R., Lorette, G., and Poulin, G., "On the intrinsic theoretical and practical difficulties of designing an automatic signature verification system", *Proc.3rd Int. Symp. on Handwriting and Computer Applications*, Montréal 1987b, pp. 154-156.

Plamondon, R., and Maarse, F., "A neuron-oriented representation to compare biomechanical handwriting models", *Proc. 3rd International Symposium on Handwriting and Computer Applications*, Montréal 1987, pp. 2-4.

Plamondon, R., Suen, C.Y., Poulin, G., and Deschenes, J.G., "*Proc. 3rd Int. Symp.on Handwriting and Computer Applications*", Editions de l'Ecole Polytechnique de Montréal, 1987a, 196 p.

Plamondon, R., and Lamarche, F., "Modelization of handwriting: a system approach", *Graphonomics: Contemporary Research in Handwriting*, H.S.R. Kao, G.P. Van Galen, R. Hoosain (Eds.), Elsevier Sci. Pub., North-Holland 1986, pp.169-183.

Plamondon, R., "La vérification automatique des signatures", Recherches et dévelopements, *Interfaces* **5**, no3, 1984, pp. 16-21.

Plamondon, R., and Brault, J.J., "A system for signature analysis and verification based on an accelerometric pen", *Proc. of the 1983 Inter. Carnahan Conf. on Security Tech.*, Zurich 1983, pp. 157-164.

Plamondon, R., Brault, J.J., and Robillard, P.N., "Optimizing the design of an accelerometric pen for signature verification", *Proc. of the 1983 Conf. on Crime Countermeasures and Security Tech.*, Lexington 1983, pp. 35-40.

Raphael, D.E., and Young, J.R., "*Automated personal identification*", Rept. Long Range Planning Service, Stanford Research Institute, 1974, pp. 1-28.

Reitter, R., Andre, C., Revillet, M.J., and Achemlal, M., "Biomac, a versatile biometric verification machine for access control", *Proc. IAPR Workshop on Comp.Vis.: Special Hardware and Ind. Applic.*, Tokyo 1988, pp. 120-123.

Requier, J.P., "*Utilisation d'algorithmes de comparaison entre graphes pour la conception d'opérateurs microprogrammés de reconnaissance de formes - application à la reconnaissance de signatures manuscrites*", Thèse de doctorat d'état, Université Paris XII, 1979.

Requier, J.P., and Lorette, G., "*Reconnaissance de signatures manuscrites*", Actes du 2e congrès AFCET-IRIA, Reconnaissance des formes et Intelligence Artificielle, III, 1979, pp. 298-307.

Riganati, J.P., "An Overview of electronic identification systems", *WESCON Technical Papers*, Los Angeles, CA 1975.

Rosenfeld, A., and Kak, A.C., "*Digital picture processing*", Academic Press, New York 1976, pp. 442-444.

Sabourin, R., and Plamondon, R., "Steps toward efficient automatic handwritten signature verification systems using grey level images", *Proc. Vision Interface'90*, Halifax, Nova Scotia, May 1990a, pp. 94-104.

Sabourin, R., and Plamondon, R., "Progress in the field of automatic signature verification systems using grey-level images", *Proc. Int. Workshop on Frontiers in Handwriting Recognition*, Montréal 1990b, pp. 1-12.

Sabourin, R., and Plamondon, R., "Segmentation of handwritten signature images using the statistics of directional data", *Proc. 9th Int. Conf. on Patt. Recog.*, Rome 1988, pp. 282-285.

Sabourin, R., and Plamondon, R., "On the implementation of some graphometric techniques for automated signature verification systems: a feasibility study", *Proc. 3rd Int. Symp. on Handwriting and Computer Applications*, Montréal 1987, pp. 160-162.

Sabourin, R., Plamondon, R., "Preprocessing of Handwritten Signatures from Image Gradient Analysis", *Proc. of 8th Intern. Conf. on Pattern Recognition*, Paris 1986, p.576-579.

Sakai, T., Kanade, T., and Ariki, Y., "Multi-feature display of on-line signature bycolor TV display", *Proc. of the Second International Joint Conference on Pattern Recognition*, Copenhagen 1974, pp. 303-304.

Salvage, C.C., "Dynamic signature verification: An effective and economical systems security control", *Proceedings of the European Computer Systems Security Forum*, 1984, pp. 141-148.

Sato, Y., and Kogure, K., "On-line signature verification based on shape, motion and handwriting pressure", *Proc. of the 6th Int. Conference on Pattern Recognition 2*, 1982, pp. 823-826.

Saudek, R., "*Experiments with handwriting*", Book for Professionals, Sacramento CA 1978, 395 p.

Schlosser, P., and Kable, R., "A new high-performance digitizer for computer graphics and display applications", *1985 SID Int. Symp. Digest of Tech. Papers,* 1985, pp. 42-43.

Schomaker, L.R.B., Thomassen, A.J.W.M., and Teulings, H.L., "A computational model of cursive handwriting", *Proc. 3rd Int. Symp. on Handwriting and Computer Applications,* Montreal 1987, pp. 5-7.

Shaw, D.F., "Proof of identity - a review", *1980 International Conference: Security through Science and Engineering*, Berlin 1980, pp. 31-45.

Simpson, Abromovitch, Bar-Ilan, Ramat-gan, and Israel, "A diagnostic approach to writing disability", *Academic Therapy* **16**, 1980, pp. 83-92.

Smith, K.U., Schappe, R., "Feedback analysis of the movement mechanisms of handwriting", *J. of Experimental Education* **36**, no 4, 1970, pp. 61-68.

Stelmach, G.E., *"Information processing in motor control and learning"*, Academic Press, New York 1978, 315 p.

Sternberg, J., "Automated signature verification using handwriting pressure", *1975 WESCON Technical Papers*, no.31/4, Los Angeles 1975, pp. 1-6.

Stockolm, E., "Recognition of a writer as a function of his method of writing", International Graphoanalysis Society, Chicago, IL, *Perceptual & Motor Skills* **49(2)**, 1979, pp. 483-488.

Strangio, C.E., *"Numerical comparison of similarly structured data perturbed by random variations, as found in handwritten signatures"*, Technical Report, Dept. of Elect. Eng., MIT, 1976, 53 p.

Stuckert, P.E., "Magnetic pen and tablet", *IBM Tech. Disclosure Bull* (USA) **22**, no3, 1979, pp. 1245-1251.

Suen, C.Y., Berthod, M., and Mori, S., "Automatic recognition of handprinted characters: the state of the art", *Proc. of the IEEE* **68**, no4, 1980.

Sykes, D., "Personal identity verification", Honeywell Information Systems Inc., Pheonix Al. USA *Inf. Privacy (GB)* **1**, no.5, 1979, pp. 222-225.

Sykes, D.J., "Positive personal identification", *Datamation (USA)* **24**, no11, 1978, pp.179-186.

Tappert, C.C., Suen, C.Y., and Wakahara, T., "On-Line handwriting recognition: a survey", *Proc. 9th Int. Conf. Patt. Recog.*, Rome 1988, pp. 1123-1132.

Tappert, C.C., Fox, A.S., Kin, J., Levy, S.E., and Zimmerman, L.L., "Handwriting recognition on transparent tablet over flat display", *SID86 Digest,* 1986, pp. 308-312.

Telmosse, P., *"Reconnaissance de nombres manuscrits à l'aide d'un crayon accélérométrique"*, Mémoire de M.Sc.A., Ecole Polytechnique de Montréal 1986.

Telmosse, P., and Plamondon, R., "Caractérisation et classification de nombres manuscrits par méthode statistique", *Proc. RAI/IPAR 86, Identification and Pattern Recognition*, Toulouse 1986, pp. 865-877.

Teulings, H.L., Mullins, P.A., and Stelmach, G.E., "The elementary units of programming in handwriting", in *Graphonomics: Contemporary Research in Handwriting*, H.S.R. Kao, G.P. van Galen, R. Hoosain (Eds.), North-Holland 1986, pp. 21-32.

Teulings, H.L., and Maarse, F.J., "Recording and processing of handwriting movements", *Human Movement Sciences* **3**, 1984, pp. 193-217.

Thomassen, A.J.W.M., Keuss, P.J.G., and Van Galen, G.P., "Motor aspects of handwriting", *Acta Psychologica* **54**, North-Holland 1983, 354 p.

Thomson-Titn, "*La technologie de reconnaissance dynamique de signatures par leprocédé Battelle / Thomson-Tinn*", Notice d'information, 1984, 23 p.

Totty, R.N., "Automatic signature comparison", *Annual Meeting of the American Soc. of Questioned Document Examiners*, Montreal 1986.

Totty, R.N., and Hardcastle, R.A., "A preliminary assessment of the SIGNCHECK system for signature authentication", *J. Forensic Science Soc.* **26**, 1986, pp. 181-195.

Van Galen, G.P., "Structural complexity of motor patterns: a study on reaction time and movements of handwritten letters", *Psychological Research* **46**, 1984.

Van Galen, G.P., and Teulings, H.L., "The independant monitoring of form and scale factors in handwriting", *Motor Aspect of Handwriting, Acta Psychologica*, North-Holland 1983, pp. 9-22.

Van Galen, G.P., "Handwriting and drawing: a two-stage model of complex motor behavior", in *Tutorials in Motor Behavior*, G.E. Stelmach and J. Requin (Eds.), Amsterdam, North-Holland 1980, pp. 567-578.

Wallner, T., "Hypothesis of handwriting psychology and their verification", *Professional Psychology* **6**, 1975, pp. 8-16.

Warfel, G.H., "*Identification technologies*", Charles C. Thomas Publisher, Springfield 11, 1979.

Warfel, G.H., "Input security in electronic funds transfer system", *IEEE Compcon* Fall, Computer Technology to Reach the People, 1975, pp. 169-171.

Watson, R.S., "Automatic signature verification (VERISIGN)", *Proceedings of a Conference on Science and Security*, paper no.13, Brighton 1979.

Weinstein, S.B., "Smart credit cards: the answer to cashless shopping", *IEEE Spectrum*, 1984, pp. 43-49.

Wing, A.M., "Response timing in handwriting", in G.E. Stelmach (Ed.) *Information processing in motor control and learning*, New York: Academic Press, 1978, pp. 469-486.

Wolf, W., "*Diagrams of the unconscious*", New York, Grune & Stratton, 1948.

Wood, H.M., "The use of passwords for controlling access to remote computer systems and services", *The Information Series*, **III**, Computers and Security, C.T.,Dinardo, Editor, AFIPS Press, Montvale, NJ 1978, pp. 137-143.

Wood, H.M., "*The use of passwords for controlled access to computer resources*", NBS Special Publication 500-9, U.S., Department of Commerce, National Bureau of Standards, GPO SN 003-003-01770-1, 1977.

Woodard, J., Nelson, J., Ferrante, J., and Maier, J., "Automated entry control: RADC technology development results and future plans", *Proc. Carnahan Conf.on Crime Countermeasures*, Lexington 1981, pp. 77-82.

Woodard, J.P., Maier, J.J., "Automatic entry control for military application", *Proceeding 1979 Carnahan Conference on Crime Countermeasures*, Lexington 1979, pp. 65-76.

Worthington, T.K., Chainer, T.J., Williford, J.D., Gunderson, S.C., "IBM dynamic signature verification", *Computer Security, IFIP 1985*, Elsevier Science Publ, North-Holland 1985, pp. 129-154.

Yasuhara, M., "Identification and decomposition of fast handwriting system", *IEEE Trans. on Circuits and Syst.* **CAS-30**, no.11, 1983, pp. 825-832.

Yasuhara, M., "*Experimental studies of handwriting process*", Kep. Univ. Electro-Comm. 25, 1975, pp. 233-254.

Yasuhara, M., and Oka, M., "Signature verification experiment based on nonlinear time alignment: a feasibility study", *IEEE Trans. Systems, Man. Cybernetics*, **SMC-7**, 1977, pp. 212-216.

Zimmerman, K.P., and Varady, M.J., "Handwriter identification from one-bit quantized Pressure Patterns", *Pattern Recognition* **18**, no1, 1985, pp. 63-72.

Zimmerman, K.P., and Werner, C.L., "SIRSYS - a research facility for handwritten signature analysis", in *Proceedings 1978 Carnahan Conference on Crime Countermeasures*, J.S. Jackson and R.W. De Vore, (Eds.), University of Kentucky, Lexington 1978, pp. 153-155.

Zweighenhaft, R.L., and Marlowe, D., "Signature Size: Studies in Expressive Movement", U. California, Santa Cruz, *Journal of Consulting & Clinical Psychology* **40**, 1973, pp. 469-473.

PART 2

ON-LINE SYMBOLS AND SCRIPT RECOGNITION

Overview

This section contains eight chapters several of which deal with aspects of the design of an electronic pen pad. This new type of computer interface should mimc the conventional pen-paper interface and is expected to become an efficient alternative to keyboards in applications where handwriting will prevail. For example, in applications where both pointing and data entry is necessary, in applications where both thinking and sketching is required and in applications where note-taking and reformatting is necessary and more naturally achieved using handwritten input.

The first chapter, by Meeks and Kuklinski, investigates the dynamic variability of some digitising tablets in terms of residual error in position, speed and acceleration. The second chapter, authored by Higgins and Duckworth, describes a prototype of an electronic pen pad. Emphasis is made both on hardware and software of the design problems.

The third chapter, by Kadirkamanathan and Rayner, focusses on the fundamental problem of segmentation of cursive script into strokes. The chapter by Teulings, Schomaker, Ritsen, Drexler and Albens discusses some of the modules designed for on-line cursive script recognition while the following chapter, by Oulhadj, Petit, Lemoine, Gaudaire and Lorette describes a practical implementation of a cursive script recognition system based on a prediction-verification strategy.

The chapter by Kondo analyses specific features of hand-printed characters using principal component analysis. The following chapter, by Welbourn and Whitrow, reports on a text editor to be incorporated in a penpad system; it uses hand-drawn gesture commands to mimic the usual pen and paper type of editing.

Finally, the last chapter of this section, by Leedham, expands the potential applications of an electronic pen pad to other forms of script. It describes the current state of research in the automatic recognition and transcription of Pitman's handwritten shorthand.

Computer Processing of Handwriting
Eds. R. Plamondon & C. G. Leedham

MEASUREMENT OF DYNAMIC DIGITIZER PERFORMANCE

M. L. MEEKS and T. T. KUKLINSKI
Meeks Associates, Inc., Lincoln, MA 01773, USA

The dynamic performance of three digitizers was investigated by recording data as a pen moved along a straight line across each tablet. A test bed was constructed to obtain accurate rectilinear motion with a constant acceleration. Residual errors in position, speed, and acceleration were measured by subtracting the controlled values of these variables. Although all three tablets performed with the specified position accuracy, two tablets produced instrumental effects that severely limited their accuracy as instruments for measuring dynamic handwriting variables. We analyzed the effect of asynchronous coordinate sampling and measured the sampling delays of the tablets. The resulting position errors in all cases were relatively small.

1. Introduction

The basic instrument for recording handwriting is the digitizer or graphics tablet. Handwriting applications place demands on the accuracy and performance of digitizers that were not anticipated when these instruments were designed. Applications such as on-line character recognition, signature verification, and measurements of human motoric performance require that pen velocity and acceleration be determined as well as accurate pen position. An extensive review of the errors associated with the recording and processing of handwriting movements has been published by Teulings and Maarse (1984). Maamari and Plamondon (1985) have described off-line processing techniques for coping with some problems encountered in using digitizers to measure pen velocity and acceleration. Tappert (1982) and Kim and Tappert (1984) have discussed various constraints such as the trade-off between retaining small angular features and smoothing to reduce noise.

In this chapter we report the results of dynamic measurements made on three commercially available instruments. We find that it is not sufficient to regard a digitizer as a black box that superimposes noise on unbiased measurements. Our measurements illustrate a variety of errors that occur in practice and indicate how investigators might examine the detailed performance of their own digitizers.

2. Digitizer Requirements and Specifications

Digitizers may be classified on the basis of manufacturer's specifications. These specifications typically state the *active area*, the *resolution*, and the *accuracy* of the instrument. For handwriting applications the digitizer is operated in the so-called *stream mode* wherein a sequence of coordinate pairs are transmitted at a specified rate while a pen or stylus is in proximity of the tablet, usually within 2 cm of the tablet surface. When each sample is reported, a bit is set to indicate whether or not the pen is pressed against the tablet; output terminates when the pen is moved out of the proximity of the tablet surface. Manufacturer's specifications state the maximum sampling rate of the digitizer in stream mode; lower rates can usually be selected.

Digitizers are essentially analogue-to-digital converters operating in two dimensions; internal logic interpolates data and performs additional digital processing prior to output. This processing is controlled by firmware in the instrument, and digitizers permit selection between several modes of internal processing.

Currently available instruments for handwriting research might have the following manufacturer's specifications: resolution 0.00254 cm, accuracy ±0.025 cm, and maximum sampling rates from 100 to 205 Hz. Sampling rates in this range are suggested by Teulings and Maarse (1984) as a compromise between accuracy and data-storage requirements. Blesser & Ward (1987) have discussed the digitizer requirements for handwriting capture and analysis. In this volume, Lorette and Plamondon describe several digitizers in the section on signature verification. Diehl and Apiki (1989) have compared the performance of 21 high-quality digitizers for drafting applications. Hearn and Lankford (1989) list 15 manufacturers and 30 different types of digitizers which are now commercially available. Unfortunately manufacturer's specifications are incomplete; they describe neither the method of measurement nor the nature of the error. Many types of error are ignored.

One must look at digitizer technology to understand the problems that may arise when these instruments are used to collect handwriting data. The random noise associated with analogue-to-digital conversion in the instrument is but one of a variety of errors that may be present. Ward & Phillips (1987) have investigated the performance of a variety of digitizers, and they specify fifteen problem areas to be considered. The impact of each of these problem areas will depend upon the user's application. The following problems cited by Ward & Phillips are particularly relevant to handwriting research:

1. *Missing coordinates* or dead bands, that is, specific values of x or y that the tablet consistently fails to report.

2. *Monotonicity failure*, a characteristic of some digitizers whereby a small jump backward is sometimes reported because of failure in the digitizer's interpolation firmware.

3. *Output continuity failure*, coordinate samples or groups of samples are incorrectly reported as "pen-up" in the output.

4. *Differential nonlinearity,* systematic errors which are correlated with position within small areas on the tablet.

5. *Boundary errors*, predictable position errors that appear near the edges of the tablet.

6. *Hysteresis errors*, failure of the digitizer firmware to report changes in pen position until the pen has moved a minimum amount.

7. *Pen-tilt error*, position errors caused by sensitivity to the angular position of the pen axis.

8. *Velocity-dependent errors*, position errors which vary with pen velocity as a result of non-simultaneous sampling of x and y.

In addition to the above problems, there is another potential source of error that requires consideration in handwriting research, although it is unimportant in drawing and drafting applications. In order to recover the x and y components of pen velocity and pen acceleration, the sampling period must be constant. All of these potential sources of error require investigation before instrumental errors can be understood and taken into account.

3. The Classification and Measurement of Digitizer Errors

The errors that may appear in digitizer output can be classified as follows: *spatial errors,* which are errors in the position coordinates; *temporal errors,* which result from irregularities in the sampling period and from non-simultaneous sampling, and *intrinsic errors* caused by changing pen tilt and by inaccurate firmware processing. In the sections that follow we focus on spatial and temporal errors and methods by which they can be measured.

3.1 Spatial Errors

Spatial errors may result from systematic defects and irregularities in tablet structure as well as from the finite resolution of the instrument. A distinction must be drawn between spatial errors which are correlated with specific positions on the tablet and those which are random and not reproducible. The types of error, which were cited by Ward and Phillips (1987) as due to differential nonlinearity, boundary effects, and missing coordinates, are clearly correlated errors. Random noise from a variety of sources is thus superimposed on the correlated errors.

The problem is to obtain measurements of the spatial errors in a form that will allow separation of the random and the correlated components. To the extent that spatial errors can be correlated with position, they may be corrected by calibrating the instrument. However calibration is feasible only when the corrections can be entered as functions of x or y. On a typical tablet there are some 1.2×10^8 different locations, so in practice corrections would need to be expressible in relatively simple functional form.

One can directly measure spatial error by the following method: first digitize a straight line drawn accurately across the tablet; then fit a straight line to the recorded xy-data by the method of least squares; and finally calculate the residual errors after fitting. Graphs of the residual error plotted as a function of position will show the spatial errors along the line. In the absence of correlated spatial errors one would expect the residual errors to be random with a normal distribution about zero and with a standard deviation roughly characterized by the resolution and quantization noise of the instrument. Teulings and Maarse (1984) have discussed this case in detail. Correlated errors will be superimposed on the random distribution, and when correlated errors are appreciably larger than noise, the pattern of these errors will show up clearly when residuals are plotted against position. Examples of plots showing correlated-error patterns are given in Sec. 6.3.

3.2 Temporal Errors

When pen velocity and pen acceleration are to be calculated from digitizer output, temporal error becomes important. Variations in the inter-sample period will directly lead to errors in the time derivatives of x and y even though the effect on the recorded pattern of the writing may be negligible. A second type of temporal error results when x and y are not sampled simultaneously. These errors and general methods for their correction are considered by Teulings and Maarse (1984). We discuss these sources of error in the following sections and describe how the errors can be measured.

3.2.1 Variations in inter-sample period

One can measure deviations from uniformity in the sampling period as follows. Record data with the digitizer as the pen moves across the tablet with accurately controlled velocity. (We controlled velocity using the test bed described in Sec. 5.) Then calculate the velocity components v_{xi} and v_{yi} from the recorded x- and y- components of position and the inter-sample period τ as follows

$$v_{xi} = (x_i - x_{i-1})/\tau$$
$$v_{yi} = (y_i - y_{i-1})/\tau \qquad (1)$$

If the objective is to measure motoric performance in handwriting, one would not use Eq. 1 to calculate velocity. The best estimate of velocity would be obtained by processing the data with a digital filter which combines a differentiator with a low–pass filter (Parks & Burrus, 1987). We used Eq. 1 because the resulting data reflect directly any irregularities in tablet performance.

One can then calculate velocity residuals by fitting the controlled-velocity function to the calculated velocity components v_{xi} and v_{yi} derived from measurement. Examples of these calculations are shown in Sec. 6.4. In the absence of variations in inter-sample period, one can put an upper limit on the standard deviation of these residuals on the basis of the measured errors in the position coordinates. Specifically, if there are no correlated position errors and if the sampling period remains constant, the velocity errors will result only from random errors in the position measurements. The standard deviations of the residual velocity components can be shown to be independent of pen speed in this case. This result is derived in Sec. 4, which describes how random errors in position coordinates and sampling period propagate to calculations of velocity.

3.2.2 Non-simultaneous sampling

The results of non-simultaneous sampling are shown in Fig. 1, which illustrates a case where y is sampled after x with a time delay $\delta = t_2 - t_1$. The pen is moving at speed v in a direction at an angle θ with respect to the x-axis. From this geometry one can see that the reported position P will be displaced parallel to the y-axis by an amount Δy given by

$$\Delta y = v\delta \sin \theta \tag{2}$$

The reported position P in Fig. 1 will be displaced perpendicular to the line of motion by an amount p given by

$$p = \frac{v\delta}{2} \sin 2\theta \tag{3}$$

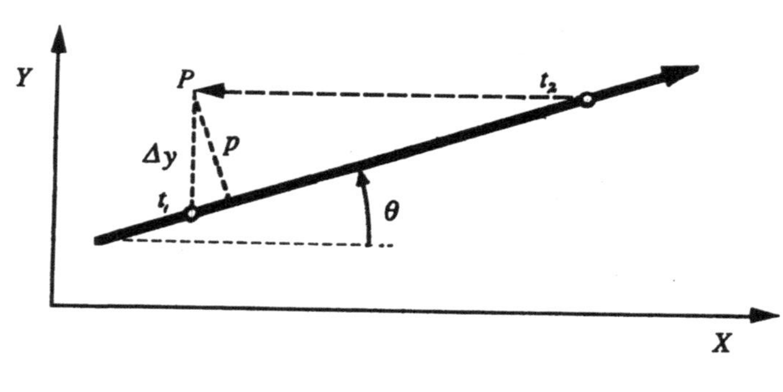

Figure 1. Position Error Caused by Asynchronous Sampling. Here x is sampled at t_1 and y is sampled at t_2, which is delayed relative to t_2 by δ, so the output coordinates correspond to an incorrect position P. The dark line represents pen motion.

The perpendicular offset is thus proportional to the pen speed, and it is a function of direction, being a maximum when the pen moves at a 45-degree angle with either axis and zero when the pen moves parallel to an axis. The size of the time delay and the maximum writing speed determine the maximum offset produced by sampling delay. If δ can be measured and shown to be independent of position, Eq. (3) can be used to calculate the errors caused by non-simultaneous sampling. Measurements of δ are discussed in Sec. 6.5.

4. Error Propagation in Digitizer Measurements

Correlated and random digitizer errors propagate differently when one calculates the pen position, velocity, and acceleration from the recorded digitizer output. In the present context we calculate these variables from the recorded digitizer output without smoothing or filtering in order to observe instrumental effects directly. In practice one would use digital signal processing to filter handwriting data by removing frequency components that are beyond the capabilities of the human motor system (Teulings and Maarse, 1984). The influence of noise on time derivatives computed from filtered data has been described by Lanshammar (1982). The numerical methods for smoothing and filtering to reduce the effect of noise are described in detail by Press *et al* (1986). But we know of no systematic method for quantitatively assessing the influence of correlated errors in the case of either direct or filtered processing.

If correlated errors are absent, one can predict the way that errors in position and sampling period propagate into errors in velocity and acceleration. The calculation of error propagation is straightforward and is well described in texts on error analysis such as Bevington (1969). If we assume that errors in x, y and τ are the result of a gaussian random processes, then the propagation of errors from x_i to v_{xi} goes as follows, where N is the sampling rate, $\tau = 1/N$, and

$$v_{xi} = N(x_i - x_{i-1})$$

For the variance of x_i and the standard deviation of x_i we write var[x] and $\sigma[x]$, respectively, and we use this notation also for the other variables. Now we define $\Delta x_i = x_i - x_{i-1}$ so that we have var[Δx] = 2 var[x]. The variance of v_{xi}, will be result of errors in x_i and errors in t, which are independent and which contribute in amounts proportional to the derivatives of v_x with respect to Δx and to t. The variance of v_x, taking into account gaussian errors in x and τ, is thus given by

$$\mathrm{var}[v_x] = 2N^2\mathrm{var}[x] + v^2_x\mathrm{var}[\tau]/\tau^2 \tag{4}$$

Hence from Eq. 4 we expect that *errors in the inter-sample period will produce velocity errors that increase with increasing velocity* and that the influence of variations in τ will depend on $\sigma[\tau]/\tau$, where $\sigma[\tau]$ is the standard deviation of τ. In the absence of variation in τ, the variance in v_x becomes independent of the magnitude of v_x, and the standard deviation is

$$\sigma[v_x] = \sqrt{2}N\sigma[x] \tag{5}$$

As a numerical example, let us assume that there is no variation in the inter-sample period; then with a sampling rate of 200 Hz and a random distribution of position errors, $\sigma[x] = 0.0025$ cm, we find from Eq. 5 that $\sigma[v_x] = 0.71$ cm/sec.

Clearly it is necessary to consider the correlated error as well as the random error in estimating the accuracy of calculated velocities. However, correlated position errors will tend to produce smaller errors in the calculated velocity than random errors. This will be true because when one calculates $\Delta x_i = x_i - x_{i-1}$ the correlated error common to x_i and x_{i-1} will be subtracted out. For this reason we expect that Eq. 5 may overestimate the measured standard deviation of velocity error; if one assumes for $\sigma[x]$ a value that includes correlated error as well as random error. In Sec. 6.4 we will see this expectation confirmed.

We can go on to calculate the acceleration from differences in successive values of the velocity as follows

$$a_{xi} = (v_{xi+1} - v_{xi})N \tag{6}$$

and we can predict the variance of a_{xi} by the same method used to obtain Eq. 4. Again assuming that τ is constant, we obtain the standard deviation of a_x in terms of velocity error

$$\sigma[a_x] = \sqrt{2}N\sigma[v_x] \tag{7}$$

and in terms of position error in the x coordinate

$$\sigma[a_x] = 2N^2\sigma[x] \tag{8}$$

The position error in x is thus amplified by a factor N in calculating the velocity error and by a factor N^2 in calculating the acceleration error as Eqs. 5 and 8 show. Of course one should not conclude that a low sampling rate should be used to reduce velocity and acceleration errors; Landshammar (1982) has described how one can use higher sampling rates combined with low-pass digital filtering to reduce errors in velocity and acceleration.

5. The Test Bed

In order to measure instrumental errors in position and pen speed, one must have a means of controlling pen motion while data are recorded. A test bed was constructed for this purpose; it constrained the pen to move along a straight line at controlled speeds. Figure 2 shows a drawing of the test bed and tablet. The pen is supported on a bridge which can move freely across the tablet on parallel tracks, 120 cm long, which are secured to a rigid wooden frame. We used model-railroad cars and track (HO scale) to implement the design. A grindstone was placed on the bridge to increase the inertial mass; the pen was secured at the center of a grindstone with modeling clay. The combined weight of bridge and grindstone was 3.3 kg. The bridge, spanning 61 cm, is wide enough for the digitizers to be positioned so that the pen could move diagonally across the surface as shown in Fig. 2.

The speed of the pen across the tablet could be manipulated by tilting the table on which the test bed rests. By this means, the forces of friction and gravity could be arranged to control the pen acceleration. For reproducible speeds one could tilt the test bed and let the bridge start from rest, or to produce higher speed, one could increase the angle or give the bridge a push before it reaches the tablet.

The tip of the pen did not touch the surface of the digitizer during these measurements; it was set to move at a constant height of 1 mm above the surface. The digitizers that we measured were designed to operate with paper between the pen tip and the digitizer surface so that pen contact was not necessary for optimum performance. However, to rule out the possibility that performance might be influenced by pen height, we made measurements with pen heights from 0.1 to 3.0 cm. For heights less than 1 cm we found negligible effect on measurement accuracy; above this height we found increased error.

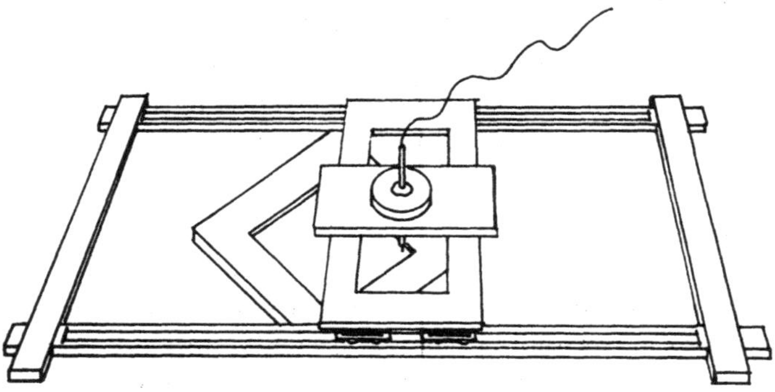

Figure 2. The Test Bed and Digitizer. The test bed is shown here with the digitizer in position for the measurements described in the text.

6. Performance Measurements

6.1 Tablets Used in the Experiment

Using the test bed described in Sec. 5, we performed a series of experiments on three commercial digitizing tablets. We will refer to these digitizers as Tablets A, B, and C. Table I provides the relevant performance specifications for the digitizers. Typically digitizers have many modes of operation. We made measurements only with the sampling rates and resolutions listed in the table. Accuracies are quoted for these tablets when used with a pen stylus; better accuracies may be obtained when a puck or cursor is used instead of a pen. The measurements were made with a single digitizer of each type.

Tablet	Manufacturer	Model	Active Area (cm)	Sampling Rate (Hz)	Resolution (cm)
A	Wacom	SD-421	30 x 30	205	0.0020
B	GTCO	Digi-Pad 5A	28 x 28	200	0.0025
C	Pencept	Penpad 300	28 x 28	160	0.0025

Table I. Digitizer Characteristics.

The experiments described here were performed with each tablet connected to an IBM PC compatible computer via an RS-232 serial link. Software was developed to read the raw data in the appropriate format from the tablet and to write these data to disk in real time. The high sampling rates (up to 205 samples per second) of the tablets required a higher than normal serial baud-rate and an interrupt-driven software driver capable of buffering the data while it is being written to disk. For example, if each sample requires 8 bytes of data, then a tablet sampling rate of 200 Hz generates some 12,800 bits of data per second, which is in excess of 9600 baud. In our measurements we used a 19,200 baud-rate for Tablets A and B and a 9600 baud-rate for Tablet C, which generated 160 samples per second.

6.2 The Experimental Measurements

Dynamic measurements can be useful in evaluating both spatial and temporal errors. Our principal experiments used the test bed to move the pen across the main diagonal of the tablet from low x and y (lower left-hand corner) to high x and y (upper right-hand corner). The bridge on the test bed was started from rest at a fixed position with the entire test bed sloped at a fixed angle. This resulted in a constant acceleration and a linearly increasing velocity of the pen. By changing the slope of the test bed we controlled the range of velocity variation across the tablet. Each tablet was placed so that the pen would cross the lower left-hand corner (0,0 coordinate) at about 20 cm/sec. This resulted in a velocity profile that is much more linear than one can obtain with hand motion.

One might be concerned that the test bed itself could have introduced errors in the measurement. However, we obtained quantifiably different errors for the different tablets. A cross check of error measurements indicates that the errors introduced by the test bed are negligible compared to the errors measured with individual tablets.

Each tablet reported its position data in tablet coordinates. For our graphs the x and y tablet coordinates were converted to actual distance, with coordinate pair 0,0 representing an x value of 0.0 cm and y of 0.0 cm. In general the reported distance was the tablet-coordinate value multiplied by the tablet resolution. We assumed a uniform sampling rate to assign each coordinate pair to a time t. The time origin ($t = 0$) was assigned to the first reported pair of tablet-coordinates in which both x and y are non-zero, and successive coordinate pairs are incremented in time by the inter-sample period τ, the reciprocal of the sampling rate N. A sequence of x and y velocity components could then be calculated by Eq. 1. In the following sections we examine in detail plots of position error *vs* distance across the tablet and tangential velocity vs time, which were derived from the recorded data.

6.3 Measurements of Spatial Error

The most direct method of examining spatial data is to plot the xy coordinates of each sampled point. Since the path of the pen was virtually a straight line, the points when plotted should fall along a straight line. More complete measurements could be done along various diagonals on the tablet, but our measurements were confined to the main diagonal. Figure 3 shows a magnified portion of xy data for each of the three tablets. These plots illustrate well the differing characteristics of the three tablets. Tablet A produced evenly spaced points along a visibly straight line. Tablet B produced points along a straight line, but some of the points are not evenly spaced. The points from Tablet C appear to be evenly spaced, but there is some side-to-side variation from a straight line. In the sections that follow we examine in more detail the origin and consequences of these variations.

In Sec. 3.1 we described how spatial errors can be measured by using the method of least squares to fit a straight line to the xy data. Let the slope of the fitted line be a and let the intercept of the line on the y-axis be b, then define the residual error in y_i to be y_{ir} where

$$y_{ir} = y_i - (ax_i + b)$$

Knowing y_{ir}, the residual error in the direction parallel to the y-axis, we can easily compute d_i the distance of the recorded point from the best–fit straight line. From the geometry of Fig. 4 we can see that $d_i = y_{ir} \sin \theta$, where $\theta = 45°$.

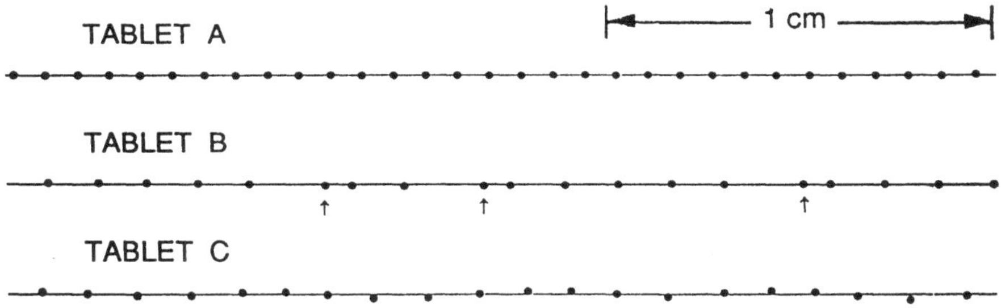

Figure 3. Enlarged Plots of Data Points Reported by the Digitizers. The pen was moving along a straight line at constant speed when these data were recorded. Deviations from linearity and from equal spacing show up at this magnification. Delayed samples recorded from Tablet B are indicated with arrows; their position along the line of points suggests that they were sampled late by about half the regular inter-sample period. The data samples from Tablet C show a sinusoidal displacement from the straight line.

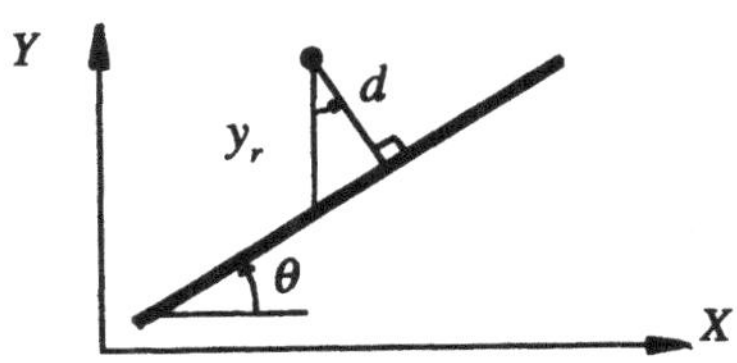

Figure 4. The Relationship Between Residual Error in the y-Coordinate and Distance from the Best–Fit Straight Line.

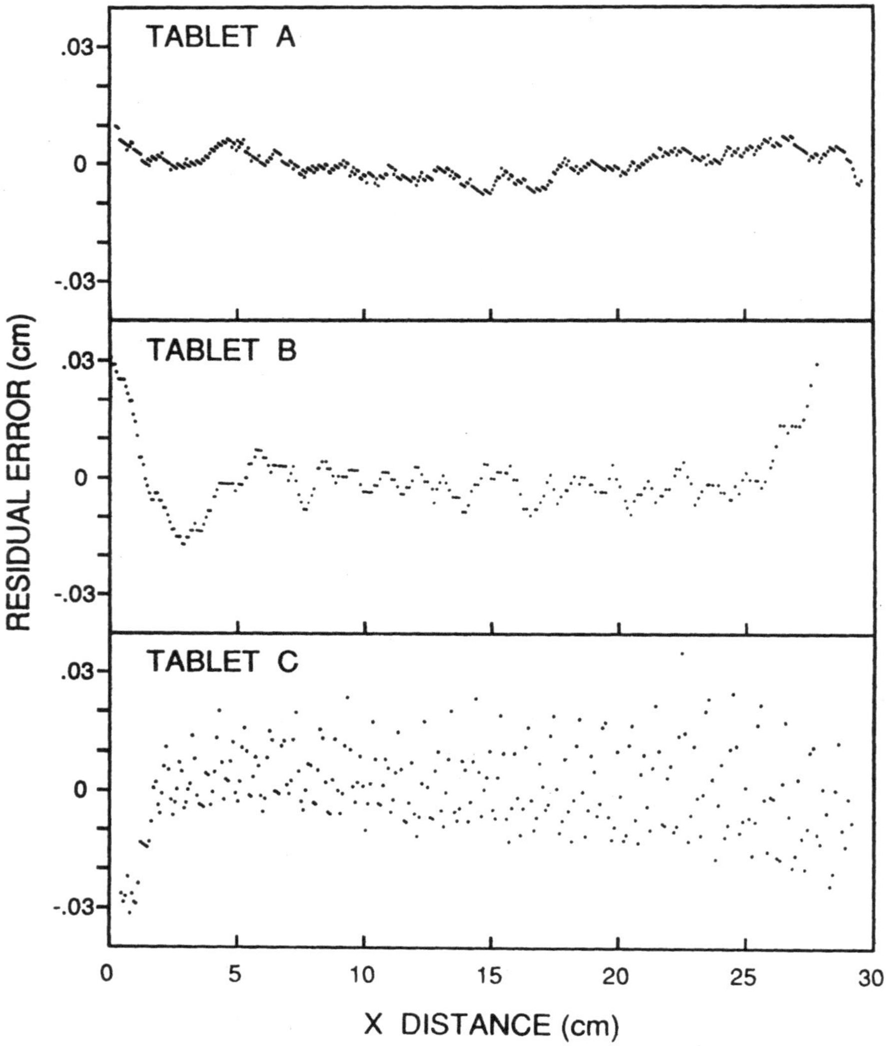

Figure 5. Perpendicular Distance of the Measured Points from the Best–Fit Straight Line Plotted vs the x-Coordinate: Tablets A, B, and C. The relative position–accuracy of the three tablets may be compared in this figure. The larger errors at both ends of these plots are examples of the boundary effect. See Fig. 6 for an enlarged plot of the data shown here for Tablet C.

Figure 5 shows the position errors d_i plotted as a function of distance along the main diagonal of the three tablets. Larger errors appear on the right and left sides of these plots at positions corresponding to the lower left and upper right corners of the digitizers. This is evidently a *boundary effect* because the errors are smaller elsewhere along the main diagonal. Table II lists the measured standard deviations of position error d_i; we omitted data distorted by boundary error for these calculations. Also included in Table II are the manufacturers' specifications for the accuracy of these tablets. The measurements in Table II show that all three tablets perform within specifications.

Tablet	$\sigma[d_i]$ (cm) *	Accuracy (cm) **
A	0.0034	±0.025
B	0.0090	±0.025
C	0.0110	±0.013

* Standard deviation of the distance of the measured points from the best–fit straight line from $x = 5$ to 25 cm

** Manufacturer's specification of the position accuracy

Table II. Position Error.

The patterns of position error for Tablets A and B in Fig. 5 do not appear to be entirely random. Errors correlated over distances of the order of one centimeter show up in these plots. In addition the effects of quantization show up in the diagonal pattern of the points from Tablet A and the horizontal pattern of some of the points from Tablet B.

Tablet C exhibits the largest residual errors in Fig. 5. But contrary to the impression given by this figure, the errors of Tablet C are not random. We need to look at the enlarged plot to see this. Figure 6 shows a segment of these data, and one can see that the errors are in fact periodic in x with a spatial period of about 0.5 cm.

According to the manufacturer this periodicity corresponds to a pattern in the structure of Tablet C. The technology employs grids of parallel wires on a printed circuit board extending in both the x and the y directions; the spacing of these wires is 0.508 cm. The errors are evidently the result of inaccuracy in interpolating positions which fall between parallel wires. Because the errors are periodic, it might be possible to develop a correction table which could be used to remove some of the measurement errors from Tablet C.

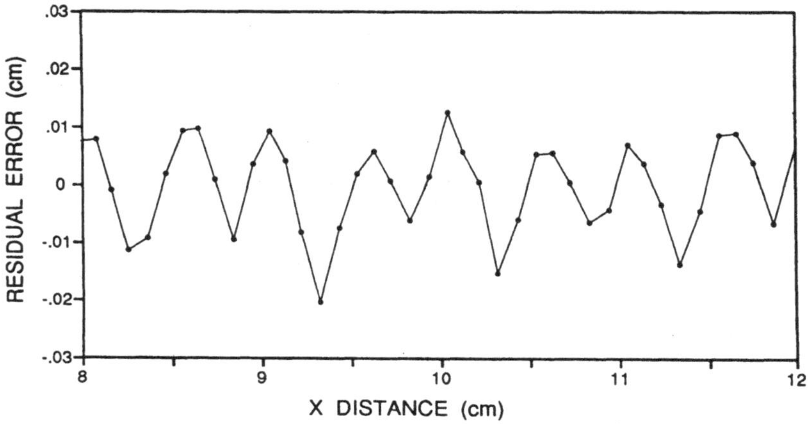

Figure 6. An Enlargement of the Position–Error Data from Tablet C Plotted *vs* Distance. This sample of the data in Fig. 5 clearly shows the sinusoidal nature of Tablet–C errors; these errors may be seen to have a spatial period of about 0.5 cm.

6.4 Measurements of Velocity Error

In handwriting studies it is important to know the effects of instrumental error on velocity measurements. We collected data on the test bed with the pen moving with nearly constant acceleration so that we could remove the slowly varying trend in velocity and obtain the residual velocity error. In this section we discuss measurements of pen velocity derived from the same data used to calculate the position errors shown in Fig. 4.

We calculated the components of pen velocity using Eq. 1, and we found negligible difference between the data for v_x and v_y. Figure 7 shows the x component of velocity plotted against time for the three digitizers. In this figure the velocity curve for Tablet A appears fairly linear with the low variance. This is a consequence of the comparatively small position error and the uniform sampling rate of this tablet, which was evident in Fig. 3. The data for Tablet B exhibits unusual behavior; velocity "outliers" show up aperiodically in Fig. 7. The underlying cause of these aberrant velocity points can be understood by again looking at the data plotted in Fig. 3, where we can see evidence of a non-uniform sampling period. The manufacturer reported that in some instances the position samples on Tablet B may be *delayed* by one–half of the normal sample period; as a consequence the speeds calculated sequentially across an aberrant point are alternately too large and too small by one–half. The pen–speed data for Tablet C in Fig. 7 shows the largest scatter. These residuals are periodic because of the spatial periodicity of the position error, and here the periodicity of the residuals was modulated by the changing pen speed. With the exception of the singular points for Tablet B, there appears to be no evidence in Fig. 7 for an increase in velocity error with increasing velocity, and excluding these singular points, we can conclude from the analysis in Sec. 4 that the sampling rates were uniform on these tablets.

We calculated the residual errors in the x and y components of velocity by removing the slowly varying speed of the pen which is evident in Fig. 7. Table III lists the measured standard deviations of residual error in the velocity components. For each of the three tablets the residual velocity errors of the x and y components have nearly the same standard deviation. Table III also lists the standard deviation predicted for each tablet by Eq. 5, which assumes that the position errors were the result of a gaussian random process and that the sampling periods were uniform. In Table III we find that the predicted value of $\sigma[v_x]$ is larger than the measured value for Tablets A and C, which we might expect because $\sigma[x]$ includes correlated as well as random errors. On the other hand for Tablet B, where we know there were variations in sampling period, the predicted value of $\sigma[v_x]$ is *smaller* than the measured value because of the variations in τ.

Tablet	$\sigma[v_x]$ (cm/sec)	$\sigma[v_y]$ (cm/sec)	Predicted $\sigma[v_x]$* (cm/sec)
A	0.19	0.20	1.0
B	6.1	6.2	2.6
C	1.7	1.8	2.5

* Predicted from the measured position error by Eq. 5, which assumes purely random position error

Table III. Velocity Error.

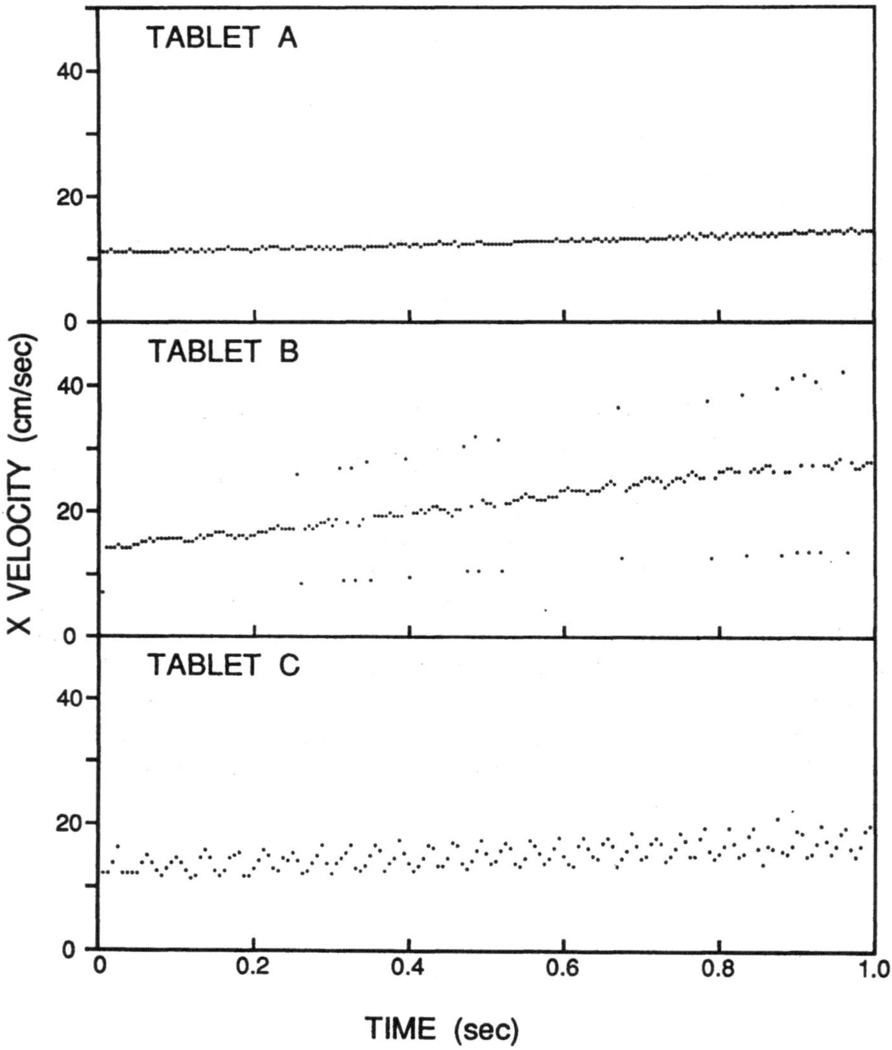

Figure 7. The Velocity Component v_x Plotted vs Time: Tablets A, B, and C. The plotted points for Tablet A show the effects of velocity quantization not the resolution of the plotter. The anomalous points in the data for Tablet B are a result of irregularities in the inter-sample period like those shown in Fig. 3. The sinusoidal position errors in Tablet C produce corresponding errors in the velocity component.

From the velocity data shown in Fig. 7 for Tablet A we computed the acceleration component a_{xi} using Eq. 6, and we removed a small, slowly varying trend to obtain residual errors which had a standard deviation of 61.4 cm/sec. This standard deviation is in close agreement with a value of 60.0 cm/sec predicted by Eq. 7, which was derived by assuming a constant sampling rate and a gaussian random distribution of velocity error 6.5. Asynchronous-Sampling Errors

The position error produced by non-simultaneous sampling of the x and y coordinates depends on the direction and speed of pen motion. Equation 3 in Sec. 3.2.2 specifies this functional dependence. Teulings and Maarse (1984) have described a simple method for observing the effect of this sampling delay. "(One) samples pen movements along a ruler in two opposite, diagonal directions. When the reported x- and y-coordinates are plotted, the up and down movements will appear as oppositely curved instead of straight lines." When we tested our three tablets in this way, we found clear evidence of sampling delay only in Tablets A and B. The largest displacement of these lines was only about an order of magnitude greater than the tablet resolution, making accurate measurement difficult.

We measured the time delay on the test bed by pushing the bridge diagonally across the tablet at relatively high speeds and by remeasuring this path at low speed. We fitted a straight line to the slow-motion data and calculated the y-displacements of the high-speed points from this fitted baseline. To obtain accurate estimates of the pen speeds corresponding to the y–displacements, we fitted a straight line to calculated values of v_{xi}. Finally we used Eq. 2 to calculate δ. Table IV gives the results of these measurements and calculations. We have implicitly assumed that the time delay δ is constant and independent of position on the tablet. Our results, of course, depend on the validity of this assumption.

Tablet	AverageΔy (cm)	Average v_y (cm/sec)	δ (msec)	$\sigma[\delta]$ (msec)
A	0.045	46.5	1.0	0.2
B	0.057	30.0	1.9	0.3
C	0.0008	29.0	0.0	0.3

Table IV. Measurement of the Sampling Time-Delay δ.

Based on these measurements of δ, we can use Eq. 3 to estimate the maximum position errors that would be produced by a pen speed of 20 cm/sec, which is a relatively high speed for cursive writing. The maximum errors, measured normal to the direction of pen motion, would be 0.010 and 0.019 cm respectively for Tablets A and B. For Tablet C the error would be negligible. Although the resulting position errors for Tablets A and B are relatively small, it should be noted that these sampling delays are not small compared with the 5-msec interval between samples.

7. Conclusions

The measurements reported here uncovered details of digitizer performance that were not previously recognized. Although we found that the accuracy of all three digitizers was within the manufacturer's specifications, serious deficiencies appeared when we used two of the instruments to measure pen velocity. Only one of the digitizers was able to sample pen positions at a constant rate *and* without serious correlated instrumental errors. Using a test bed we were able to record data with controlled pen speeds which allowed us to measure the residual errors in position and velocity. We found that defects in digitizer performance showed up clearly when we calculated components of pen velocity from first differences and plotted these data versus time. Two types of instrumental error were found: (1) breaks in the uniformity of the sampling period which appeared aperiodically with Tablet B and (2) systematic errors in interpolation between grid lines which were found with Tablet C. We subtracted out the controlled variations in pen motion to obtain the residual errors, and we found a wide variation in the standard deviations of the residuals among the three tablets. With the most accurate tablet we measured a standard deviation of 0.21 cm/sec for the residual errors in v_x and v_y. An error–propagation analysis showed that velocity error is independent of magnitude of the velocity provided that the sampling period is constant.

We investigated the effects produced by asynchronous sampling of the x and y coordinates and found that sampling delay produced errors in the reported positions that were perpendicular to the direction of pen motion and proportional to pen speed. These errors were largest when the pen moved along the diagonal and were zero when the pen moved parallel to the x– or y–axes. For Tablets A, B, and C we measured sampling delays of 1.0, 1.9, and 0.0 msec, respectively. With the largest sampling delay the position errors are small, amounting to an offset of 0.02 cm from a diagonal line drawn with a pen speed of 20 cm/sec. The analysis and measurement of the effects of sampling delay show these effects to be small.

Acknowledgements

The authors wish to thank Elaine Briggs for her assistance in programming the data analysis and producing the computer–generated graphs for this article. This research was supported by the National Institute of Child Health and Human Development under grant HD22343.

References

Bevington, P. R., *Data Reduction and Error Analysis for the Physical Sciences*, McGraw Hill, New York, 1969, pp. 56-65.

Blesser, B., Ward, J. R., "Design criteria for an electronic tablet technology with acceptable performance for handwriting capture and analysis," *Proc. 3rd Int. Symp. Handwriting and Comp. App.*, Montreal, 1987, pp. 178-180.

Diehl, S., Apiki, S., "Graphic details," *BYTE*, **14** 1989, pp. 162-174.

Hearn, E. D., Lankford, L. G., *Digitizer Source Book*, Geocomp Ltd., Golden, Colorado, 1989, 137 pp.

Kim, J., Tappert, C. C., "Handwriting recognition accuracy versus tablet resolution and sampling rate," *IEEE 7th Int. Conf. Patt. Rec.*, Montreal, 1984, pp 917-918.

Lanshammar, H., "On precision limits for derivatives numerically calculated from noisy data," *Journal of Biomechanics,* **15** 1982, pp. 459-480.

Maamari, F., Plamondon, R., "Extraction of the analog pentip position, velocity and acceleration signals from a digitizer," in *Neuronal and Motor Aspects of Handwriting*, J. S. R. Kao, (Ed.), North Holland Publ., 1985.

Parks, T. W., Burrus, C. S., *Digital Filter Design,* John Wiley & Sons, 1987, pp. 286-287.

Phillips, M. J., "Several simple tests can help you choose the correct digitizer," *Computer Technology Review,* **7** 1987.

Press, W. H., Flannery, B. P., Teukolsky, S. A., Vetterling, W. T., *Numerical Recipes,* Cambridge University Press, 1986, pp. 381-443.

Tappert, C. C., "Cursive script recognition by elastic matching," *IBM Journal of Research and Development*, **26**, 1982, pp. 765-771.

Teulings, H. L., Maarse, F. J., "Digital recording and processing of handwriting movements," *Human Movement Science*, **3**, 1984, pp. 193-217.

Ward, J. R., Phillips, M. J., "Digitizer technology: performance characteristics and the effects on the user interface," *IEEE Computer Graphics and Applications*, April 1987, pp. 31-44.

Computer Processing of Handwriting
Eds. R. Plamondon & C. G. Leedham

THE PAD (PEN AND DISPLAY) - A DEMONSTRATOR FOR THE ELECTRONIC PAPER PROJECT

COLIN A HIGGINS
Department of Computer Science, University of Nottingham, Nottingham. NG7 2RD, United Kingdom (email c.higgins@uk.ac.nott.cs)

R JAMES DUCKWORTH
Worcester Polytechnic Institute, Department of Electrical Engineering,,Worcester. MA 01609 USA (email rjduck@ee.wpi.edu)

The Electronic Paper Project (EPP) of the National Physical Laboratory, United Kingdom, has been in progress for five years. The project involves using the natural gestures and strokes of writing and drawing to manipulate data presented on a flat panel display/digitizer. It thus requires an electronic mimic of pen and paper. A hardware device was required to test and demonstrate the project software. The PAD has been designed and constructed to support the EPP. The PAD consists of a flat AC plasma display panel with a transparent digitizer overlay, controlled by a microprocessor. The prototype is at present desk-bound but it is hoped that future designs should be light enough to be carried or used on the lap. The size and weight is very dependent on current flat screen technology. The prototype operates in a *stand-alone* sketch mode, or it can be driven by a host computer via a serial line, or linked directly to a SUN workstation via the VME bus. The PAD acts as a display device and 'mouse' to the SUN window manager enabling it to be used with all SUN facilities. Initial observations indicate that the PAD is a useful prototype device for investigating the EPP concept. Due to the technology level available on construction it has several limitations such as high parallax and large size.

1. Introduction

Electronic paper will consist of a flat panel display that can be written on with a stylus, see Figure 1 (Pobgee (1988)).

The movement of the stylus is traced onto the screen, and the gestures interpreted and understood. *Electronic paper* attempts to overcome many of the problems of present day human computer interfaces by presenting itself to the user as though it is a piece of paper. However, *electronic paper* will be more than just an input/output device (Worksheet (1988)) it will also include intelligence at a functional level (Ford (1987)). An interaction with the machine will involve pointing to an object, drawing a symbol or character, or writing. This leads to a very *user-natural* interface, since compound actions can be expressed very simply with a few handwritten symbols. The user's hand-drawn/hand-written message is recognized by the interface and appropriate action is taken. Due to the simplicity of the interface there is the potential to dispense with complex user manuals placing the burden of understanding onto the machine.

A hardware device called the PAD was designed and constructed in 1986 to test and demonstrate the project software available at the time and developed since. This paper describes the design of this hardware and explains the choice of particular components. It later describes the *Electronic Paper Project (EPP)*. At the time of design, no hardware systems were available, although IBM had a prototype on which some test experiments were performed (Tappert 1986) and several others have been developed since. Kankaanpaa (1988) describes a similar, smaller device, built in 1988, based on an electroluminescent display, rather than a plasma display. Several commercial systems are now available, all based on liquid crystal displays with resolution limited to a maximum of 640 by 480 pixels.

2. Requirements

In the simplest sense the hardware must mimic pen on paper, that is as the stylus moves over a surface the display beneath illuminates the path taken. In a wider sense *electronic paper* includes the software to increase the functionality of the system, such as text and diagram editors, sketch packages, diaries and the ability to flick through 'pages' of an electronic note-book. The hardware has to support high speed communications with a powerful processor and lower speed communications for data transfer to other machines.

The hardware for *electronic paper* consists of a flat panel display over which is placed a 'transparent' digitizer (Figure 2). The guide-lines suggested by the NPL for an ergonomically acceptable system at the time of design are shown in Table 1. Unfortunately these are somewhat vague since more precise specifications can only be given after a prototype has been tested. Moreover, the technology available is a major limiting factor. Blesser (1987) has subsequently analysed handwriting recognition requirements and gives more concrete specifications for opaque digitizers.

In the first instance the PAD worked in a stand-alone mode whereby movements of the stylus were simply reproduced by the display. It also had a serial interface to enable it to communicate with a host machine.

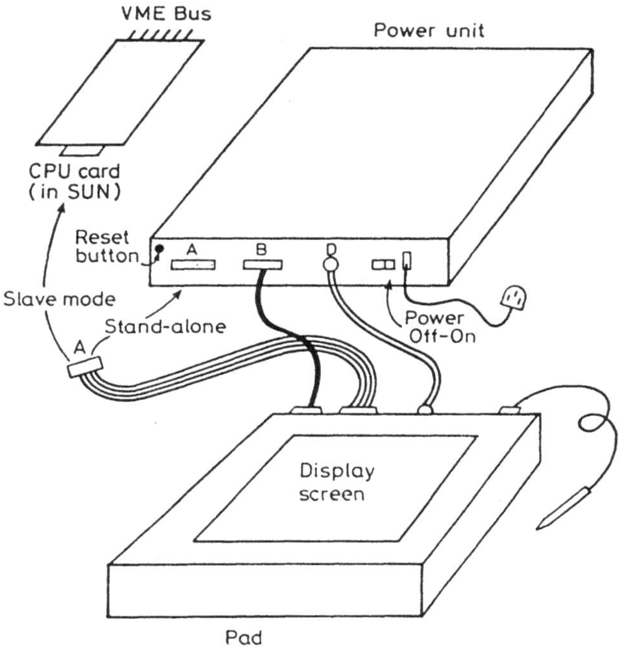

Figure 1. The PAD demonstrator.

Resolution:	display	Minimum 80 pixels per inch (0.3 mm pitch) in both horizontal and vertical directions, or as close to this as possible.
	digitizer	Minimum 1/100th of an inch (0.25 mm pitch) in both horizontal and vertical directions or as close to this as possible.
Speed:	digitizer	Minimum 100 points per second or as close to this as possible.
Size:		As large as possible or practical, at least A4 size (11 inches by 8 inches) was desirable.
Feel:		The stylus must not be too unwieldy or awkward. Action must be natural, for example the user must be able to rest his/her hand on the surface.
Display:		The brightness and contrast of the display should be as high as possible. Reflectance minimized. Parallax between stylus point and display should be a minimum.

Table 1. Component guide-lines.

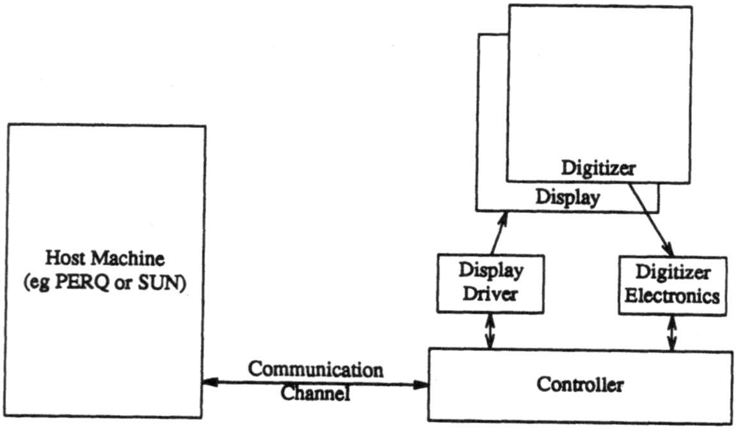

Figure 2. The PAD connected to a host machine.

3. The PAD Components

The following component parts were required for the construction of the PAD.

i) A host computer or workstation to run the recognition and intelligence software.

ii) A display device to illuminate the path taken by the stylus.

iii) A digitizing device that enabled the coordinates of the stylus to be detected as it moved over the writing surface.

iv) A microcomputer controller to synchronise the data capture and display elements and also provide interfaces to the host machine.

v) On-board software to control the system and provide external interfaces.

vi) Host software to communicate with the PAD.

Item i) initially consisted of a Perq 2 workstation which was later replaced by a more powerful SUN workstation. The criteria for the selection of the remaining elements are considered below.

3.1 Display

There were four main alternative technologies available for the display; the cathode ray tube (CRT), liquid crystal display (LCD), plasma and electroluminescent (EL). The features of these types of display are considered in Table 2.

AC plasma displays made by Thomson-CSF seemed to offer the best, large area display at the time and the Thomson TH 7617 was chosen for the prototype. It had a high resolution (0.3 mm pitch), large size (300 by 300 mm) with 1024 by 1024 pixels, and had good brightness and contrast. Although plasma displays use relatively old technology there are still few displays that combine a large display with high resolution. The cost of such a large, high resolution display is therefore high.

One feature of this type of display is inherent memory, i.e. illuminated pixels stay lit until they are explicitly turned off. This would be an advantage in some applications but for our product it meant that a more complicated controller had to be designed.

CRT		Inexpensive, but too thick, flat screen versions too small, too high energy (which might affect the digitizer), not ergonomical, and not robust (danger of implosion).
LCD		Barely adequate size or resolution, far too dull, poor viewing angle. (Some of these characteristics have recently been improved).
EL	AC thin-film	e.g. manufactured by Sharp. High resolution, bright, good contrast, relatively low cost and thin. Maximum size 640 by 400 pixels (120 by 192 mm.).
	DC thick-film	e.g. manufactured by Phosphor Products. Most suited to the construction of custom displays requiring special legends or symbols. Poor brightness and contrast.
Plasma	DC	Dull, poor contrast and suffer from coating of the electrodes. Limited size and generally inferior to AC plasma displays.
	AC	High resolution, large size, bright with good contrast, but are high energy, have inherent memory and are expensive.

Table 2. Flat panel display types.

3.2 Digitizer

The selection of a suitable input device for the PAD proved to be the most difficult part of this project. The authors feel that a combined display and digitizer, purpose built, is the only ideal solution. The problem was that the input method used should not effect the quality of the display in any way. A digitizing tablet placed on top of the display must therefore be totally transparent, not too thick (otherwise parallax may be a problem), and should not introduce unwanted reflections. Unfortunately these criteria eliminated many of the most common digitizers that were available.

Another consideration when choosing the digitizer was whether the stylus should be active or passive. Digitizers that use a passive stylus rely on the pressure applied by the stylus to bring two resistive membranes into contact. With the application of voltages to the membranes the position of the stylus can be determined. Because this method uses pressure, it is also sensitive to anything else resting on the surface and since many people rest their hand or wrist on the writing surface this method is clearly unacceptable. This membrane type of input also has poor resolution and accuracy and is usually limited to applications where a finger is used to select an option from a menu.

On the other hand, using an active stylus can achieve far better resolution and accuracy but it must be connected to the system by a lead of some kind which may detract from its ease of use.

An acoustic digitizer initially seemed to offer the ideal solution. This type uses a point source of sound in the stylus and by measuring the time it takes to reach two microphones it can determine, using trigonometry, the x and y coordinates of the stylus. This method allows high resolution and accuracy to be achieved and it also keeps the surface of the display completely clear. The problem, however, is that the spark gap used as the point source in existing acoustic digitizers makes a buzzing noise which is unacceptable. Future devices may be able to overcome this problem by the use of ultrasound.

A transparent glass digitizer made by Scriptel was chosen as the digitizer for the prototype. Scriptel offered a range of models with active areas ranging from 300 mm by 300 mm to 600 mm by 900 mm. They used an active stylus that detects AC voltages transmitted from a resistive layer sandwiched between two pieces of glass. This method gave excellent resolution (0.025 mm) but the accuracy was only ±0.64 mm. Also the thickness of the digitizer was 7 mm which caused parallax problems. This was, however, the best solution available at the time.

3.3 Controller

The controller selected for the PAD was the Plessey PME 68-1B single board controller with a VME bus interface. This board has a Motorola 68000 microprocessor running at 10 MHz. One of the reasons this board was selected was its extensive I/O capability. It has three serial I/O interfaces and a very versatile 24-bit parallel interface. One serial interface was connected to the digitizer and another was connected to the Host Computer. The parallel port was used to control the plasma display.

3.4 Software

The software for the system was developed on a VAX computer and then cross compiled and placed in PROM allowing it to function as a completely stand-alone system without the need for user intervention.

The majority of the software for the single board controller was written in 'C' supplemented by some assembler routines. Software was provided for driving the display (which was not a memory-mapped device), capturing and transforming the digitizer coordinates, system integration and serial communications with a host machine. The digitizer's coordinate system was used as a base as this had the highest resolution. Much additional software was added later to integrate the PAD with a SUN workstation.

3.5 The PAD System

The actual system (shown in Figure 3) consisted of a Thomson TH7617 plasma display, a Scriptel transparent digitizer (300 mm by 300 mm, 1024 by 1024 pixels) and a Motorola 68000 single board computer with associated electronics as the controller.

3.6 SUN Integration

The PAD was integrated with a SUN workstation during 1987. A block diagram of this arrangement is shown in Figure 4. For this integration a replacement CPU board was installed on the SUN's VME bus with dual port RAM to allow access by both the PAD controller board's CPU and the SUN's CPU. The SUN operating system kernel was adapted to allow it to pass messages to the replacement PAD controller board as well as writing to its usual memory mapped display buffer. The replacement board converted the bit-mapped information from the SUN kernel into the form required to drive the plasma display. At this stage a separate hardware digitizer controller was also integrated into the system to provide more flexible control over the digitizer output. This new system meant that the PAD could then act as if it were a SUN display and mouse, replacing or augmenting the normal SUN screen and mouse. Thus all new software could be written simply as a SUN application and any software that executed on a SUN could be used on the PAD/SUN system.

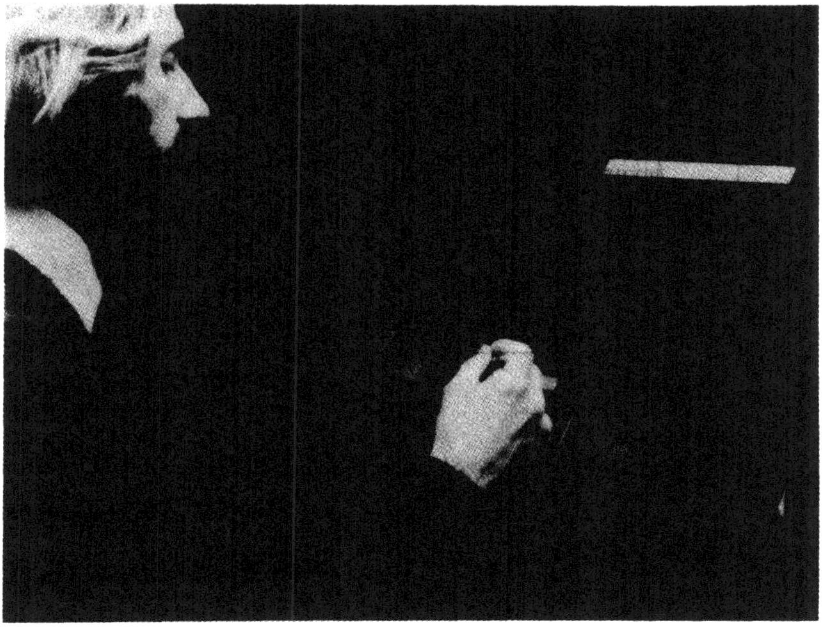

Figure 3. The PAD (photograph by kind permission of the National Physical Laboratory).

4. *Electronic Paper* Software Modules

Electronic paper as a concept requires the enhanced functionality available from software modules that will execute on the PAD (Brocklehurst (1988)). To demonstrate the potential of the new interface and to encourage others to visualise novel applications, many software modules are envisaged. All will run under the control of a generic 'note-pad' controller. To fully utilise these modules, both single separated handwritten character recognition and more importantly connected cursive script recognition (CSR) will be needed. Both of these are under investigation. The software modules so far developed are: a script editor, a text editor, a graph drawing program, a table drawing module, and a histogram (bar-chart) drawing program. These are described below. Other envisaged software included a diary/calendar, a pie-chart program and an interactive command line interpreter based on the use of an iconic interface and CSR.

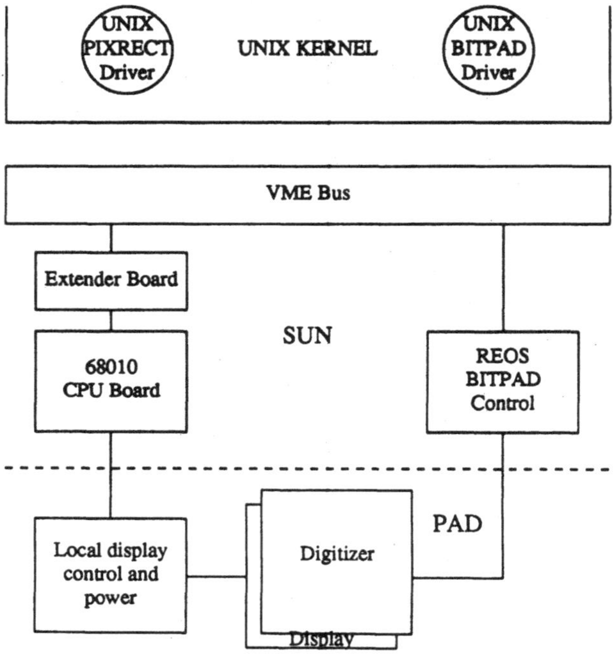

Figure 4. The PAD/SUN system.

The programs currently in existence all operate in a 'stand-alone' mode. They can, however, be run from within a simple controlling program which allows the user to select the required module from a menu and combine the output into a single document. The algorithms NPL used have not been published, but more extensive functional descriptions can be found in Brocklehurst (1988) and Stevens (1988).

4.1 The Text Editor

This is a powerful interactive general text editor for correcting typescript which may have been generated from manuscript text by the CSR sub-system. The writer uses the stylus to make corrections and additions just as he/she would on a typewritten paper draft. Natural gestures are used to specify the corrections to be applied, based on commonly used correction marks as specified by BSI standard BS5261 (1976), 'Copy Preparation and Proof Correction.' For instance a word may be deleted simply by striking it out with a line. The editor then adjusts the text interactively in real time. It can run in two modes, the default, text mode and a table mode which allows simple manipulation of columns of text.

4.2 The Script Editor

This is a simple editor for manipulating and correcting unrecognized cursive script, for use either as a simple 'note-pad' or prior to recognition. For example pre-processed script ready to be input to the recognizer could be edited to join incorrectly segmented words say, or the editor could be used for moving bit mapped text and sketches around a note-pad page. Editing symbols available are a superset of the Text Editor symbols.

4.3 The Table Drawing Module

This module allows the construction of tables and forms. The user roughly sketches a table and on pausing, the sketch disappears and is replaced by the computer's interpretation. Further lines can then be sketched. The system straightens these lines and makes 'intelligent' guesses about where the user intended end points to be located. The lines must be rectilinear. Lines can be toggled between two thicknesses simply by touching them. Text can be inserted into a box by touching in that box (although until the CSR system is implemented this must be done by typing). The text can then be centred or left or right adjusted by pointing to it. Lines can be deleted by drawing a cross through them. A 'snap to' grid is also available. As with all these modules, ease of use can only be fully appreciated by actually using the system.

4.4 The Graph Drawing Module

Here two axes are roughly sketched and the value of the end points is entered. The system redraws these axes and makes sure they intersect at a point appropriate to the given end points. It also inserts intermediate annotation at regular points along the axes. Graph points are next inserted by moving the stylus around on the PAD and lifting it at the desired position (the position being indicated by the numbers in a small box). Straight and curved line fits can be requested and a title, etcetera added. An example graph produced by the system is shown in Figure 5.

4.5 The Histogram Drawing Module

The sequence of steps performed by the user in this module is firstly to sketch the bar-chart's axes (Figure 6) which the program then tidies up. Next the user marks in the horizontal spacing and enters the minimum and maximum values of the vertical axis, along with the value at which the horizontal axis is to be placed. The system re-adjusts the diagram. Individual bars are then 'pulled out' to the required height, the height indicated by a value in a box above the bar, Figure 7. Next each bar may be shaded by selecting an intensity and touching inside the relevant bar. A title, key and dimensional units are requested by the system which the user currently enters via a keyboard but will later use CSR. These can then be moved to the desired position simply by drawing a line from them to that location, Figure 8. At any time the overall size of the chart can be adjusted using 'thumb-bars', and at most points in the construction it is possible to go back and redo previous operations. A completed bar chart is shown in Figure 9.

5. Experiments on the PAD

Two experiments carried out on the PAD are described below. These were performed by a human factors group at the National Physical Laboratory.

5.1 Text Editor *vs* MacWrite

In the first experiment the task was to edit a piece of text using both the NPL text editor and MacWrite (Thomas (1986)). A repeated measures design was used for carrying out the experiment. This design is such that all people carrying out the experiment do all of the tasks, and comparison is made between the results of the different tasks by grouping all the subjects' results together. This design was employed in order to elicit as many comments as possible from the subjects and to get subjective data about which system was preferred. Apparatus used consisted of a Perq 2 workstation for the text editor tasks, and an Apple Macintosh for the MacWrite tasks.

The task to be carried out by the subjects was to amend a piece of text consisting entirely of 'paragraphs' of letters arranged in the same way as words, as shown in Table 3. The aim of the task was to rearrange the letters

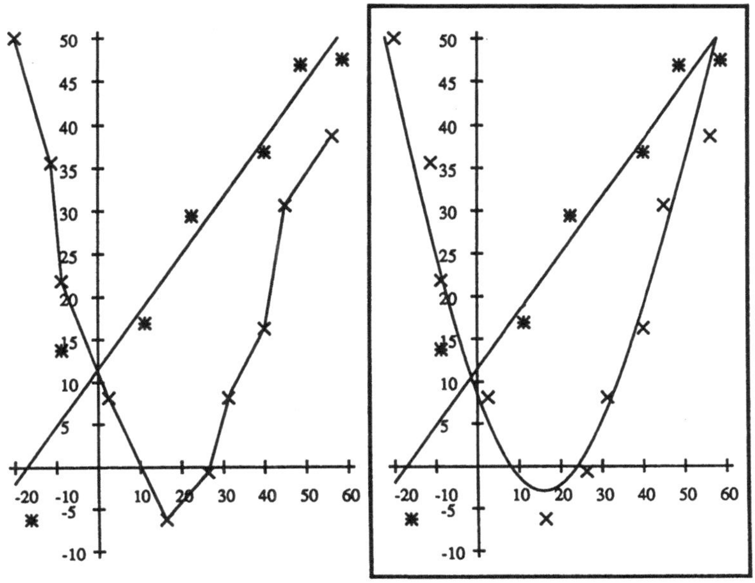

Figure 5. Output of the graph drawing module.

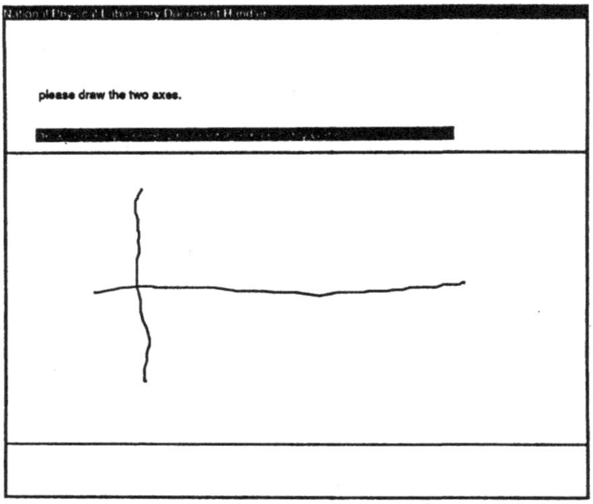

Figure 6. Histogram display (a).

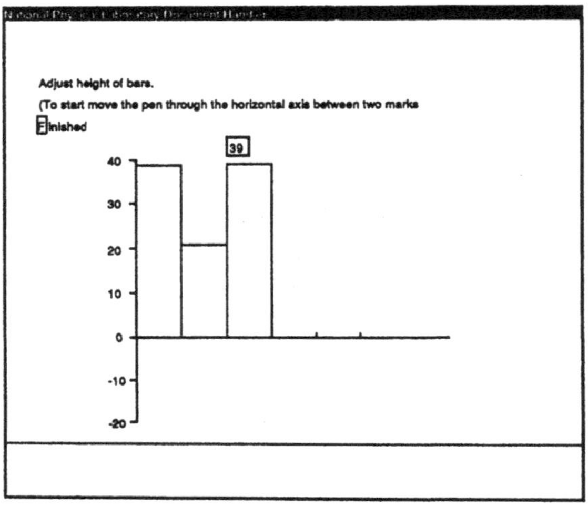

Figure 7. Histogram display (b).

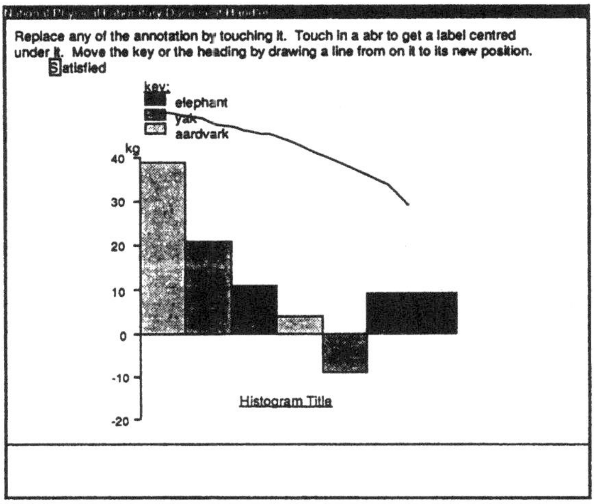

Figure 8. Histogram display (c).

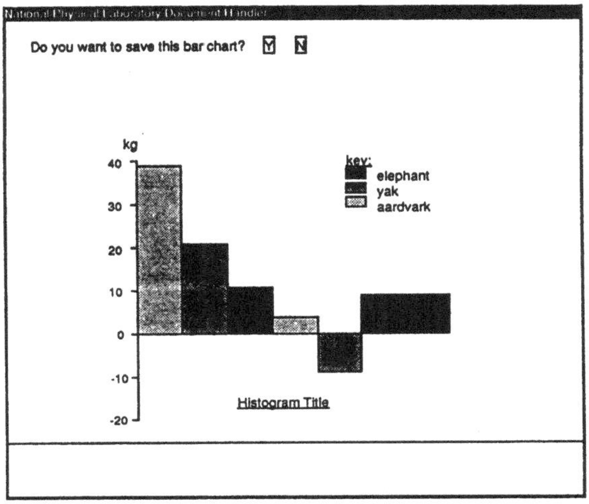

Figure 9. Histogram display (d).

into five paragraphs, each consisting entirely of unique letters (a, b, c, d and e), ordered alphabetically. This design makes it easier to identify the operations that need to be carried out, and factors such as interest level and reading speed do not interfere as they may if normal text were used. The exercise was designed to require all the tasks that are common to both systems.

Various measurements were taken and comments solicited from the participants. Conclusions are tentative as only twelve subjects, all employees of the NPL, participated in the experiment. Basically no differences were found, although in general the text editor came out marginally better. The MacWrite task involved many fewer references to the manual. Might this be due to the ease with which the text editor manual can be referred (it is a single sheet of paper)?

Bbbb bb bbb bbbb bb b bbbbbbb bb bbbbbbbbbb bb bbbbb bb bbbb bbbb bbb bbbbbbb bb bbbbbbbb bb bbbbbbbbb bbb bbbb bbbbbb bbbb bb bbb bbbbbb bbbbb bbbb bbbb bbbbbb bbbbb b.

Aa aaaa aaaa aaaaaa aaaa aa aaaa aa aaa aaa aaaaaa aaaaa aaa aaa aaaa aaaaaaa aa aaa. Aaaaa aaaa aaa aaa aaaaaaaa aaaa.

Xxx xxxxxxx xx xxxxxxxx xxx xx xxxxxxx xxx xxxxxxxxx xxxxx xxxxx xx xxxx xx xxxxx xxxxxxx xxx xxxxxxxxxxx. Xxxxxxxxx xxx xxxx xxxx xx xxxxx xxxxxxxxxx.

Cccc ccccc cc c ccc ccc ccc cccc xxxxxxx cc cccccccc ccc ccc cccc ccccc ccccccccc. Ddd dddd ddddddd ddddd dd aaaaaa dddd dd ddddd d ddddd ddd. Cc c ccccc cc cccc cccc c ccccx cc ccccccc, cccc cc ccccc ccc cccc ccc cc ccc cccc cccccc.

Dd ddddd dddd ddd dddddddd ddd dddddddd dddd ddd ddd dd dddddddd dd ddd ddd dddddd dd dddd ddd ddddd dddddd.

Dddd dddd dddddd dddd ddddd ddd dd ooooooo ddd dddd ddd dd dddddd dddddd dddd ddddd dd. Eeeeeee ee eeeeeee eeee eeeeeeee eeeeeee eee eeeeeeeee. Ee eee ee eee ee eeeee eeeee, eeeee, eee.

Table 3. Text used for the editing task.

5.2 *Electronic Paper* Usability

In this experiment fourteen users were given two tasks: to edit some text using the text editor and to produce a histogram (Rengger (1989)). The former was chosen as an example of a structured task, where all the instructions on how to use the module could be presented to a user as they were needed, on the screen. The latter was chosen as an example of an unstructured task, which required a user be given a written instruction manual, at the start of an editing session, in order to be left free to generate his or her own editing strategy.

The subjects were video recorded for later assessment using unobtrusively mounted cameras. The evaluation was designed to generate the maximum amount of data in order to enable a wide variety of usability metrics and methodologies to be investigated. The tasks were performed about two months apart and each took about thirty minutes to accomplish. A questionnaire was given to the subjects at the end of each task to ascertain how easy the system was to learn, how easy the operations were to perform, how many difficulties were encountered and how annoying it was to overcome these difficulties. The video tapes were later reviewed by the experimenter and various parameters measured and documented.

Various results were reported. In the context of the current paper the most interesting are related to the PAD physical characteristics. Part of the questionnaire related specifically to the PAD and subjects indicated their answers by marking a radius on analogue dials, Figure 10. These dials were later divided into segments for analysis. The results are shown in Figure 11 and summarised in Table 4.

Figure 10. Dials presented to users and segmented for analysis.

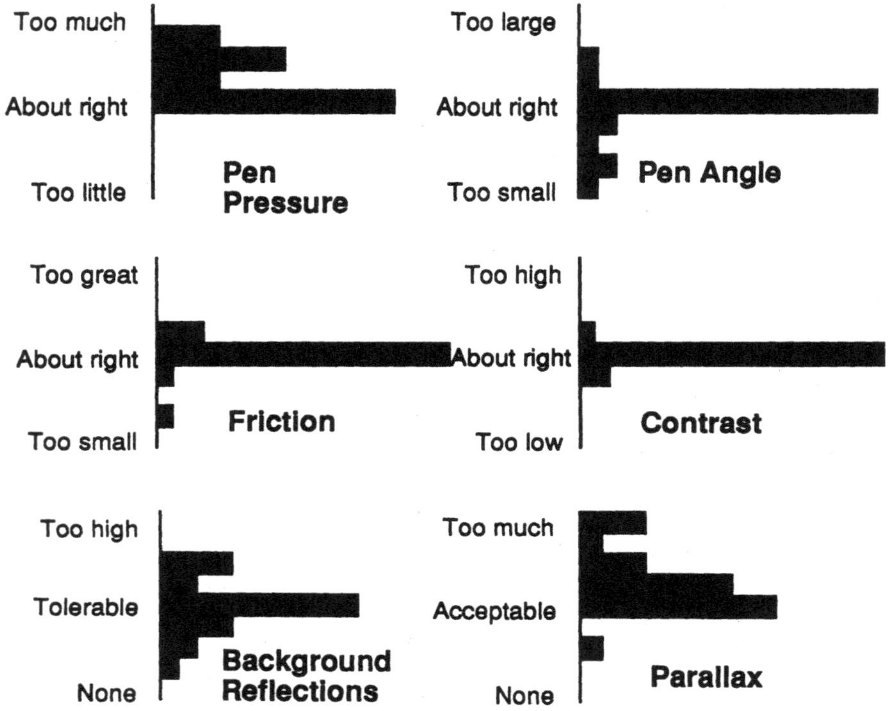

Figure 11. Physical characteristics.

6. Future Developments

The display technology most likely to be suitable in the future will be LCD or possibly electroluminescent. These are getting larger, but are still too small. A minimum size of 1000 by 800 pixels at 100 per inch should be achieved within the next few years.

Transparent digitizers are becoming more widely available. They are still not thin enough, nor does the stylus for these systems yet feel 'natural'.

Several combined units are now available 'off the shelf' which plug into a PC and have active dimensions roughly 220 by 140 mm (640 by 400 pixels). For example the Photron FIOS-6440 has a pixel pitch of 0.33 mm and digitizer resolution of 0.1mm. Again the parallax and feel of these is poor.

A complete unit which includes the processor (Intel 8088), floppy disk drive, smart card reader, RAM, batteries etc is made by Linus Technologies. This is basically an IBM PC compatible device with a CGA (640 by 200 pixels) display.

Pen pressure:	In general it was felt that too much was required to operate the stylus.
Pen angle:	The amount of tilt allowed was not quite sufficient, although a large number of users were happy with the system.
Stylus friction:	Surprisingly for nylon on glass, the feel of the pen on the surface was very acceptable.
Display contrast:	This was also acceptable. This is not a surprise since the main criteria for the selection of the display was its brightness and contrast.
Background reflections:	These were tolerable, perhaps because of the favourable location of the experimental device.
Parallax:	As we expected this is the worst aspect of the system. The parallax between the stylus tip and the display is about one half inch. Amazingly some subjects actually found it acceptable!

Table 4. Reported results of PAD physical characteristics.

7. Conclusions

The *electronic paper* style of interface has many interesting properties and could well become a major style of interface in the future. The PAD has been a useful demonstrator of the possibilities for this type of interface, but is being superseded by improvements in technology. When the problems of connected handwriting recognition have been solved, pocket diaries, note-books and personal organisers can be replaced with an electronic device which will create as large an increase in functionality as the PC provided for the desk top.

Acknowledgement

We would like to thank the National Physical Laboratory and Central Computer and Telecommunications Agency of the British Government for their support of this work. Thanks are also due to David Ford for his assistance with the project.

References

Blesser, B. and Ward, J.R., "Design criteria for an electronic tablet technology with acceptable performance for handwriting capture and analysis," *Proceedings of the Third International Symposium on Handwriting and Computer Applications*, Montreal, Canada, July 1987, pp. 178-180.

Brocklehurst, E.R. and Stevens, M.J., "Software modules for electronic paper," *National Physical Laboratory Report DITC* 127/88, October 1988.

"*Copy Preparation and Proof Correction*," BSI standard BS5261, 1976.

Ford, D.M., Higgins, C.A. and Brocklehurst, E.R., "The electronic paper project," *Proceedings of the Third International Symposium on Handwriting and Applications*, Montreal, Canada, July 1987, pp. 194-196.

Kankaanpaa, A, "FIDS A Flat-Panel Interactive Display System," *IEEE Computer Graphics & Applications*, Vol. 8, No. 2, March 1988, pp. 71-82.

Kim, J., "On-line gesture recognition by Feature Analysis," *Vision Interface '88*, Edmonton, Canada, June 1988, pp. 51-55.

Mel, B.W., Omohundro, S.M., Robinson, A.D., Skiena, S.S., Thearling, K.H., Young, L.T. and Wolfram, S., "Tablet: Personal Computer in the year 2000," *Communications of the ACM*, **31** No.6, June 1988, pp. 638-646.

Phillips, M. and Ward, J., "Digitizer technology: performance characteristics and the effect on the user interface," *IEEE Computer Graphics and Applications*, April 1987, pp. 31-44.

Pobgee, P.J., "A prototype system for interactive input of cursive information," *National Physical Laboratory Report DITC* 125/88, September 1988.

Rengger, R.E., "Exploring experimentally derived usability metrics by a laboratory evaluation of electronic paper," *National Physical Laboratory Report DITC* 119/88, July 1988.

Stevens, M.J., "A text editor driven by hand-drawn symbols," *National Physical Laboratory Report DITC* 124/88, September 1988.

Tappert, C.C., Fox, A.S., Kim, J., Levy, S.E. and Zimmerman, L.L., "Handwriting recognition on a transparent tablet over flat display," *SID International Symposium*, May 1986, pp. 308-312

Thomas, C.M., "A pilot evaluation of an electronic paper text editing system by comparison with MacWrite," *National Physical Laboratory Technical Memorandum DITC* TM5/86, 1986.

"The Worksheet. The personal computer in the year 2000," *Personal Computer World*, **11** No.10, October 1988, pp. 155-162.

Computer Processing of Handwriting
Eds. R. Plamondon & C. G. Leedham

A SCALE-SPACE FILTERING APPROACH TO STROKE SEGMENTATION OF CURSIVE SCRIPT

M. KADIRKAMANATHAN and P. J. W. RAYNER
Cambridge University Engineering Department, Trumpington Street, Cambridge CB2 1PZ, England

In cursive script writing, variations occur in many sizes from small to large. It is therefore preferable to analyse and compare these variations at their respective sizes and hence the method of scale-space filtering is pursued. A scale-space based stroke segmentation algorithm, which takes as input a sequence of XY coordinates sampled at equidistant points along the path of the cursive script, is proposed. A suitable smoothing function is also proposed to scale the data for analysis at different resolutions. The segmentation points obtained at a range of scales are suitably combined to perform the final segmentation. An experiment is carried out to compare the consistency of segmentation of the proposed algorithm with that of the conventional pen-speed minima picking algorithm. The design of a cursive script recognition system with the proposed segmentation algorithm as its front end is also briefly discussed. A second experiment is carried out to evaluate the overall recognition performance of the recognition system.

1. Introduction

Stroke segmentation of cursive script has received a lot of attention for applications in cursive script recognition, signature analysis, child behavioural studies, motor dysfunction studies and handwriting assessment. However the stroke segmentation obtained by most methods proposed in the on-line cursive script recognition literature are different from the hypothetical stroke, defined by psychophysicists as the period of motion between two successive pen-speed minima during a continuous pen movement (Morasso, 1986). This discrepancy arises partly due to the data representation used by such algorithms, a time sequence of geometric features/parameters, used for segmentation in most cursive script recognition systems (Berthod and Ahyan, 1982; Ehrich and Koehler, 1975). One of the main reasons for not using the pen-speed information is to allow the possibility of extending the on-line system to recognize off-line data (data for which pen-speed information is not available).

In this paper, two conventional stroke segmentation algorithms, one of which picks troughs in the pen-speed data and the other of which picks peaks in the local curvature, are reviewed. As the strokes and other related properties of handwriting have a wide range of variations, a scale-space filtering approach to segmentation is pursued. The proposed algorithm produces a multidimensional representation of cursive script by scale-space filtering and then compares the different representations obtained at a range of scales to arrive at a set of segmentation points. The input to the proposed algorithm is an XY coordinate description of the process sampled at equidistant points along the curve in the temporal order of construction. An experiment is carried out to measure the consistency of segmentation of cursive letters as they occur in words by the proposed algorithm and by the pen-speed based segmentation algorithm. Wann and Kadirkamanathan (1989) have compared the proposed segmentation algorithm with the conventional segmentation algorithm with childrens' handwriting data.

An author-dependent on-line cursive script recognition system is built with the proposed stroke segmentation algorithm as its front end. The design and performance of this system is also briefly discussed. A more detailed description of the recognition system may be found in Kadirkamanathan and Rayner (1989) and Kadirkamanathan (1989).

This paper is organised into two parts; the first part, from sections one to six, describes the theory of the scale-space based segmentation algorithm. The segmentation consistency experiment appears at the end of this section. The second part of this paper, from sections seven to nine, discusses the layout of the recognition system and outlines the recognition theory. The recognition experiment is included at the end of the second section.

2. Curvature and Pen-Speed

Fig. 2 shows a typical pen-speed plot consisting of many cycles, of peaks and troughs, obtained from the cursive script shown in Fig. 1. The period between two successive pen-speed minima is attributed to a single motor action. The pen-speed plot is therefore split into a series of profiles where each profile corresponds to the duration of a single stroke.

Figure 1. Handwritten word "husk".

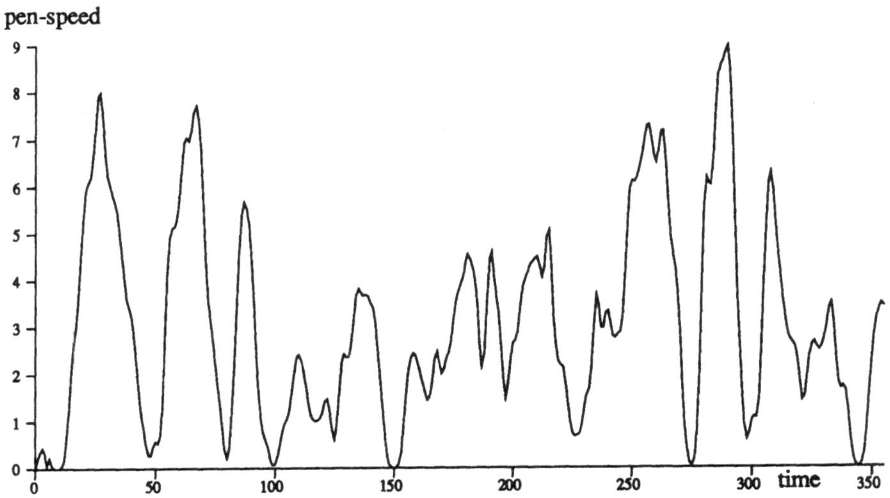

Figure 2. The pen-speed profile of "husk".

A negative correlation between the pen-speed and the local curvature was first observed many decades ago. Recently Viviani *et al.* (1985) claimed that the relationship may be modeled as a power law

$$K = K_T P^{-\gamma} \qquad \text{where } K_T,\ \gamma > 0 \tag{1}$$

K_T is an author-dependent parameter, K is the local curvature and P the pen-speed. The value of γ is observed to be between 1/3 and 2/3. It therefore seems natural to perform segmentation at locations of maximum curvature as an alternative to the conventional pen-speed based method.

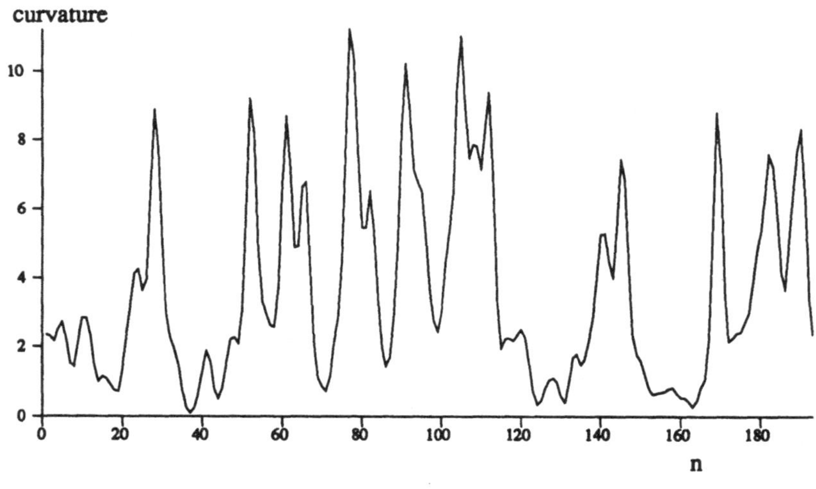

Figure 3. The curvature profile of "husk".

2.1 Stroke Segmentation by Pen-Speed Minima

The conventional stroke segmentation method locates the pen-positions in cursive script which correspond to a local minimum of pen-speed. These are designated as points of segmentation. In an alternative approach, Morasso (1986) uses curvature peaks for segmentation, exploiting the relationship between curvature and pen-speed.

Although the curvature, or the pen-speed variations, for simple hand-drawn patterns, show cycles consisting of maxima and minima, corresponding with the strokes (Viviani 1985), the same "nice" behaviour is not present in the majority of handwritten words. The curvature and pen-speed plots for a handwritten word "husk" are shown in Figs. 2 and 3. The absence of this "nice" behaviour is generally attributed to effects of coarticulation, discrete quantisation owing to poor resolution and the noise characteristics of the digitiser, especially when considering handwriting of normal sizes. It has been observed (Morasso, 1986) that excessive coarticulation may totally destroy some of the peaks, which would otherwise be present in the data. In addition, discrete quantisation and the digitiser noise have been observed to introduce additional spurious peaks. These spurious peaks are, however, smaller in size as compared with the average velocity profile.

In addition to various signal enhancement processes, it is quite common to use a peak size threshold to separate the spurious from the genuine peaks. Some stroke boundaries may be lost if the threshold is too large while extra segmentation points may be gained if the threshold is too small. For naturally constrained handwriting, a threshold to achieve good separation either does not exist or is not a known *a-priori*. It appears that the conventional method based on a predetermined fixed threshold may not yield a very high performance.

3. Scale-Space Filtering

The size, or duration, of strokes within cursive script writing, varies widely both within and across words, and so do the pen-speed and curvature profiles. It is therefore preferable to look at these variations at many different scales or resolutions. The term scale, here and in the remainder of the chapter, implies a measure of spectral scaling.

In scale-space filtering, the information signal is convolved with a family of smoothing functions, each with a different `bandwidth parameter'. The resulting smoothed signals are then further processed to obtain a multi-dimensional description of the signal. The process involved here is that of stroke segmentation. At a very fine scale (large bandwidth parameter), the signal description will have extraneous segmentation points, while at a very coarse scale (small bandwidth parameter), only a few prominent stroke boundaries will be present (see Fig. 5). Computer vision has seen many applications of this method (Asada and Brady 1984; Babaud *et al.* 1986; Yuille and Poggio, 1986).

It is a requirement in scale-space filtering that the signal description obtained at one scale must be related to the description obtained at another scale (Babaud *et al.* 1986). For example, a segmentation point detected at a coarser scale must also be detected a finer scale, but not necessarily at the same temporal location in time. The locations of the segmentation points detected, however, are likely change with smoothing. If the locations of the segmentation points detected are plotted against the scale and the points which correspond to the same stroke boundary are connected together, the resulting graph known as the scale-space plot, would appear as shown in Fig. 12. In general, the smaller the variations are, the easier it is for the algorithm which analyses the scale-space map.

If it is assumed that an optimum 'scale' of analysis exists (this is similar to assuming that an optimum threshold exists for pen-speed segmentation), the appropriate scale may be found very efficiently with the aid of the scale-space map. The criterion used is based on higher level constraints or minimisations. A typical application may be found in Mokhtarian and Mackworth, 1986. However, this approach is not pursued by the proposed segmentation algorithm, presented in this paper.

3.1 Properties of Smoothing Functions

The choice of the smoothing function, which determines the scale of measurement, is central to scale-space filtering. The smoothing functions should satisfy the properties stated below. The traditional assumptions made on the properties of the smoothing function are (Babaud *et al.* 1986):

A1 The filter is linear time invariant. This implies that the smoothing process may be estimated as a convolution. The smoothed signal may be expressed as

$$f(s) \otimes h(.,.) = \int f(s - \zeta).h(., \zeta)d\zeta \qquad (2)$$

where $\underline{h}(.,.)$ is the smoothing function and $f(.)$ the information signal.

A2 The filter has no preferred scale and hence expressed in the form,

$$h(\sigma,\zeta) = 1/\sigma \; h(1,\zeta/\sigma), \tag{3}$$

where σ is the *bandwidth parameter*. $1/\sigma$ ensures that the filter is normalised for unit gain.

A3 The position of the center of the filter is independent of except at boundaries.

A4 The filter recovers the whole signal at sufficiently small scales as

$$\lim_{\sigma \to 0} h(\sigma, s - \zeta) \otimes f(s) = f(s)$$

thus implying $$\lim_{\sigma \to 0} h(\sigma, \zeta) = \delta(\zeta), \tag{4}$$

where $\delta(.)$ denotes the Dirac delta function.

A5 The filter goes to zero as $|x| \to \infty$ and as $\sigma \to \infty$. Assumptions A4 and A5 ensure that the filter covers the entire range of scales.

4. Scale-Space Filtering of Stroke Boundaries

4.1 The Sampling Process

The pen coordinates are sampled as the curve is being drawn, such that successively sampled coordinates are equidistant along the curve in xy the plane. The cursive script process is then treated as a time series of two dimensional vectors,

i.e. $$\mathbf{p}(n) = (x(n), y(n)) \qquad \text{where } s = nS. \tag{5}$$

S is the distance between any two adjacent samples and $\mathbf{p}(n)$ corresponds to an instantaneous position vector on the xy plane.

4.2 The Smoothing Process

Choosing an isotropic smoothing process the smoothed script at a scale σ is denoted by $\{\mathbf{p}(i,.)\} = (\mathbf{x}'(\sigma, n), \mathbf{y}'(\sigma, n))$ where

$$\mathbf{x}'(\sigma, n) = \sum_{\tau = -\omega}^{\omega} h(\sigma, \tau)\, x(n - \tau)$$

$$\text{and} \quad \mathbf{y}'(\sigma, n) = \sum_{\tau = -\omega}^{\omega} h(\sigma, \tau)\, y(n - \tau) \qquad (6)$$

$h(\sigma,i)$ is the impulse response of the filter and ω the window size.

4.3 The Smoothing Function

The Gaussian function, which exhibits excellent smoothing properties, is the conventional choice of smoothing function. However, in this paper, a first order symmetric Infinite Impulse Response function is preferred to the Gaussian, since the former appears to provide more consistent stroke boundary information. This is illustrated in Figs. 11 and 12.

The symmetric Infinite Impulse Response function may be derived by considering smoothing as having a "contextual" influence of the pen process. The term "context" here refers to the immediate past and the immediate future pen-positions with respect to the current pen-position. If the influences are modelled as linear functions, the process could be described as

$$\mathbf{p}(\sigma, n) = \frac{\mathbf{p}(\sigma, n+1) + k\mathbf{p}(n) + \mathbf{p}(\sigma, n-1)}{k+2} \qquad (7)$$

and the z-transform of the smoothing function is

$$H(z) = \frac{k}{z - (k+2) + z^{-1}} \qquad (8)$$

where $k = k(\sigma)$ is the ratio of the influence of the guiding process to that of the unbiased "context". Intuitively, it can be observed that decreasing values of k would introduce more smoothing, as the "context" has more influence on the process than the guiding signal.

4.4 A Physical Interpretation of the "Context" Function

Consider a process in which the smoothed pen process at position $\mathbf{p}(k, n)$ is influenced by the guiding $\mathbf{p}(n)$ and the "context" $\mathbf{p}(k, n-1)$, $\mathbf{p}(k, n+1)$, in such a way that a force proportional to the distances from $\mathbf{p}(k, n)$ is exerted on $\mathbf{p}(n)$ due to each influence. If $\mathbf{p}(k, n)$ is in equilibrium under these three forces, then

$$k_e(\mathbf{p}(k, n) - \mathbf{p}(n)) + k_c(\mathbf{p}(k, n) - \mathbf{p}(k, n - 1)) + k_c(\mathbf{p}(k, n) - \mathbf{p}(k, n + 1)) = 0. \tag{9}$$

Manipulation of this equation leads to Eq. 7 with the substitution $k = k_e/k_c$, where k is the ratio of the spring constants. These forces are equivalent to extensions of linear springs as illustrated in Fig. 4.

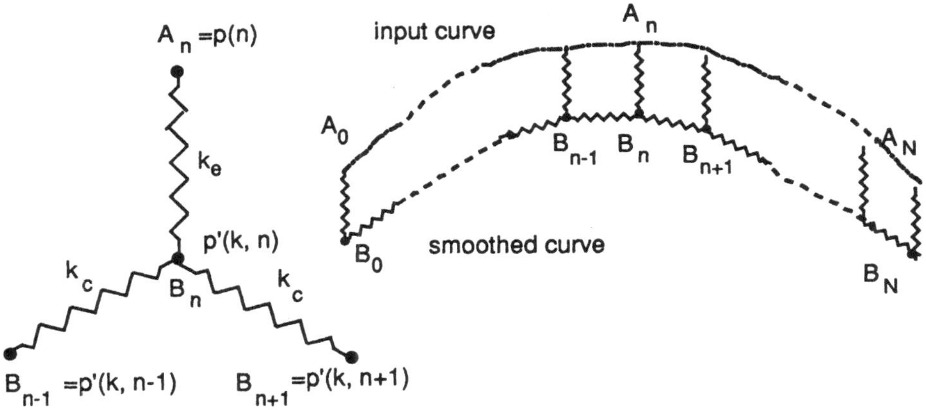

Figure 4. A process in which the Pen at **p** is in static equilibrium under the three forces exerted by springs. The complete structure is on the right.

4.5 The Boundary Conditions

The influence of either the "past" or the "future" does not exist at the boundaries n = 0 and n = N, hence the boundary conditions are

$$k_e(\mathbf{p}(k, 1) - \mathbf{p}(1)) + k_c(\mathbf{p}(k, 1) - \mathbf{p}(k, 2)) = 0$$

$$k_e(\mathbf{p}(k, n) - \mathbf{p}(n)) + k_c(\mathbf{p}(k, n) - \mathbf{p}(k, n - 1)) = 0$$

which leads to

$$\mathbf{p}(k, 1) = \frac{\mathbf{p}(k, 2) + k\mathbf{p}(1)}{k + 1}$$

$$\mathbf{p}(k, n) = \frac{\mathbf{p}(k, n - 1) + \mathbf{p}(n)}{k + 1} \tag{10}$$

4.6 The Impulse Response Function

The "context" smoothing function described here does not satisfy the traditional assumption A2, that the filter has no preferred scale. The consequence of this violation is that the filter has a preferred scale and that the shape of the filter does change as the parameter k is varied. The smoothing performance of this filter will therefore depend on the quantisation interval. The use of a fixed quantisation interval is easily justified for handwriting since the height of most handwriting is within the range of a couple of millimetres. All the other assumptions A1 to A5 are satisfied by the function.

The proposed function has zero phase characteristics with an impulse response of infinite duration which is derived below for later analysis. The two poles of the function are α and $1/\alpha$ which are the roots of the equation $y^2 - (k + 2).y + 1 = 0$. One of the roots would lie inside the unit circle while the other lies outside. Considering the z-transform of the function

$$H(z) = \frac{k}{z - (k + 2) + z^{-1}}$$

$$= -\frac{k}{\alpha - 1/\alpha}\left\{\frac{1}{1 - (1/\alpha)z} - \frac{1}{1 - \alpha z}\right\}$$

where

$$\alpha + \frac{1}{\alpha} = k + 2, \quad |\alpha| > 1, \tag{11}$$

the inverse z-transform, by integrating inside the unit circle for positive time response and integrating outside the unit circle for negative time response, is obtained as

$$h(n) = \frac{k}{\alpha - 1/\alpha} r(n)$$

$$\text{where } \mathbf{r}(n) = \begin{cases} \alpha^{n} & n < 0 \\ \alpha^{-n} & n \geq 0. \end{cases} \tag{12}$$

4.7 Filter Implementation and the Range of k for Convergence

The filtering is implemented as an iterative scheme, the procedure for which is described below.

1. Initialise. $\hat{\mathbf{p}}(\sigma, j)$, $k = k(\sigma)$

$$\mathbf{p}(\sigma, j) = \mathbf{p}(j) \qquad \text{for all } 1 \leq j \leq N.$$

2. Obtain new estimate. $\hat{\mathbf{p}}(\sigma, j)$

$$\hat{\mathbf{p}}(\sigma, j) = \frac{\mathbf{p}(\sigma, j+1) + k\mathbf{p}(j) + \mathbf{p}(\sigma, j-1)}{k+2} \tag{13}$$

3. Compare the maximum distance between points of the new and the old estimate.

$$D_{max} = \max_{1 \leq j \leq N} \left| (\hat{\mathbf{p}}(\sigma j) - \mathbf{p}(\sigma, j)) \right| .$$

4. Substitute $\mathbf{p}(\sigma, j) = \hat{\mathbf{p}}(\sigma, j)$ for all $1 \leq j \leq N$.

5. Jump to 2 if $D_{max} > \epsilon$.

The estimated values of $\mathbf{p}(\sigma, .)$ are also the locations of $\{B_j\}$ in the spring structure under static equilibrium when the energy stored in the springs is a minimum. This equilibrium position is stable, as a perturbation of B_j only increases the total energy stored in the structure, and hence will always reset itself back to the equilibrium position.

If k is negative and in the range of $-2 < k < 0$ any perturbation decreases the energy stored in k_e. However, there is an increase in the energy stored in the set of springs k_c, and the increase in net stored energy would still be positive in which case the equilibrium is still stable. If $-2 > k > -\infty$ the system is unstable since any perturbation produces a decrement in the net stored energy. The range of values considered for smoothing are $0 \leq k \leq +\infty$ and hence the iterative technique is applicable.

The smoothed curves for the word "husk", obtained using this technique, are shown in Fig. 5 .

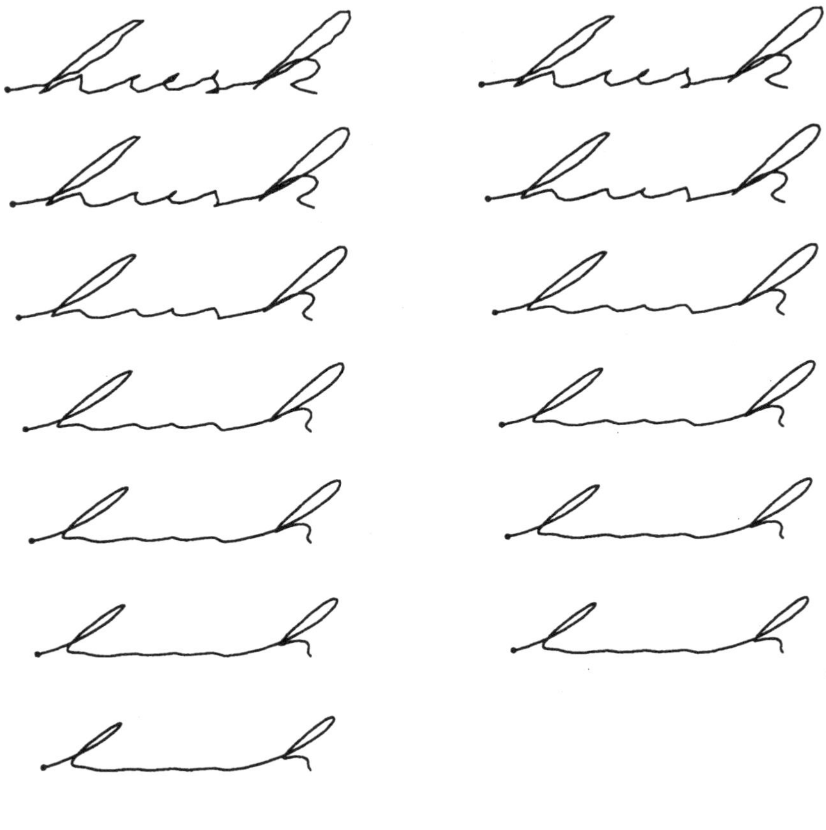

Figure 5. Smoothed script for the word "husk" using the "context" filter. The top left curve shows the unsmoothed script. Fine to coarse scaling at uniform intervals produce curves as shown from left to right, top to bottom.

5. Detection of Stroke Boundaries

5.1 Curvature Maxima

The stroke boundaries in the smoothed script may be detected by picking peaks in the curvature variations. For a discrete process, the local curvature ρ is,

$$\rho_n = \frac{\theta_n - \theta_{n-1}}{S} \tag{14}$$

where θ_i is the tangential angle along the curve

$$\theta_i = \tan^{-1}\frac{y'(.,i+1) - y'(.,i)}{x'(.,i+1) - x'(.,i)} \tag{15}$$

and S is a fixed constant for the sampling scheme.

The local curvature estimated for the words shown in Fig. 5 is illustrated in Fig 6.

It can be observed in Fig. 6 that the peak curvature drops significantly, from a high value for the unsmoothed script, to a very low value for the heavily smoothed script. In practice, the curvature variations become very small, even for moderate smoothing. The noisy peaks for low smoothing appear to be bigger than the peaks of significant stroke boundaries for moderate smoothing. Direct estimation of stroke boundaries by local curvature may therefore need different thresholds for different scales in order to achieve best results. In the following sections, a different scheme based on the rate of smoothing with respect to scale, is considered as data for segmentation.

5.2 Maximum Rate of Smoothing

The low frequency properties of the smoothing function affect the corners in the script, where the tangential angle variations are higher than the straight parts. The displacement of each pen-position due to smoothing should therefore increase with the local curvature at that point. The proposed stroke boundary detector is based on the rate of smoothing with respect to the scale. The equations of the composite smoothing filter/detector and its properties are derived below.

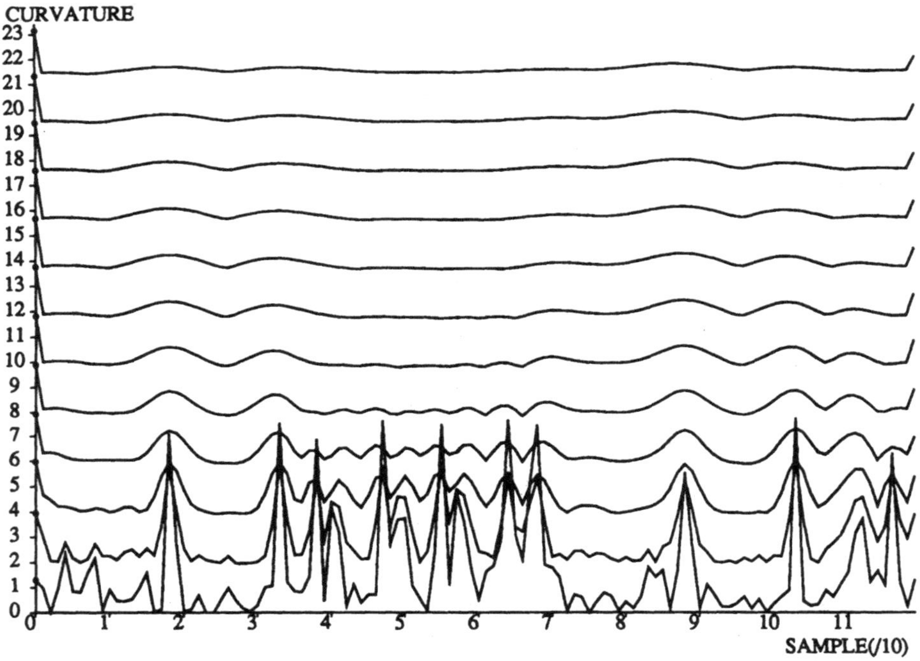

Figure 6. Curvature estimate of the word "husk". The graphs are plotted with a Y-offset for each smoothed curve. The fine scaled variations are at the bottom and the coarse scaled variations are at the top.

The rate of smoothing *w.r.t.* the scale σ in both the x and y directions along the curve, $\Psi(\sigma, .)$ may be expressed as

$$\Psi(i, \sigma) = \frac{\partial}{\partial \sigma} \Phi(\mathbf{p}(n), \sigma) \tag{16}$$

$$= \frac{\partial}{\partial \sigma} \sum_{\tau} \mathbf{p}(n) h(n - \tau)$$

$$= \sum_{\tau} \mathbf{p}(n) \frac{\partial}{\partial \sigma} h(n - \tau).$$

Considering the impulse response $h(n)$ for $t \geq 0$

$$\frac{\partial}{\partial \alpha} h(n) = \frac{\partial}{\partial \alpha} \frac{k}{\alpha - 1/\alpha} \left(\frac{1}{\alpha}\right)^n$$

$$= \frac{\alpha + 1/\alpha - 2}{\alpha - 1/\alpha} \frac{1}{\alpha} \left(\frac{2}{\alpha - 1/\alpha} - n\right) \left(\frac{1}{\alpha}\right)^n . \tag{17}$$

If σ is the point at which the function crosses the x axis then

$$\sigma = \frac{2}{\alpha - 1/\alpha} \tag{18}$$

$$\frac{\partial}{\partial \alpha} h(n) = \frac{k\sigma^2}{4} (\sigma - n) \left(\frac{1}{\alpha}\right)^n \tag{19}$$

therefore $$\frac{\partial}{\partial \sigma} h(n) = \frac{\partial}{\partial \alpha} h(n) \frac{d\alpha}{d\sigma}$$

$$= \frac{-k}{k+2} \frac{\sigma - n}{\sigma} \left(\frac{1}{\alpha}\right)^n \tag{20}$$

and the complete function for $-\infty < n < \infty$

$$\frac{\partial}{\partial \sigma} h(n) = \frac{-k}{k+2} \frac{1}{\sigma} r(n)$$

where $r(n)$ is

$$r(n) = \begin{cases} (\sigma + n)\alpha^n & n < 0 \\ (\sigma - n)\alpha^n & n \geq 0. \end{cases} \tag{21}$$

The changing shape of $r(n)/\sigma$ for the "context" filter plotted over a range of values for σ is shown in Fig. 7, illustrating the shapes of the detector operater for different 'bandwidth parameters'. The change in shape is attributed to the violation of property A2. If a Gaussian function be used, the shape of $r(n)$ would be independent of the scale σ.

It can be easily shown that $r(n)$, and hence the impulse response, have the property that both their first and second integrals are bounded, *i.e*

$$\left| \sum_{n'=-\infty}^{n} r(n') \right| \leq M_1 \quad \text{for all } n \tag{22}$$

$$\left| \sum_{n'=-\infty}^{n} \sum_{n''=-\infty}^{n''} r(n'') \right| \leq M_2 \quad \text{for all } n \tag{23}$$

This property ensures that the displacements, which are estimated from the discrete convolution of $r(n) \otimes \mathbf{p}(n)$, are also bounded.

$$|r(n) \otimes \mathbf{p}(n)| = |\sum_{n'=-\infty}^{n} r(n') \otimes \Delta\mathbf{p}(n)| \tag{24}$$

$$\leq |S \otimes \sum_{n'=-\infty} r(n')|$$

$$= S\,|\sum_{n'=-\infty}^{n} \sum_{n''=-\infty}^{n''} r(n'')|$$

$$\leq \mathbf{M}, \tag{25}$$

where **M** is an upper bound Δ in Eq. 24 is used as a difference operator. It can be shown that if the smoothing function is Gaussian, then **M** is independent of the scale σ. In the remainder of the section, a displacement function is defined as the discrete representation of the rate of smoothing function to be used for segmentation.

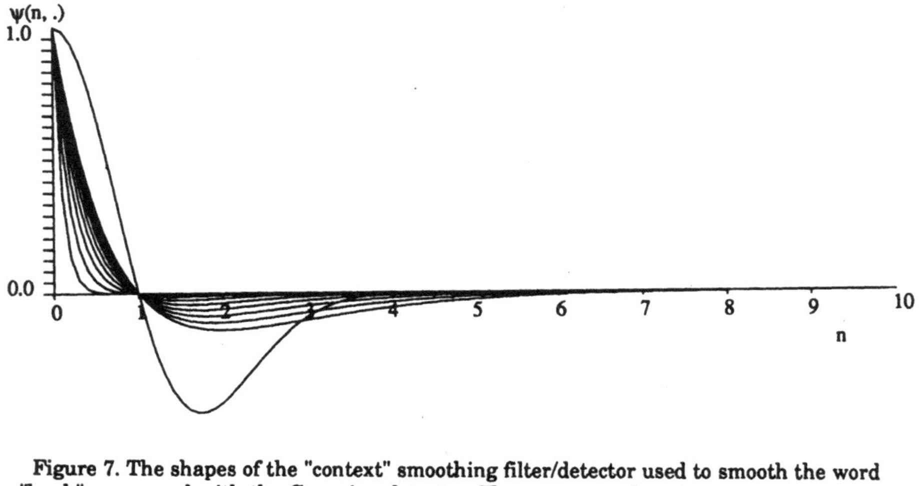

Figure 7. The shapes of the "context" smoothing filter/detector used to smooth the word "husk", compared with the Gaussian detector. The narrowest function correcsonds to the finest scaling.

Consider the case in which the scale σ is uniformly increased by equal sized steps. Let $D_i(n)$ be the displacement at the nth sample point after applying the ith scaling and let be the largest of these displacements over a word or a sentence. $D_i(n)$ is the discrete representation of the continuous function $\partial\Psi(i,\sigma)/\partial\sigma$ described previously, *i.e*,

$$D_i(n) = |\mathbf{p}(\sigma_{i-1}, n) - \mathbf{p}(\sigma_i, n)|, \quad i > 0, \quad 1 \le n \le N \qquad (26)$$

$$\mathcal{D}_i = \max_{1 \le n \le N} |\mathbf{p}(\sigma_{i-1}, n) - \mathbf{p}(\sigma_i, n)|, \qquad i > 0 \qquad (27)$$

where

$$\sigma_i = i\,\sigma_{step}. \qquad (28)$$

N is the number of samples in the data.

A plot of the maximum displacement against σ is shown in Fig. 8. This indicates that the largest displacements are rather uniform with scale. These patterns are typical for almost all cursive script data.

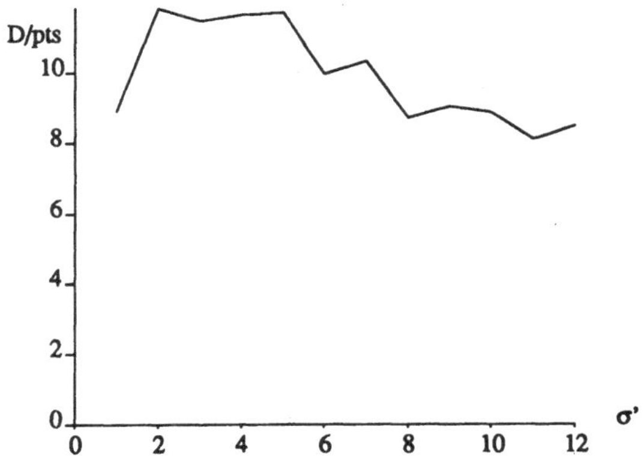

Figure 8. Peak Displacements againt σ of the "context" filter for the word "husk".

The displacements for the word "husk" are shown in Fig. 9. Unlike the curvature plots, the size of the peaks in the displacement plots, obtained over the entire range of scales, appear to be uniform. This property allows direct comparison of the displacements across scales. The peak picking algorithm may now be designed with a single threshold to analyse the entire range of scales.

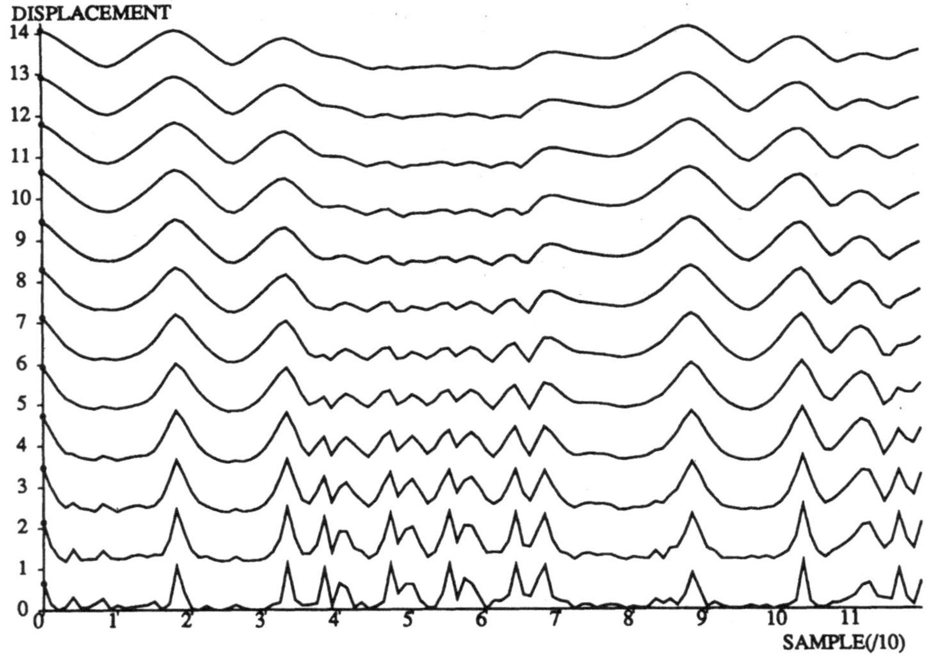

Figure 9. Displacement for word "husk" due to "context" smoothing. The graphs are plotted with a y-offset for each smoothed curve. The fine scaled variations are at the bottom and the coarse scaled variations are at the top.

5.3 The Segmentation Algorithm

If it is generally assumed that the displacement at a coordinate point would be the largest across scale only when the scale of the analysis is the appropriate one for the corresponding part of cursive script, then these largest displacements could be regarded as being the optimally filtered displacements. A peak picking algorithm applied to this data would obtain a set of optimally filtered segmentation points.

Let $\pi(n)$ be the largest distancement at sample point n over the range of scales, *i.e*,

$$\pi(n) = \max_{\sigma_1 \le \sigma \le \sigma_{ns}} D_\sigma(n). \tag{29}$$

The step between successive scales may be set depending on the quantisation interval of the sampling scheme and the range up to a value at which most of the strokes in the script are totally smoothed out. The largest displacements for the word "husk" are shown in Fig. 10. Compare this curve with the pen-speed and curvature profiles of Figs. 2 and 3 for the clarity of peaks and troughs. The strokes of the word "husk" obtained by the proposed segmentation method are illustrated in Fig. 11 with some more examples of segmentation from the other subjects.

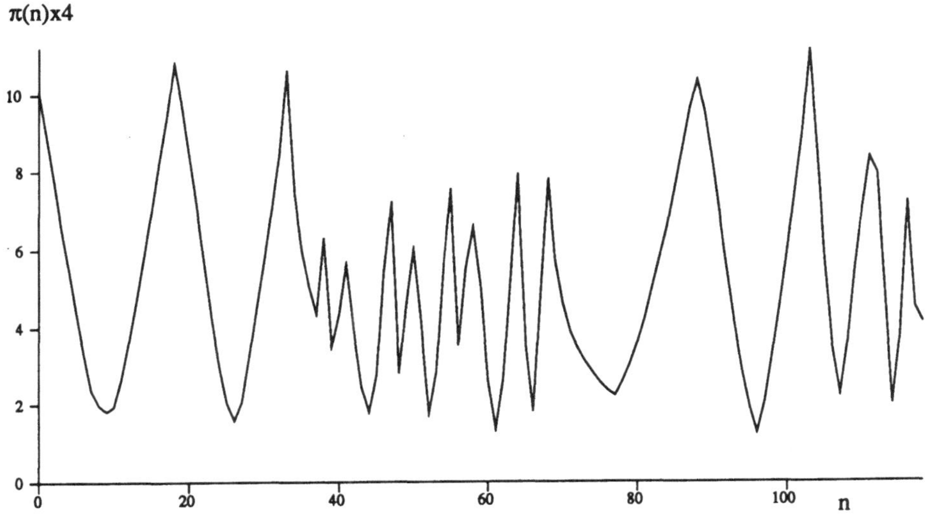

Figure 10. The combined displacements π_i for the word "husk".

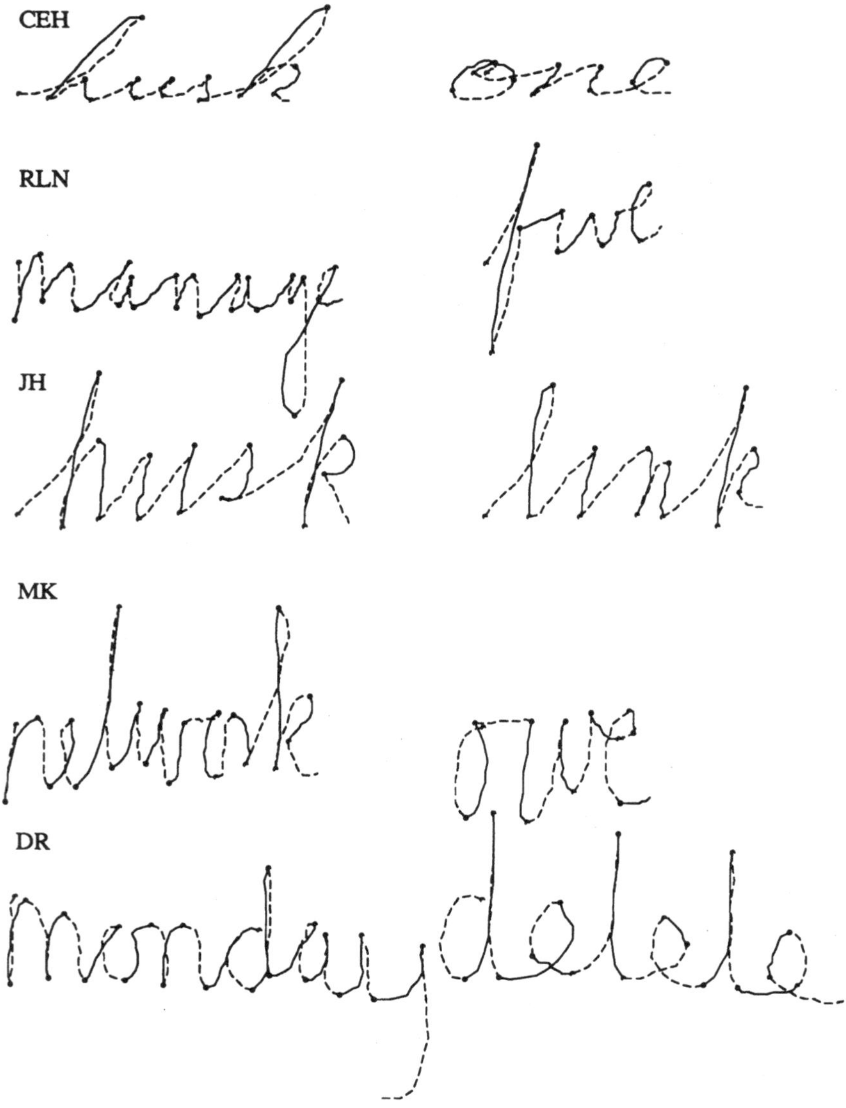

Figure 11. Some examples of the segmentation by the proposed algorithm. Alternating strokes are drawn with solid and dashed lines.

5.4 Construction of Scale-Space Map and the 'Prominence' of Stroke Boundaries

For a given cursive script word, the scale-space map is constructed as follows. The finest scale of measurement is chosen first and the stroke boundaries significant at that scale are detected and plotted on the map against the appropriate scale. The next scale is then selected by increasing the 'bandwidth parameter' by one step. Each significant stroke boundary detected at this scale is associated with one of the stroke boundaries detected at the previous scale. The associations are made based on proximity in sample position. A part of a contour is formed by joining the two associated points on the map. The contours constructed are extended in the same way for the entire range of scales.

The scale-space map of stroke boundaries, plotted for the word "husk" appears as shown in Fig. 12 . The larger the scale of measurement, the more prominent the detected stroke boundary is likely to be. The height of the trace which belongs to a stroke boundary in the scale-space map may therefore be treated as a measure of the 'prominence' of the stroke boundary. Compare the scale-space map obtained with the context function to that obtained with the gaussian illustrated in Fig. 13. Observe the shifts in the location of segmentation points.

The prominence of a segmentation point may used to assist the matching process in considering the possibility of a segmentation error during recognition.

6. Experiment I: Consistency of Segmentation

In order to compare the performance of the proposed segmentation algorithm with that of the conventional pen-speed based algorithm, the consistency of segmentation for data was evaluated for both methods. The experiment involved five subjects in which each subject wrote about forty-five words, each word twice. The data was recorded on a Micropad interfaced to an Acorn BBC microcomputer. The resolution of the Micropad was such that the word "husk" illustrated had a range of 145 x 43 divisions. The subjects wrote using ordinary pens on a sheet of single lined paper placed on the pressure sensitive pad. There were no restrictions placed on the size of the writing but the subjects wrote most words about two or three times bigger

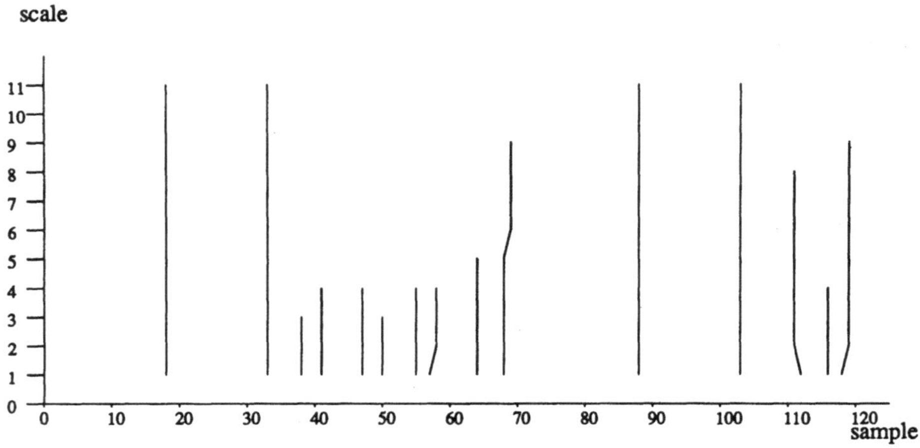

Figure 12. Scale-space map of Stroke Boundaries for the "context" filter.

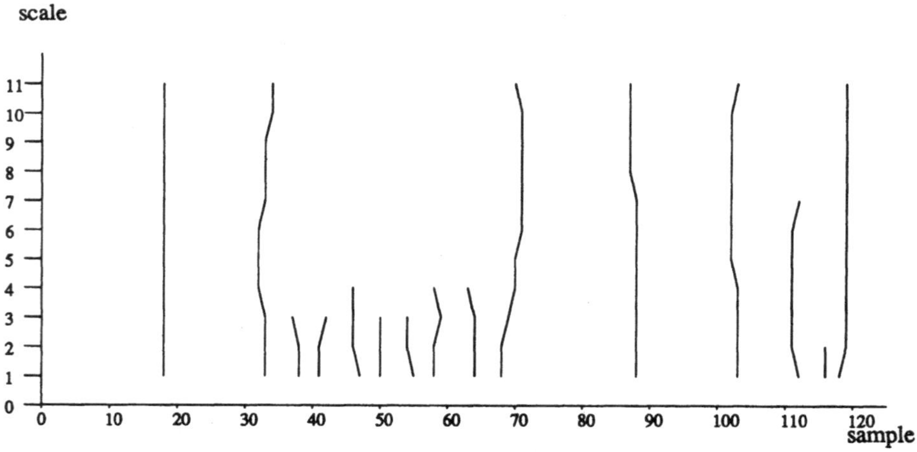

Figure 13. Scale-space map of Stroke Boundaries for the "Gaussian" filter.

than their normal size. All five subjects wrote very cursively, among whom two subjects wrote most words in more than one pen continuous movement. The data was sampled every 20 ms.

6.1 Penspeed Estimation

As the average height of handwriting was between a half and one centimetre, the quantisation errors owing to poor resolution of the graphic tablet and the tablet noise were significantly high. A kernel estimation approach was therefore used in estimating the pen-speed. A gaussian kernel was chosen, as it exhibits excellent properties for estimation of signals in white noise, which is a good approximation to the quantisation error distribution. The velocity components and are estimated as

$$V_x(n) = \sum_{\tau=-\omega}^{\omega} g(\sigma,\tau)x(n - \tau)$$

$$\text{and } V_y(n) = \sum_{\tau=-\omega}^{\omega} g(\sigma,\tau)y(n - \tau) \tag{30}$$

$g(\sigma,.)$ is the gaussian function of variance σ^2

$$g(\sigma, n) = \frac{1}{\sqrt{2\pi}}\frac{1}{\sigma} \exp(-1/2(n/\sigma)^2) \tag{31}$$

and ω is the window size. The estimated pen-speed υ (.) is therefore

$$\upsilon(n) = \sqrt{V_x^{\,2} + V_y^{\,2}} \tag{32}$$

Typical values of σ were 2.0 and 3.0

The handwritten data was resampled at a quantization interval of 0.4 units on the MicroPad. The 'bandwidth parameter' σ was increased in 12 steps from 1.0 to 12.0 in obtaining the displacements.

6.2 The peak/trough picking algorithm

The peak/trough picking algorithm used on both pen-speed and displacement data was implemented by dividing the data into profiles of trough-peak-trough. In the case of pen-speed segmentation, additional conditions were imposed on the maximum pen-speed at a trough and the minimum height of the peak from both its adjacent troughs. The thresholds required adjustment for different subjects. Typical values were 1.0 and 0.125 units per sampling interval, respectively. For the displacement data, additional conditions were again imposed on the minimum height of a trough and on the minimum height of the peak from both its adjacent troughs. The thresholds were 0.8 and 0.05 units respectively. The average largest displacement was about 2.5 units.

The average computation time for the scale-space filtering and the segmentation was about 40s per word on a MicroVax II workstation.

6.3 The consistency criterion

An estimate of the inconsistency of segmentation for the two algorithms is calculated as the number of segmentation errors, such as the inclusion of an extra segmentation point and the omission of a stroke boundary. The decisions were based on the most common segmentation of each letter in a subject's database. However, two acceptable forms of segmentation were considered for letter 'S', as most of the subjects seemed to use more than one construction method. Contextual variations of cursive letter formation are incorporated into the criterion by allowing the use of different single segment ligatures as connections between two letters, if necessary. Up to ten types of ligatures were assigned for each author.

Table 1. shows the number of segmentation errors per 100 letters. About 450 to 500 letters were present in the dataset of each author.

7. The Cursive Script Recognition System

An author-dependent on-line cursive script recognition system based on stroke segmentation has been designed to recognize any cursive word in a word dictionary, written with lower case characters. The system is illustrated in Fig. 14.

Cursive script writing is modelled as a sequence of strokes, and the stroke matching algorithm employs a suitable distance metric to measure the distance between an input and a template stroke. The template set consists of stroke descriptions for each of the 26 letters and about 10-12 ligatures. The dissimilarity between the input stroke sequence and synthesized stroke sequences, which are constructed from the letter templates, is considered in order to find a number of good sequences with low overall distances. The search strategy to obtain these good solutions is based on the "Connected word algorithm" applied originally to continuous speech recognition, J. Bridle *et al.* (1983), and later developed by Ney (1984). The proposed algorithm (Kadirkamanathan *et al.* 1989) also takes into account any isolated errors due to missing stroke boundaries or extra segmentation points.

	Displacement Method			Pen-Speed Method		
Author	Extra	Omitted	Total	Extra	Omitted	Total
ceh	11.9	2.1	14.0	13.9	17.9	31.5
rln	5.3	1.6	6.9	9.6	11.3	20.9
jh	2.9	1.4	4.3	9.2	12.0	21.2
mk	6.4	1.6	8.0	12.6	4.6	17.2
dr	15.8	1.5	17.3	20.4	16.0	36.4

Table 1. Number of inconsistent segmentations per 100 letters.

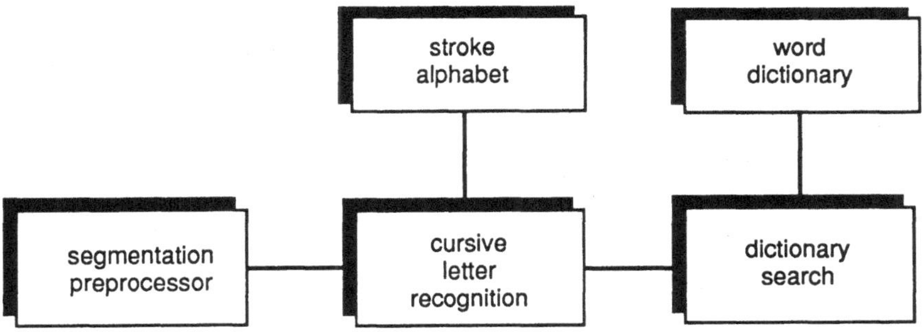

Figure 14. The cursive script recognition system.

The postprocessor performs a dictionary search to filter out any incorrect answers or character strings, which are not present in the dictionary.

8. The Recognition Model

Let the input, or test pattern, consist of $i = 1, \ldots, N$ strokes, which are indexed as i. The input pattern is known to be composed of individual letters and ligatures which are chosen from a prespecified alphabet $\mathcal{A}$. The letters and ligatures of the alphabet, $\mathcal{L}_k$, are distinguished by the index $k = 1, \ldots, K$,

$$\mathcal{A} = \{\mathcal{L}_1, \ldots, \mathcal{L}_K\}. \tag{33}$$

The strokes of a letter template k are denoted as $j = 1, \ldots, J(k)$, where $J(k)$ is the number of strokes of template k.

Suppose the ultimate goal of the algorithm is to determine that sequence $q(1), \ldots, q(R)$ of templates that best matches the input pattern, where the matching criterion needs further specification. The concatenation of the templates $q(1), \ldots, q(R)$ may be referred to as the "super" reference pattern.

The matching of template stroke sequences against input stroke sequences is similar to the matching of template speech pattern vector sequences against unknown input speech pattern vector sequences, in connected word recognition. The dynamic time warping of speech frames is analogous to the merging and splitting of strokes for "optimal" alignment. The splitting and merging of strokes may be done freely as the locations of the stroke boundaries detected are rather consistent, while the more common types of errors are the introduction of an extra segmentation point or the missing of a stroke boundary.

If it is assumed that:

1. No two consecutive stroke boundaries are both undetected by the segmentation preprocessor,
2. No two consecutive segmentation points detected are both extra and in error,

with the additional requirements that

3. The time order of the strokes be preserved,
4. No stroke may be omitted in the sequence $i(1), \ldots, i(l), \ldots, i(L)$,,

then the dynamic programming optimization with three types of branching can be employed.

Any violation of these assumptions at the segmentation stage would lead to a total failure to recognize the input pattern. On the other hand, such assumptions significantly reduce the computational cost of the algorithm. In order to define the optimization function, the following are defined:

The ith segmentation point of the input pattern and the jth segmentation point of template k define a grid point (i, j, k). Segment i ends at the ith segmentation point of the input pattern and stroke (j, k) ends the jth segmentation point of template k. Each grid point (i, j, k) is associated with three local distance measures $d_{1/2}(i, j, k)$, $d_1(i, j, k)$, and $d_2(i, j, k)$, defining a measure of dissimilarity between the corresponding strokes as described below.

1. $d_{1/2}(i, j, k) = \mathbf{d}(i - 1 \oplus i, (j, k))$ is the distance between the concatenated strokes $i - 1$ and i and stroke j of template k. This distance is associated with the case when the i - 1th segmentation point is an error. $\oplus$ denotes a concatenation of strokes.
2. $d_1(i, j, k) = \mathbf{d}(i, (j, k))$ is the distance between the stroke i and stroke j of template k.
3. $d_2(i, j, k) = \mathbf{d}(i, (j - 1, k) \oplus (j, k))$ is the distance between the stroke i and the concatenated strokes $j - 1$ and j of template k. This distance is associated with the case when a segmentation point, between the i-1th and the ith segmentation points, is undetected.

The local distances across letter templates could be defined in a similar way. The distance measure used is defined in the next section.

The connected letter problem can be regarded as one of finding the path through the set of grid points (i, j, k) which provides the "best match" between the input pattern and the unknown sequence of letter templates.

The path $\mathcal{W}$ is given as a sequence of grid points

$$\mathcal{W} = (w(1), w(2), \ldots, w(l), \ldots, w(L)) \qquad (34)$$

where $w(l) = (i(l), j(l), k(l))$ and l is the path parameter for indexing the ordered set of path elements. The criterion for the matching is the global distance, *i.e* the sum over the local distances along a given path. The problem of connected letter recognition can now be stated as the minimization problem

$$\min_{w} \sum_{l} d_\eta(w(l)) \tag{35}$$

where $d_\eta(w(l))$ is one of the three distance measures $d_{1/2}(.,.,.)$, $d_1(.,.,.)$, $d_2(.,.,.)$, depending on the previous path point $w(l-1)$ such that

$$w(l-1) \overset{T_\eta}{\rightarrow} w(l) \tag{36}$$

where $\overset{T_\eta}{\rightarrow}$ denotes a transition. The complete solution may be found in Kadirkamanathan and Rayner(1989) and Kadirkamanathan(1989).

8.1 The Distance Metric

Let S_A and S_B be two strokes, the distance between whom is to be computed. Both S_A and S_B are equally divided into a fixed number of n parts, and labeled as illustrated in fig. 15. The distance measure is defined as

$$\mathbf{d}(S_A, S_B) = \sum_{i=1}^{n} |(a_i - a_{i-1}) - (b_i - b_{i-1})|, \tag{37}$$

where a_i and b_i are the position vectors of points A_i and B_i respectively.

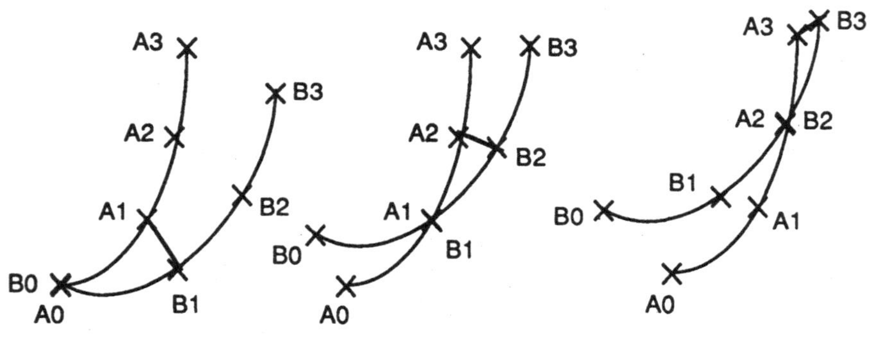

Figure 15. The distance metric.

9. Experiment II: The Recognition

9.1 Generation of Stroke Alphabet

The data collected in experiment I was again used for the recognition experiment described here. For each subject, two separate databases were constructed, one consisting of all the first examples of each word, and the other consisting of the second examples.

With the aim of obtaining the most common shape of each letter and ligature, the first database was segmented using the proposed segmentation algorithm. The segmented words were then hand-labeled and grouped into similar letters/ligatures and strokes. The most common form of each stroke in each letter was picked using a nearest-neighbour clustering algorithm, employing the distance measure proposed in the previous section. The prototypical strokes chosen for each letter and ligature were used to create the stroke alphabet for the each subject. Figs. 16 and 17 show some of these strokes, obtained from one particular author.

	DB I		DB II		Overall	
Author Let/Lig	Words	Correct	Words	Correct	Accuracy	Errors
ceh 26, 12	45	43	45	37	90	1
rln 27, 11	43	40	47	44	93	0
jh 26, 11	43	41	44	43	98	0
mk 26, 13	42	36	50	42	85	0
dr 26, 12	45	40	44	41	91	2

Table 2. Recognition performance.

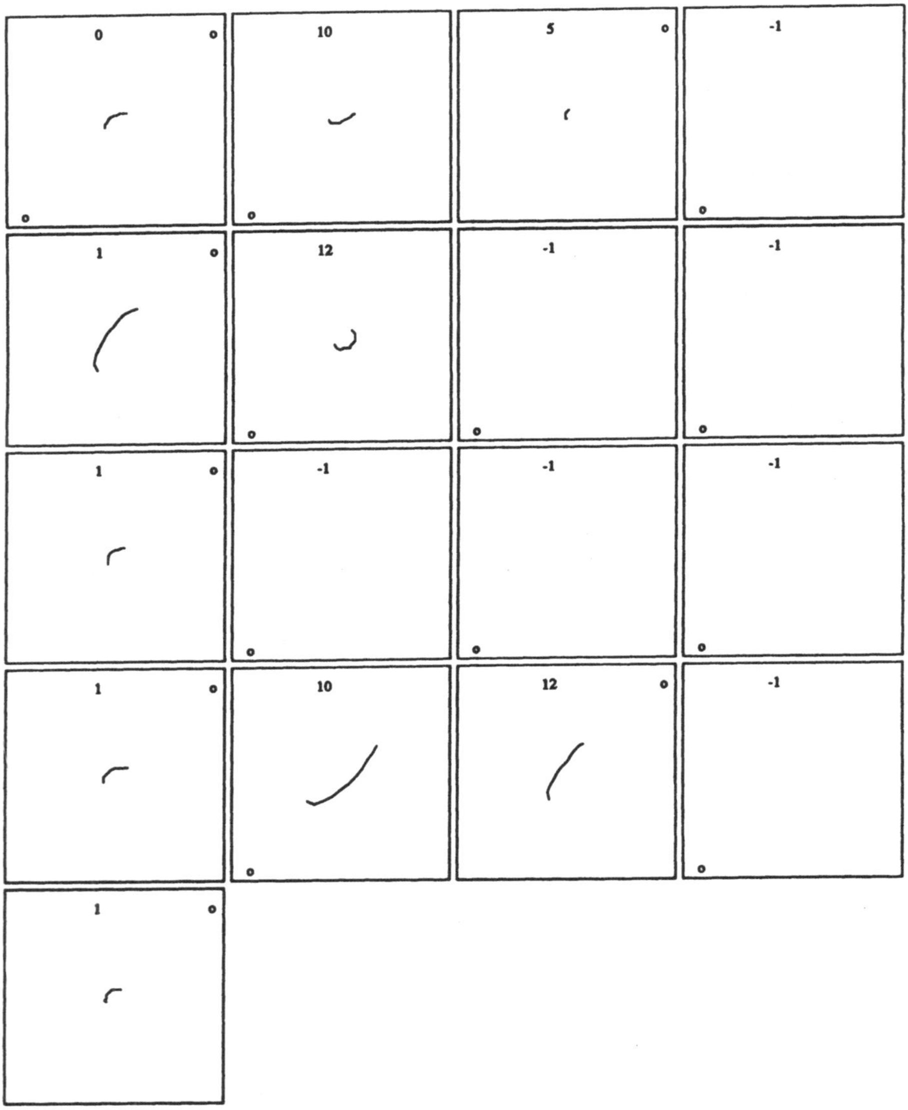

Figure 16. The first five letters in the stroke alphabet of the author "ceh". The circle in the corner of the box indicates the beginning end of the stroke.

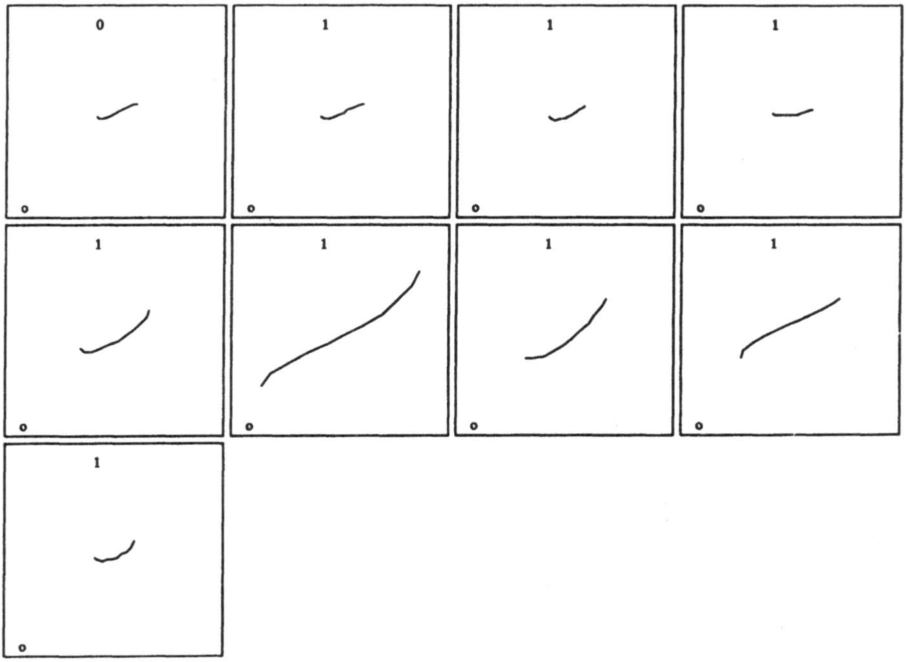

Figure 17. Some types of ligatures used by author "ceh".

9.2 Results

Both the datasets obtained were tested for correct recognition by the system using the stroke alphabet of the subject who created the databases. The time taken by the dynamic programming algorithm, running on a MicroVax II workstation was an average of 90s per word. The performance results are listed in Table 2.

10. Conclusion

A multi-scale segmentation algorithm has been described, which exploits the displacement data obtained using scale-space filtering. The displacements are suitably combined by choosing for each sample point, the largest displacement recorded over the range of scales. When the peaks of this combined displacement are located as segmentation points, the resulting stroke description appears to be much more consistent than the conventional pen-speed minima method of segmentation. The segmentation inconsistency measure of the proposed method is, for all subjects, less than half of that obtained with the conventional method for the limited dataset used in the experiment. The consistency criterion, however, considers the contextual variations of cursive letter formation only to a very limited extent. The strokes obtained by the proposed algorithm also appear to be very close to the `strokes' described by the psychophysicists.

The performance of the on-line author-dependent cursive script recognition system illustrates the robustness of the proposed stroke segmentation algorithm and the robust modelling achieved with the strokes of cursive script. The performance appears to be comparable to the performance of the significant designs reported in the literature, even though simplifying assumptions were made in the design. Unlike most other systems, the proposed system has no *ad hoc* features.

However, it must be pointed out that the choice of the smoothing function, used in the scale-space filtering, was an intuitive one and is not optimal. The chosen smoothing function was preferred over the gaussian function since the deviations introduced in the location of the segmentation points with scaling were smaller. The iterative "context" filtering procedure could be speeded up by introducing a momentum term in the new estimates obtained at the end of each iteration.

One significant feature of the proposed segmentation algorithm is that it could be extended to process static data in the form of an image, after applying the necessary preprocessing steps to obtain the required data representation. The only temporal information that is required by the algorithm is the order in which the different parts of the cursive script were constructed. It is a general belief that the temporal order of construction of the different parts of the off-line cursive script may be deduced from the static pattern, although no algorithm has yet been proposed to do this. Berthod (1982) discusses some of the required preprocessing steps. However, a semi-automatic system could be designed to process prewritten documents in which the only human interaction would be that involved in assisting the resampling process.

There are some weaknesses in the script recognition algorithm presented in this paper. The stroke sequence model is too simple and too approximate. Diacritical marks such as dots or crossbars, which are constructed as separate pen movements, are not used in recognition. A potential source of information which might enhance the performance considerably.

Acknowledgement

The authors wish to the thank D. Rainton of C.U.E.D for his comments on the final draft of this paper.

Reference

Asada, H., Brady, M.,(1984). "The curvature primal sketch", *Massachusetts Institute of Technology, Cambridge* , AI Memo 758, 1984.

Babaud, J., Witkin, A. P., Baudin, M., Duda, R. O., "Uniqueness of the Gaussian kernel for scale-space Filtering", *IEEE Trans. Patt. Analysis and Machine Intell.* , **8,** 1986, pp. 26-33.

Berthod, M., "On-line Analysis of Cursive Writing" in *Computer Analysis and Perception, Volume I: Visual Signals* , C. Y.Suen, R. De Mori (Eds.) C. R. C. Press Inc., 1982, pp. 55-81.

Berthod, M., Ahyan, S., "On-line cursive script recognition: A structural approach with learning", *Proc. 5th Int. Joint. Conf. on Patt. Rec.* , 1980, pp. 723-726.

Bridle, J., Brown, M., Chamberlain, R., "Continuous Connected Word Recognition using Whole Word Templates", *Radio & Electronic Engineering*, **53,** 1983, pp. 167-175.

Ehrich, R. W., Koehler, K. J., "Experiments in contexual recognition of cursive script", *IEEE Trans. on Comp.* , **24** , 1975, pp. 182-194.

Kadirkamanathan, M., Rayner, P., J., W., "A unified approach to on-line cursive script segmentation and feature extraction", *Proc. IEEE Int. Conf. Acoust., Speech and Signal Processing*, 1989, pp. 1659-1662.

Kadirkakmanathan, M., *Ph.D Dissertation, Cambridge University Engineering Department, Cambridge, England*, 1989, 125 pp.

Mermelstein, P., Eden, M., "Experiments on computer recognition of connected handwritten hords", *Information and Control*, **7**, 1964, pp. 255-270.

Mokhtarian, F., Mackworth, A., "Scale-based description and recognition of planar curves and two-dimensional shapes", *IEEE Tran. Pattern Analysis and Machine Intelligence*, **8**, 1986, pp. 34-43.

Morasso, P., "Understanding cursive script as a trajectory formation paradigm", in *Graphonomics: Contemporary Research in Handwriting*, H. S. R. Kao, G. P. Van Galen, R. Hoosain, (Eds.), Elsevier Science Publ., 1986, pp. 137-167.

Ney, H., "The use of a one-stage dynamic programming algorithm for connected word recognition", *IEEE Trans. Accoust. Speech and Signal Processing*, **32**, 1984, pp. 263-271.

Viviani, P., Centazo, M., "Segmentation and coupling in complex movements", *Journal of Experimental Psychology: Human Perception and Performance*, **11**, 1985, pp. 828-845.

Wann, J., P., Kadirkamanathan, M., "Variability in childrens' handwriting: Appraising spatial segmentation", in *Development of Graphic Skills: Contemporary research and educational implications*, J. P. Wann, A. M. Wing, N. Sovik, (Eds.), Academic Press: London (in print).

Yuille, A., L., Poggio, T., A., "Scaling Theorems for Zero Crossings", *IEEE Trans. Patt. Analysis and Machine Intell.*, **8**, 1986, pp. 15-19.

Computer Processing of Handwriting
Eds. R. Plamondon & C. G. Leedham

AN ON-LINE HANDWRITING-RECOGNITION SYSTEM BASED ON UNRELIABLE MODULES

HANS-LEO TEULINGS, LAMBERT R.B. SCHOMAKER, JAN GERRITSEN, HANS DREXLER and MARC ALBERS
Nijmegen Institute for Cognition Research and Information Technology (NICI) P.O. Box 9104, 6500 HE Nijmegen, The Netherlands

In automatic recognition of unrestricted handwriting the ambiguities can be solved by top-down processing. However, automatic systems never have access to the extended background knowledge available to human readers. In order to replace this higher-level information we need to improve the reliability of the bottom-up processing. A handwriting-recognition system can be split up into six discrete blocks: (1) digitizing, word segmentation, pre-processing, and segmentation into strokes, (2) normalization of global handwriting parameters, (3) extraction of features per stroke, (4) allograph recognition, (5) optional word hypothesization, and, in order to allow recognition (6) a learning phase. The present paper discusses the design of four of these processing blocks (normalization, feature extraction, allograph recognition, and learning) and briefly specifies feature extraction. Normalization concerns orientation, size, and slant. However, various alternative algorithms can be chosen and some algorithms yield more reliable results than others. A mechanism is proposed that will, sooner or later, find the most appropriate normalization algorithms. Consequently, the features extracted from each stroke in the handwriting pattern will be more uniform within a writer and even between writers. In the recognition phase, handwriting patterns are segmented into allographs using an algorithm that can handle allographs with various numbers of strokes and with optional connection strokes between them. In order to teach the recognizer allographs a method has been designed that builds non-interactively a lexicon of allographs by automatically discovering the allographs in a large corpus of cursive script.

1. Introduction

Handwriting cannot be neglected as a universal tool to store a wide variety of graphical information, such as personal notes containing non-standard-ized characters together with graphs. Although this material may sooner or later be entered into a computer for further processing, people might prefer a pen as it cannot replace mouse interaction or voice input. In order to develop methods that optimally employ the movement information of the pen we investigate the possibilities of on-line handwriting.

Thirty years ago automatic recognition of handwriting was considered an 'exercise' for automatic speech recognition (Harmon, quoted by Lindgren, 1965). Still, no cursive-script recognition system appeared appropriate in a sufficiently wide range of applications. The need for automatic handwriting recognition is high, though, also for entering non-western script into computers (e.g., Chinese or Kanji, Hiragana, Katakana, Korean, Brahmi, Devanagari, Tamil, Indian, Hebrew, Arabic). A useful recognizer should be able to deal also with digits, arbitrary character sequences in handprint and cursive script where the identification has to be based upon the bottom-up processing and in more redundant character sequences where higher-level information is required to disambiguate the stroke pattern. User friendliness requires recognition of all kinds of rapid and slow, cursive writing and handprint recognition by a wide variety of subjects.

Two methods to enter handwriting data into the computer exist. The first is to optically scan the picture of a sample of handwriting that was produced off-line. The second is to record the movements of the pen tip on-line by means of a digitizer. We employ the latter method because it is a richer source of information which allows us to exploit our basic knowledge about the motor system. The lack of practical results from traditional approaches is probably caused by the limitations of off-line recognition. Recent developments in the field of digitizer-annex-displays and microcomputers enable us to employ on-line recognition methods. This approach allows us to exploit dynamic movement information. Also promising off-line cursive-script recognition methods are under development (e.g., Srihari & Bozinovic, 1987; Aoki & Yamaya, 1986), but we discuss here only on-line recognition of handprint and cursive script.

A recent overview of on-line handwriting recognition is presented by Tappert et al. (1988). They list four experimental cursive-script recognition systems, ten experimental hand-print recognition systems, and fifteen commercial hand-print recognition systems. Most systems act in a restricted domain such as boxed handprint characters or allowing specific words only.

The usefulness of a handwriting interface is stressed in Brocklehurst (1988). To demonstrate the idea of "electronic paper" they provide a prototype handwriting-recognition system. In the frame of the Esprit Project 295, Wright (1988) designed a cursive-script recognizer with efficient low-level processing but at the cost of reliance upon a dictionary and higher-level linguistic processing. 94% of the characters of a data set of 112 people was correctly recognized, which compares well with the rate of human recognition without context. Higgins & Whitrow (1985) achieved good results using only a quadgram table based on a dictionary. In the same project Doster & Oed (1984) and Mandler (1989) developed a recognizer for handprint. Within the frame of Esprit Project 419 a recognizer for normal cursive script, based on knowledge of the motor system is under development (Teulings et al., 1987; Thomassen et al., 1988a; 1988b; Schomaker and Teulings, 1990; Morasso et al., 1990). The system consists of a parallel structure of redundant modules and includes an non-interactive learning phase. Oulhadj et al. (1989) developed a cursive-script recognizer where characters are hypothesized from left to right and are verified in parallel using a tree-like representation of a word list. Using a wordlist of 100 different words 90% of the words by 5 writers were recognized correctly.

Leedham and Downton (1986) developed a shorthand recognizer but state that the performance is still poor due to the simplified nature of the algorithms without knowledge base and due to ergonomic deficiencies of the digitizer. Recognition of on-line Arabic script is discussed by Amin (1985), who states that the bottom-up analysis is still insufficient such that complicated grammatical analyses are required due to the complexity and non-uniformity of Arab grammar. In the USA, Ward & Blesser (1985) developed a commercial handprint recognizer. Tappert (1986) at IBM Watson Research Center is developing a cursive-script recognizer using elastic matching techniques for IBM PC/AT. Plamondon & Baron (1986) used a handwriting recognizer to input computer programs. Skrzypek & Hoffman (1989) presented a neural network for cursive-script recognition. In Japan, Kondo (1989) realized a recognizer of handprinted numerals based on a noisy character generator.

Our aim is to design a system based on knowledge of the motor system which is able to recognize unrestricted handwriting (Schomaker & Teulings, 1990). In such a manner a computer can be operated without using a keyboard, nor a mouse, or voice input (e.g., Higgins & Duckworth, 1989;

Welbourn & Whitrow, 1989) as everything we want to do can be done using a pen.

In Teulings et al. (1987) we introduced a modular architecture for the low-level bottom-up analysis of handwriting of the so-called Virtual Handwriting System (VHS). A handwriting recognition system can be split up into six discrete modules, in accordance with Srihari & Bozinovic (1987):

1. Digitizing, preprocessing and segmentation of cursive script into words and subsequently into sequences of discrete strokes, which are defined here as the segments of pen movement between two successive minima of the absolute velocity,
2. Normalizing handwriting patterns,
3. Stroke feature extraction,
4. Allograph recognition,
5. Optional word hypothesization using higher-level information,
6. Learning phase in order to build the allograph lexicon required for recognition.

Of course, the higher-level subsystems of linguistic processing are essential in cursive-script recognition (e.g. Higgins & Ford, 1989; Wells et al., 1989). However we want to emphasize the importance of the optimal low-level handwriting processing because we doubt whether the higher-level linguistic processing can restore the information lost by unreliable lower-level processing as it does not have available the extended knowledge of an educated reader of a handwritten message. This holds especially if character sequences are used that do not at all occur in the lexicon of words.

The present paper discusses the current status of four of these processing blocks: normalization, feature extraction, allograph recognition, and learning.

2. Normalizing Handwriting Patterns

In order to extract the feature vector of the handwriting patterns, several normalization steps have to be performed (See Thomassen et al., 1988b, for an overview). The reason is that a sample of a person's handwriting contains various subject-specific parameters, like size or slant (e.g., Maarse et al., 1988). On top of that, the motor system is able to transform handwriting deliberately such as changing orientation, size or slant (e.g., Pick & Teulings, 1983). But all these parameters do not contain any information about the identity of the characters. Therefore, the handwriting patterns have to be normalized first.

A particular problem is that different writers may also have different cursive-script styles, and one writer may even employ various cursive-script versions of the same character. For that purpose we use the term "allograph" which is a subject-specific and context-specific topological structure of a character. By "character" we mean one symbol out of the series of upper and lower case keys that can be typed using a keyboard. We suggest that normalizing the handwriting patterns may reduce the between-writer differences to the extent that they are not caused by differences in topological structure.

The sequence of levels of normalization we find the most appropriate are: orientation, vertical position and size, and slant (Thomassen et al., 1988b). The rationale is that orientation needs to be normalized in order to estimate vertical size and position, and that vertical size and position, in turn, should be normalized in order to estimate slant. However, various algorithms for each normalization step are available and not every algorithm may be appropriate in all conditions. For instance, orientation of the word currently being processed can be estimated by a global algorithm based on a whole page or by a local algorithm based on the very word only. Although the latter algorithm bases its estimate on less data, it may yield a more relevant orientation estimate especially in writers who produce words with varying orientations. Although several estimates for the orientation are available just one orientation parameter is needed in the above example. In general, each normalization module consists of three types of submodules: Several "estimators" of a normalization parameter (e.g., the orientation of a word), a "selector" of the optimal normalization parameter, and an "operator", which finally performs the normalization using the parameter selected. The selector is actually part of one larger unit ("Scheduler") which surveys all levels and chooses the most confident hypothesis to be evaluated next.

The problem is thus which estimate to adopt if they yield different values. Choosing the wrong estimate of the orientation, for instance, may have dramatic effects upon the subsequent slant estimation (See Figure 1). Taking the average of the estimates is probably not a good choice because one of the estimators may be totally wrong. Evaluating all possible combinations of estimates forms no solution either because it would yield an explosion of computations. The solution we propose is evaluating only the most confident, non-evaluated hypothesis.

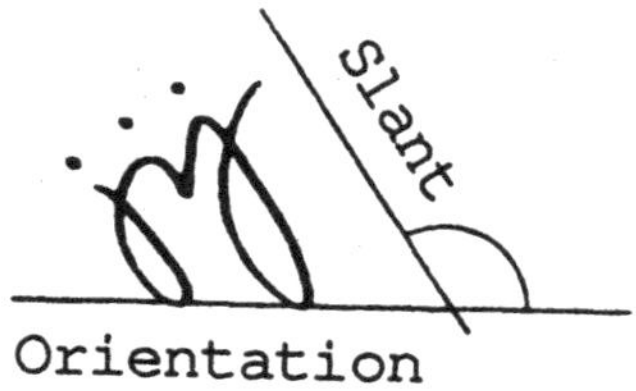

Figure 1. An example of an incorrectly estimated orientation of a word, which causes subsequently an incorrect estimate of the slant.

Figure 2. The pride of the baseline estimate of the second word is higher than that of the first word because of the additional evidence of several segmentation points consistent with the base line in the second word.

In order to quantify the compound confidence of a certain estimator in a particular case, two sources of confidence are employed. The first source of confidence is a measure for the consistency of a specific instance of an estimate ("Pride"). For instance, the pride of a specific baseline estimate increases with the number segmentation points that are on a straight line. (See Figure 2 for an example). The second source of confidence is the cumulated reliability demonstrated from the past ("Prejudice"). In previous cases, the prejudice of an estimator was increased if its estimate has contributed to a successful final solution and was decreased otherwise. The extent of the increase or decrease of the prejudice depends on the value of the pride. For instance, if an estimator reports a high pride, but turns out to produce an incorrect estimate the prejudice needs to be reduced badly.

In order to obtain the compound confidence of a specific estimate, the prejudice and pride are multiplied (yielding the "optimism"). The rationale is that the confidence should only be high if both prejudice and pride are high. In case of several alternative estimator algorithms, the algorithm producing the estimate with the highest optimism is selected to transform the data. If two or more estimators produced compatible estimates this provides additional evidence. The estimators are then combined in a group (called a "committee") and the estimators' optimisms are added up. In this way a committee may outrule a single estimator which happens to have an exceptionally high optimism.

Finally, the combined confidence score of a particular sequence of algorithms from the first processing step to the current processing step is needed. A sequence of estimates receives a confidence score which is equal to the product of all optimism scores of the individual estimates. At any time, the estimator, with the highest confidence score will be evaluated further. In this way, the system can backtrack to earlier levels where an erroneous decision was made and finally arrive at the most appropriate sequence of choices of the estimator algorithms. In this fashion, the system learns which estimators are in general reliable and which are not. Gradually, the most reliable estimators are adopted by default and the system gains efficiency by making fewer wrong choices of estimators (See Figure 3).

A problem of the present design is that it is rather conservative. If an estimator happens to be acceptable, its prejudice keeps on increasing. If at a later stage another kind of handwriting needs to be processed and if an alternative estimator yields better results, its pride may be insufficient to compete with the high prejudice of the first estimator. In fact it would have been better to require still some competition between estimators.

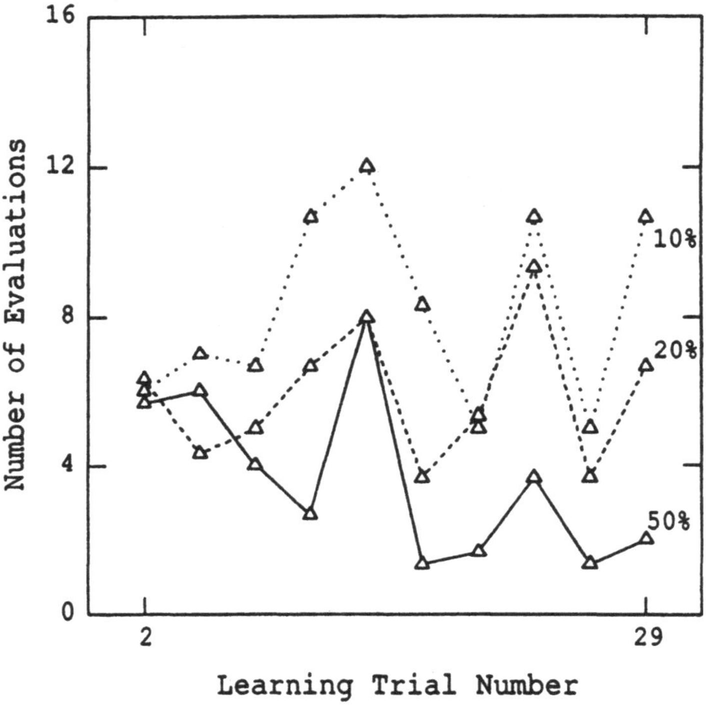

Figure 3. Simulation of the number of evaluations of an estimator (averaged across three sequential trials) as a function of number of trials. The estimators were perturbated with noise (the standard deviations of the added noise were 10, 20, and 50% of the exact estimate values, respectively). The more noise added, the less frequently an estimator was evaluated.

3. Feature extraction

Various sets of features per handwriting stroke have been used for different purposes. In the present application strokes are characterized by the following set of 14 features (Schomaker & Teulings, 1990) which make up the "feature vector":

(a) The vertical positions of the beginning (Y_b) and end of a stroke (Y_e) relative to the base line and the path length of the stroke (S) all scaled to the average body height, also called x-height, referring to the lower case x.

(b) The directions ϕ_n of the five, straight stroke segments between two subsequent points corresponding with the time moments

$$t = t1 + (n/5)*(t2-t1),$$

where t1 and t2 are the time moments of beginning and end of the stroke and n=[0, 1,..., 5], i.e., (ϕ_1, ϕ_2, ϕ_3, ϕ_4, ϕ_5). Here we explicitly use dynamic movement information. The rationale is that in equal time intervals the movement direction is changing a relatively constant amount (e.g., Thomassen and Teulings, 1985) such that each new stroke segment adds an equal amount of new information. The two previous and the two following stroke segments (respectively, ϕ_{b4}, ϕ_{b5}, and ϕ_{e1}, ϕ_{e2}) are included as well in order to capture the stroke's context.

(c) The size of the enclosed area between the end of the stroke and the subsequent stroke (λ_e) is rather a visually salient feature.

(d) A pen up indicator (P), which shows whether the pen is predominantly up or down during a stroke. It may be noted that strokes above the paper also count as strokes.

Feature vectors can also be used to define a distance measure between two strokes, called the "stroke distance". This distance is similar to the Euclidian distance of the normalized feature vector with the exception that the features referring to directions are not entered as the square difference but as the tangent of half of the difference. This is most consistent with our notion that strokes with opposite directions should have maximal distance. Finally, feature vectors are useful to define the average stroke across a set of strokes with small distances between them.

The number of features is high, indeed. This number can be reduced by building a two-dimensional self-organizing Kohonen network and then replacing the feature vector by the two network coordinates of the cell closest to the feature vector (Morasso, 1989; Schomaker & Teulings, 1990). This speeds up and improves the performance.

4. Allograph Recognition

In the recognition phase the handwriting patterns in terms of sequences of stroke features are translated into characters using an "allograph lexicon". This allograph lexicon relates specific stroke-vector sequences to the corresponding character. The building of this allograph lexicon will be described in the section on the learning phase.

Two ways to do this job are under investigation. The first method is based on a neural-network approach (E.g., see Schomaker & Teulings, 1990). The second method, which is the one described here, is an algorithmic approach. The advantage of the latter method is that it optimally splits up the input sequence of strokes into more or less adjacent sub-sequences of strokes, where each sub-sequence forms one entry in the allograph lexicon. An established procedure to segment a sequence into adjacent sub-sequences is the Viterbi algorithm. However, the standard Viterbi algorithm is not applicable in the present case because it was designed for equal-length sub-sequences. In cursive script, however, the allographs have different numbers of strokes. Moreover, the algorithm has to detect which of the strokes are the connecting strokes between the allographs. The algorithm we developed to detect the allographs, is a modified version of the Viterbi algorithm.

Apart from the stroke distance introduced earlier, the algorithm requires two other distance measures. In order to quantify the resemblance of a piece of handwriting and an allograph of the allograph lexicon, we define the "allograph distance" as the average stroke distance between the corresponding strokes. In order to quantify the confidence of a sequence of hypothesized characters, we define the "word distance" as the average allograph distances. The connection strokes are not rated in the distance scores.

The algorithm estimates first the allograph distances between each allograph in the lexicon and each part of the to-be-recognized handwritten word. By storing only the allographs having relatively small distances in a table, reduces the number of possible allograph sequences dramatically (see Figure 4). Now, the modified Viterbi algorithm is started. During the first iteration, each allograph starting at the first or at the second stroke of the input word is selected. Namely, the first stroke of the input word may be either an irrelevant connection stroke or the allograph's initial connection stroke and is not necessarily included in the allographic lexicon, because they vary with context.

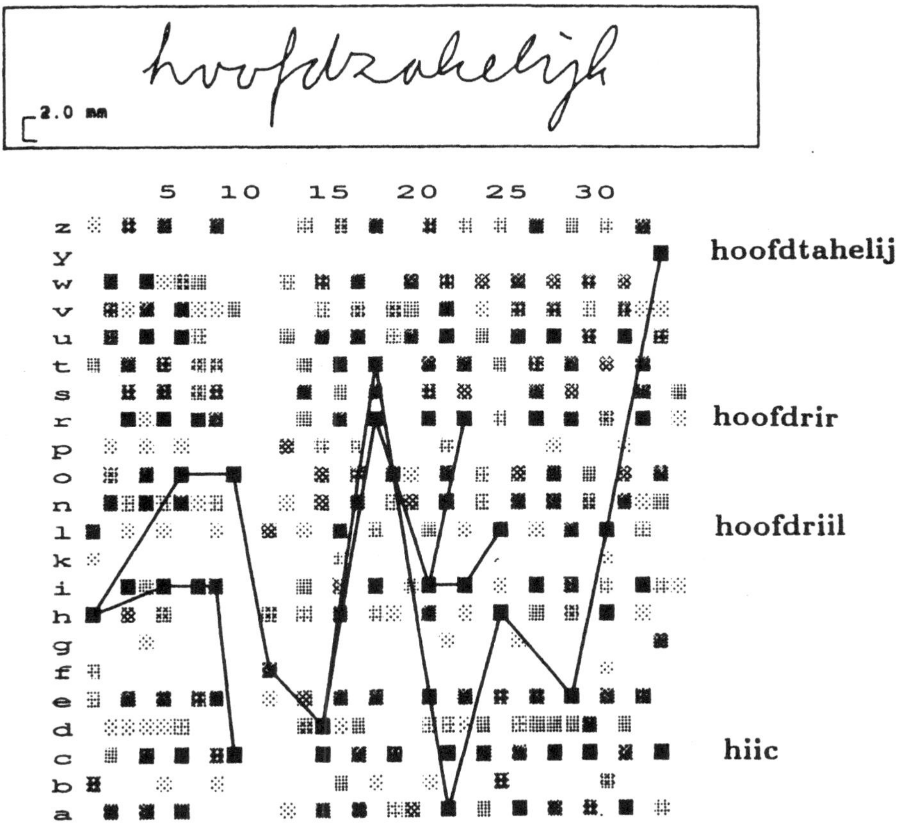

Figure 4. A representation of the allograph-distance table and a few examples of discontinued survivors and the most confident survivor. (Horizontal are all strokes of the Dutch word 'hoofdzakelijk' and vertical are all characters in the allograph lexicon. For the sake of simplicity, the distances of the allographs representing the same character have been collapsed. The darker the entry the lower the distance.)

The allographs selected constitute the "survivors" of length one. In subsequent iterations, each survivor is extended by the best fitting allograph which fits either immediately after the survivor or with one intermediate, connection stroke. If an expanded survivor covers the whole handwritten word, it is added to a list of word hypotheses. The iterative process continues until all survivors are added to this list.

A problem in the present design is that the Viterbi algorithm is not optimal for cursive handwriting patterns where allographs have varying numbers of strokes whereas the initial strokes may be highly similar. The consequence is that the system cannot decide between two survivors of similar quality without incorporating the between-word context. Too many solutions are competing at any serial position.

5. Learning Phase

Before a cursive-script recognition system is able to work, it has to learn the name of specific stroke sequences. If the written text is available, the learning module could just assign each allograph to a character in the text. However, in cursive handwriting it appears to be a rather cumbersome task to teach a system each allograph, that may occur in a person's handwriting. The reason is that the allograph boundaries in cursive script have to be specified somehow. The simplest solution would be to set the number of strokes in the characters manually. For instance, characters 'a' and 'b' count three strokes and character 'c' counts one stroke (apart from optional connection strokes, etc.). However, this may not work in natural, sloppy handwriting because the number of strokes may vary. Another method is to segment cursive script into allographs using a-priori assumptions about the shape of the connection strokes between allographs (e.g., Maier, 1986). However, such a method produces persistent errors (e.g., segmenting allographs like cursive b, v, w, u, or y into two parts). A third and promising method has been proposed by Morasso et al. (1990). First a few representative prototype allographs are identified manually. If a stroke sequence is presented together with the character sequence it is representing, the sequence of strokes can be split into sub-sequences where each sub-sequence matches the prototype allograph with minimum distance. In this way, more prototype allographs can be identified. Although the set of prototypes may form the allograph lexicon, it is wise to condense the set of prototypes. For that purpose they built "allographotopic maps" in an array of Kohonen networks for sequences of two, three, ... strokes, respectively.

In the present paper we describe an algorithmic approach which may achieve the segmentation of cursive script into allographs completely automatically, but without using prototypes, and also without specific knowledge about the structure of the allographs used, nor their numbers of strokes, nor assumptions about the properties of connection strokes. The method assumes that handwriting recordings have been collected from a specific text, containing words with all kinds of character sequences. The corpus of handwriting is segmented into words and these words, in turn, into strokes. Subsequently, the feature vector of each stroke is determined. The first step is to select the largest group of words starting with the same character. Those words starting also with the same allograph will show small mutual stroke distances for the first few strokes. In order to find out which words show relatively small stroke distances a nearest-neighbour clustering method is employed. However, an allograph may or may not start with an arbitrary connection stroke. Therefore, the stroke sequences have to be aligned first. This can be done by clustering the stroke distances of the first and the second strokes and selecting the strokes in the largest cluster.

After aligning the selected words all first strokes will be very similar. The group of similarity of the first strokes implies that the cluster analysis on the first stroke's feature vector forms at least one large cluster. The next step is to estimate the average inter-stroke distance of the second stroke, etc. As an example, consider the selected words starting with the character 'b', which has three strokes, apart from one optional connection stroke at the beginning or one at the end. Because all kinds of characters may follow the first character in the selected group of words, the cluster size will decrease markedly beyond the third stroke (See Figure 5). Using an appropriate threshold, the number of strokes that actually belong to character 'b' can be estimated. Finally, the average stroke pattern belonging to this character will be appended to the allograph lexicon.

In order to identify the second and following allographs in the corpus of handwriting, the initial character is removed from the words that were part of the resulting cluster and the corresponding feature vectors are removed from the sequence of feature vectors of the words in the corpus. Then the previous step is repeated. Processing all data in this way may yield a condensed list of prototypes, which forms the lexicon of allographs. In fact it may be possible to identify even various allographic versions of the same character. As the condensed prototypes found on the basis of the writings of different writers do not need to be inconsistent their lists of allographs may be merged. The number of times that a certain allograph is used in the recognition phase indicates how universal each prototype allograph is. These numbers enable the system to discriminate between useful and redundant allographs.

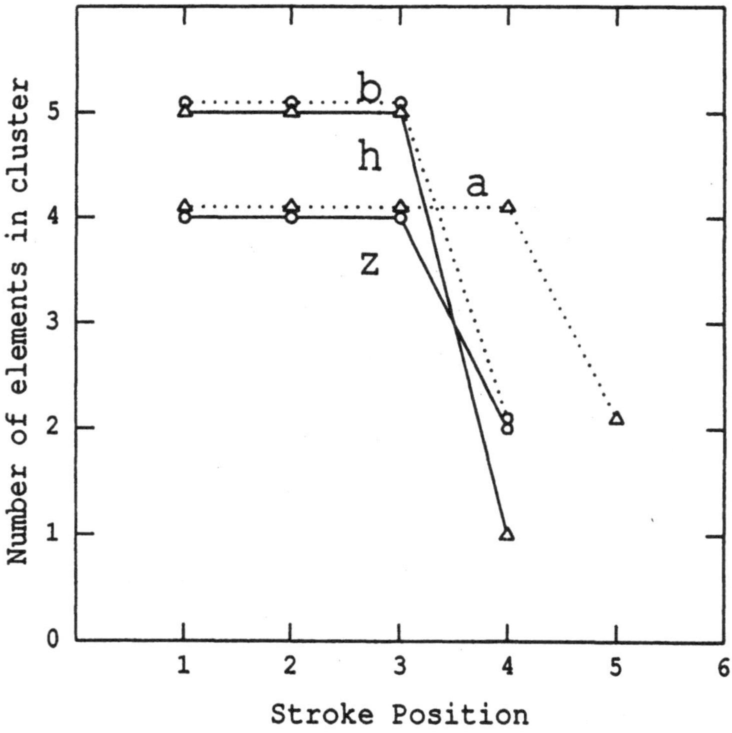

Figure 5. The cluster size of the feature vectors of the first six strokes from four subgroups of aligned words (The groups of words are words starting with an allograph 'b' (N=5), 'a' (N=4), 'h' (N=5),and 'z' (N=7). The cluster size in the words starting with 'a' decreases significantly after the fourth stroke, indicating that this allograph counts four strokes. Similarly, 'b', 'h', and 'z' count three strokes.).

Specific problems in the present design are still the necessity of tuning parameters for deciding when the cluster size drops significantly. Another problem is introduced by the artificial left-to-right scanning of the words, making the algorithm less reliable for the final characters of the words. In both respects the approach by Morasso et al. (1990) seems more appropriate.

6. Conclusion

The present paper discussed some of the modules designed for on-line cursive script recognition. However, the modules are still fragile and slow. But this is not the major problem we are dealing with: how to quantify the performance of each of the modules such that they can be compared with the performance of other designs. The four modules described above do their task but what does this tell us? In the present paper the modules were processing natural, purely cursive script of arbitrary sequences of only lower case allographs with omission of all diacritical signs such as dots on i and j. In fact the performance varies more with the properties of the sample of cursive script than with the appropriateness of the algorithms. This implies that the modules should be tested on a large number of subjects, while each subject should produce a considerable amount of cursive script. It would be wise to distinguish several "qualities" of handwriting ranging from extremely regular handwriting to the normal, sloppy handwriting, and several "character sets" ranging from only some lower case allographs to all kinds of allographs. Even if the modules are tested with large amounts of handwriting it may be hard to tell whether it performs well. The least subjective quality criterion of, for instance, normalization of the orientation, may be requiring the subjects to write horizontally on lined paper and to look for the closest-to-zero and the least-variance orientation estimator. The best known quality measure is probably still the overall percentage-of-characters correctly recognized, but this implies that the modules of a complete recognition system have been realized and run in a realistic time scale.

The second problem is the sensitivity of some modules' parameters. It would be ideal if a module could operate using only selection criteria like "the largest", or "the smallest", or "the closest to zero". But very often discrete parameters are required. They may be as harmless as "the maximum number of strokes per character is eight" to more suspect parameters as "select only strokes longer than a fixed stroke size". We feel that the micro-architecture design rules of a complex system of a cursive-script recognizer are important.

7. Acknowledgement

This research was supported by ESPRIT Project 419, Image and Movement Understanding.

References

Amin, A. *IRAC: Un système pour la reconnaissance et pour la compréhension a l'arabe écrit et imprimé,* PhD Thesis, CRIN, Université de Nancy, 1985.

Aoki, K., & Yamaya, Y. (1986). Recognizer with learning mechanism for hand-written script english words. *Proceedings of the Eighth International Conference on Pattern Recognition*, Paris, pp. 690-692.

Brocklehurst, E.R. (1988). *The NPL electronic paper project*. NPL Report DITC 133/88.

Doster, W., & Oed, R. (1984). Textbearbeitung auf Personal-Computern mit handschriftliche Direkteingabe. *PC-Praxis.*

Higgins, C.A., & Ford, D.M. (1989). A comparison of N-grams and tree structure for letter graph reduction *Proc. Fourth IGS Conference* p. 26, Trondheim, July. (Also in this volume)

Higgins, C.A., & Duckworth. (1989). The pad (pen and display): A demonstrator for the electronic paper project. *Proc. Fourth IGS Conference* p. 45, Trondheim, July. (Also in this volume)

Higgins, C.A., & Whitrow, R. (1985). On-line cursive script recognition. In B. Shackel (Ed.), *Human-Computer Interaction- INTERACT '84* pp. 139-143.

Kondo, S. (1989). A model of the handwriting process and stroke-structure of character-figures. In R. Plamondon, C. Y. Suen, M. Simner (Eds.), *Computer recognition and human production of handwriting* pp. 103-118. Singapore: World Scientific.

Leedham, C.G., & Downton, A.C. (1986). On-line recognition of Pitman's handwritten shorthand: an evaluation of potential. *International Journal of Man-Machine Studies*, **24**, pp. 375-393.

Lindgren, N. (1965). Machine recognition of human language: Part III - Cursive script recognition. *IEEE Spectrum*, **2**, pp. 104-116.

Maarse, F.J., Schomaker, L.R.B, & Teulings, H.L., (1988). Automatic identification of writers. In G.C. van der Veer & G.Mulder (Eds.), *Human-Computer Interaction: Psychonomic Aspects* pp. 353-360. New York: Springer.

Maier, M. (1986). Separating characters in scripted documents. *Eighth International Conference on Pattern recognition,* Paris, (ISBN: 0-8186-0742-4), pp.1056-1058.

Mandler E. (1989). Advanced preprocessing technique for on-line recognition of handprinted symbols. In R. Plamondon, C.Y. Suen, M. Simner (Eds.), *Computer recognition and human production of handwriting* pp. 19-36. Singapore: World Scientific.

Morasso, P. (1989). Neural models of cursive script handwriting *International Joint Conference on Neural Networks*, Washington, DC, June.

Morasso, P., Kennedy, J., Antonj, E., Di Marco, S., & Dordoni, M. (1990). Self-organisation of an allograph lexicon. *International Joint Conference on Neural Networks*, Lisbon, March.

Oulhadj, H., Lorette, G., Petit, E., Lemoine, J., & Gaudaire (1989). From primitives to letters: A structural method to automatic cursive handwriting recognition. In Pietikainen, M. & Roning, J. (Eds.), *Proceedings of the Sixth Scandinavian Conference on Image analysis*, Oulu, pp. 593-598.

Pick, H.L., Jr., & Teulings, H.L. (1983). Geometric transformations of handwriting as a function of instruction and feedback. *Acta Psychologica,* **54**, pp. 327-340.

Plamondon, R., & Baron, R. (1986). On-line recognition of handprinted schematic pseudocode for automatic Fortran code generation. *Proceedings of the Eighth International Conference on Pattern Recognition*, pp. 741-744.

Schomaker, L.R.B., & Teulings, H.L. (1990). A handwriting recognition system based on properties of the human motor system. *International Workshop on Frontiers in Handwriting Recognition*, Montreal, April.

Skrzypek, J., & Hoffman, J. (1989). *Visual Recognition of Script Characters; neural network architectures.* Technical report UCLA MPL TR 89-10, Computer Science Department University of California, Los Angeles.

Srihari, S.N., & Bozinovic, R.M. (1987). A multi-level perception approach to reading cursive script. *Artificial Intelligence*, **33**, pp. 217-255.

Tappert, C.C. (1986). An adaptive system for handwriting recognition. In H.S.R. Kao, G.P. van Galen, & R. Hoosain (Eds.), *Graphonomics: Contemporary research in handwriting* pp.185-198. Amsterdam: North-Holland.

Tappert, C.C., Suen, C.Y., & Wakahara, T. (1988). On-line handwriting recognition: A survey. *Proceedings of the ninth International Conference on Pattern Recognition*, 1123-1132, Rome, November 1988.

Teulings, H.L., Schomaker, L.R.B., Morasso, P., & Thomassen, A.J.W.M. (1987). Handwriting-analysis system. In R. Plamondon, C.Y. Suen, J.-G. Deschenes, & G. Poulin (Eds.), *Proceedings of the Third International Symposium on Handwriting and Computer Applications* pp. 181-183. Montreal: Ecole Polytechnique.

Thomassen, A.J.W.M., & Teulings, H.L. (1985). Time, size, and shape in handwriting: Exploring spatio-temporal relationships at different levels. In J.A. Michon & J.B. Jackson (Eds.), *Time, mind, and behavior* pp. 253-263. Heidelberg: Springer.

Thomassen, A.J.W.M., Teulings, H.L., & Schomaker, L.R.B (1988a). Real-time processing of cursive writing and sketched graphics. In G.C. van der Veer & G. Mulder (Eds.), *Human-Computer Interaction: Psychonomic Aspects* pp. 334-352. New York: Springer.

Thomassen, A.J.W.M., Teulings, H.-L., Schomaker, L.R.B., Morasso, P., & Kennedy, J. (1988b). Towards the implementation of cursive-script understanding in an online handwriting-recognition system. In Commission of the European Communities: D.G. XIII (Ed.), *ESPRIT '88: Putting the technology to use. Part 1* pp. 628-639 . Amsterdam: North-Holland.

Ward, J.R., & Blesser, B. (1985). Interactive recognition of handprinted characters for computer input. *IEEE, Computer Graphics and Applications*, **5 (9)**, pp. 24-37.

Welbourn, L.K. & Whitrow, R.J. (1989). A gesture based text and diagram editor. *Fourth IGS Conference* p. 48, Trondheim, July. (Also in this volume)

Wells, C.J., Evett, L.J., & Whitrow, R.J. (1989) The use of orthographic information for script recognition. *Fourth IGS Conference* p. 27, Trondheim, July. (Also in this volume)

Wright, P.T. (1988). Design and implementation of a handwriting recognizer. *Plessey New technology*, **2**, 1988.

Computer Processing of Handwriting
Eds. R. Plamondon & C. G. Leedham

A PREDICTION-VERIFICATION STRATEGY FOR AUTOMATIC RECOGNITION OF CURSIVE HANDWRITING

H.OULHADJ, E. PETIT, J. LEMOINE, M. GAUDAIRE
Laboratoire de Génie Electrique de Créteil, Université Paris Val-de-Marne, avenue du Général de Gaulle F094010 CRETEIL Cédex, FRANCE

G. LORETTE
IRISA - Université de RENNES I, F035042 RENNES Cédex, FRANCE

To date, more natural and friendly man-machine interfaces are still needed. A lot of research has been done to investigate speech recognition but less work has been done on handwriting recognition. Recently, the latter has received a new surge of interest, but man-machine interaction by means of personal handwriting remains a problem to be solved. This paper presents a new method of pure cursive handwriting recognition. In our approach, we try to recover from the noisy data the relevant information which underlies the handwriting. We assume the existence of models such as those learned in grade school and common to a lot of people. This paper describes an on-line recognition system, based on a letter prediction-verification rather than a word global recognition. This overcomes the problem encountered with an a-priori segmentation process. Recognition systems are often designed as a hierarchical set of different modules. In the present approach, the segmentation and the recognition procedures are carried out simultaneously, as in a closed-loop system with feedback. This last property, coupled with the use of various aspects of knowledge about handwriting (such as ideal letter models), results in a high performance system.

1. Introduction - State of the Art

Man-machine interaction can be achieved in various ways and new interactive tools have been designed for this purpose in the past years. To communicate with a computer two means, speech and writing, are of particular interest.

If speech has already been much investigated, it is not the same with respect to writing, which has always been challenged by typing. But it seems that recently, there has been a new surge of interest for this type of man-machine interface. This is due to the increase of micro-computing and the advent of new specific input devices such as scanners, specialized digitizers, interactive flat panel displays, and also to the progress observed in artificial intelligence techniques such as cooperation between several different knowledge sources.

More than twenty years ago Frishkopf and Harmon (1961) and Mermelstein and Eden (1964) carried out the first experiments on cursive handwriting recognition and proposed the first underlying models of the human handwriting systems [Van der Gon and Thuring (1965)] [Eden (1968)].

Dealing with handwriting, several aspects can be considered.

First of all, the handwriting can be seen from two points of view: either an external shape which is characteristic of the writer or an internal significance which is the meaning. The first point is investigated if the object of the study is to verify the author's identity or to authenticate the text [Plamondon and Lorette (1989)]. The second point is processed if the aim is to recognize what is written in the document.

A second aspect consists in distinguishing between the different types of handwriting which can be processed. A complete list has been given in a recent paper [Tappert (1984)]. The first experiments were carried out on separated handprinted characters and some reviews of this research works were published [Suen *et al.* (1980), Mori, Yamamoto and Yasuda (1984)]. Other studies dealt with pure cursive handwriting or cursive script .

A third aspect is to make a system classification according to the data input acquisition and the recognition processing method. Often systems are differentiated as "on-line" or "off-line". A handwriting recognition system is said to be "on-line" if the handwriting is digitized and recognized while it is traced [Teulings *et al.* (1989), Whitrow *et al.* (1989)]. A review of existing approaches has been carried out recently [Tappert *et al.* (1988)]. Alternatively, if the acquisition and the recognition occur from prewritten documents, the corresponding systems are known as "off-line". A few off-line word recognition systems were designed in recent years [Bozinovic and Srihari (1989)]. It may also be considered that on-line recognition systems are processing signals, while off-line systems deal with images.

The fourth aspect is that a handwriting recognition system is defined as "monoscriptor", "multi" or "omniscriptor" according to whether it is designed to process the handwriting of one, several or any scriptors.

The last parameter of classification of a recognition system is connected to the knowledge that it considers. It can include low level modules conceived to extract features in the input data (i.e. speech or writing) and/or high level modules representing various knowledge sources (e.g. semantic or syntactic) [Srihari *et al.* (1982)].

In this paper we describe an on-line system designed to recognize pure cursive handwriting of any scriptor by means of low level feature extraction. The general principles of the proposed method, the synoptic scheme and the main characteristics of the system are described in part two of this paper. A more detailed description is given in part three. The experiment and the corresponding results are presented in part four.

2. General Presentation of the Method

2.1. Principles of the method

Handwriting may be considered as a representation via a particular coding of ideas and concepts of the scriptor. In this context separated parts of words or letters are traced, using some underlying models [Teulings *et al.* (1986)]. This coding is degraded because of several noise sources which interfere with the handwriting process.

At the opposite side, reading or handwriting recognition are tasks of decoding the transmitted messages. In a way, the aim of these operations is to rediscover the underlying word models shared by both the writer and the reader. In other terms, we assume that there exists some knowledge and models of letters and words which are common to a large number of literate people. The main model of our system is composed of the letters of the latin alphabet as they are learned by children in grade school. In the case of a scriptor who uses a very different underlying model for certain letters (e.g., shapes deduced from handprinted characters), the recognition system is adapted by a training step.

The handwriting model we use is not an explanatory model of the handwriting generation process by the writer such as those used to produce natural handwriting traces [Plamondon and Maarse (1989)]. The underlying model which is assumed here is only dedicated to handwriting recognition. In this model a word or a part of a word is considered as the result of the association of different letters tied together with ligatures, each letter being formed from subparts or strokes which are themselves joined together by ligatures. Each ligature can be one of the three following ones: a cusp, a hump or a loop, depending on the letter context and the writer's habits. In ordinary handwriting, numerous changes between these different kinds of ligatures are often observed and therefore, this obviously constitutes irrelevant information which has to be removed or skipped over to achieve good recognition. Furthermore, we assume that handwriting can be considered as resulting only from two main motions: a basic motion which is approximated as being a translation from left to right and a sequence of quick motions perpendicular to the previous principal direction. Thus, the symbolic representation of letters and words consists in four direction code sequences corresponding to right, up, down and left movements. This gives, in some sense, a minimal and concise description of the handwriting. In this perspective the slant is regarded as a noise which must be removed before the symbolic description occurs.

In handwriting recognition two main approaches are encountered: the first is the global word recognition method; the second is the segmentation method, which successively consists in separating and then identifying the letters within a word. The present method does not use any attempt of a-priori segmentation. It is only based on segmentation hypothesis generation. This is performed according to a prediction-verification strategy of letters, and a final decision on the word candidates. The main interest of this principle is that all possible segmentations into letters are considered. Thus we avoid the segmentation error problem which occurs when an a-priori segmentation is carried out directly on the drawing. In the following method all possible words are considered and the final result consists of a list of candidate words ranked according to a recognition quality factor. If no other criterion is considered (such as the semantics of the sentence from which the word to recognize is issued) the first element of the list can be regarded as recognized.

Reading is done independently of the position, orientation, size and slant of handwriting. This property may also be needed from a handwriting recognition system; it can be achieved in two different ways : normalization and extraction of invariant characteristics (features or primitives) with regard to the previously mentioned geometric transformations. Here, both are used; first a normalization procedure, then extraction of primitives which represent the relevant constituents of the handwriting based on topological properties: the pen-paper contact, the local sense of rotation (clockwise or anti-clockwise), the direction of the pen tip motion in two perpendicular directions, the presence of ascenders or descenders, etc... This provides robustness to the recognition algorithm with regard to handwriting degradations and variations.

Handwriting is obviously ambiguous and can lead to various misunderstandings. In order to remove these ambiguities from these coded messages several kinds of knowledge or context levels have to be used: morphological, lexical, syntactic, semantic and pragmatic. In the following system, the syntactic and semantic parts are not considered.

2.2. General block diagram of the system

Figure 1 displays the general block diagram of the system. Two main steps, training and recognition, can be seen.

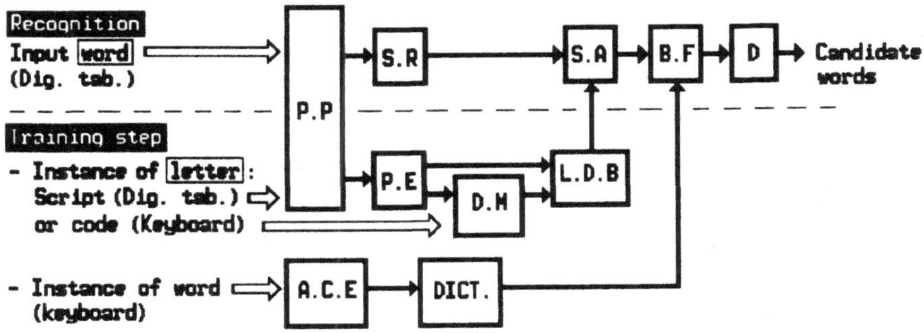

Figure 1. General Synoptic of the Cursive Script Handwriting Recognition System.
P.P. : Preprocessing P.E. : Primitive Extraction
D.M. : Degeneration Model A.C.E. : ASCII Code Extraction

During the training step, a tree-structured dictionary (DICT.) and a letter database (L.D.B.) are formed. The recognition step allows to acquire an input word and transforms its drawing into a symbolic representation (S.R.). Then a syntactic analyser (S.A.) attempts to define matches between the symbolic representation of the input word and the letter codes contained in the letter database. Finally, a decision (D) is made on the base facts (B.F.) in order to define a list of candidate words.

3. Description of the Recognition System

3.1 Data input and preprocessing

Handwriting is digitized by means of a graphic digitizing tablet. Then a preprocessing step, common to both the training and the recognition stages, reduces noise and redundancy by angular and distance filterings. Afterwards, the pen-up pen-down sequence and the temporal evolution of the drawing according to a 16-direction Freeman coding method are defined. This initial 16-direction coding is modified to obtain a 4-direction coding which corresponds to the model of handwriting that we previously described.

To this purpose, transformation rules are applied to the initial coding. These rules allow the topologic properties of the drawing to be retained (e.g. sign of Freeman differential coding, number and position of cusps,...).

It should be noted that using the rewriting rules insures some invariance with regard to the slant of the original handwriting and some suppression of irrelevant informations (Figure 2).

The preprocessing step also includes a zonal detection process [Chen and Lorette (1985)], to detect the vertical extension of letters (ascenders and descenders).

3.2 Training step

3.2.1 Dictionary formation

First, the vocabulary of the recognition system must be determined. It is limited to about two hundred words.

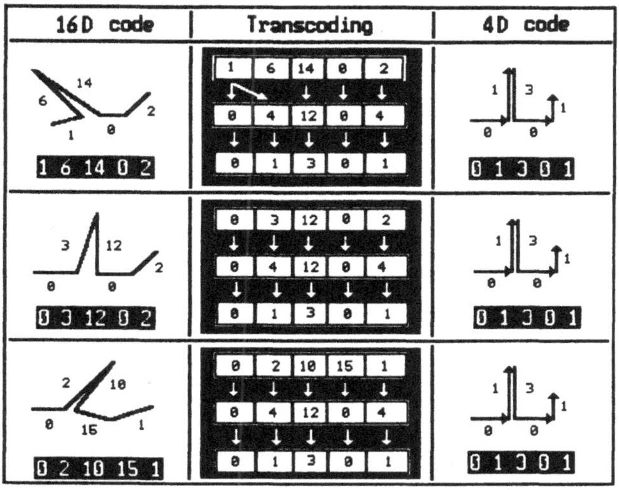

Figure 2. Examples of the use of rewriting rules.

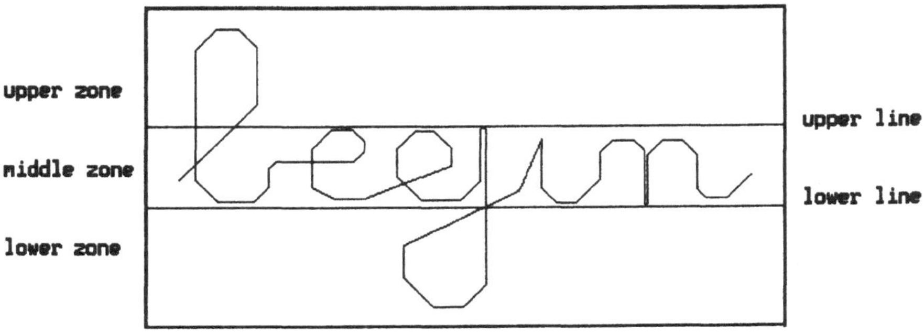

Figure 3. Middle Zone detection.

Then, the words of the vocabulary are introduced into the system. Words to be recognized are classified using a tree structure. Each node of the tree represents one letter of the alphabet. A path represents a complete word which is formed from all the encountered letters from the root to a particular leaf. Implicitly, the organisation of this word dictionary involves that the recognition algorithm consists in searching paths in this tree structure.

Figure 4 displays a reduced dictionary which we will use only as an illustration of how the system works.

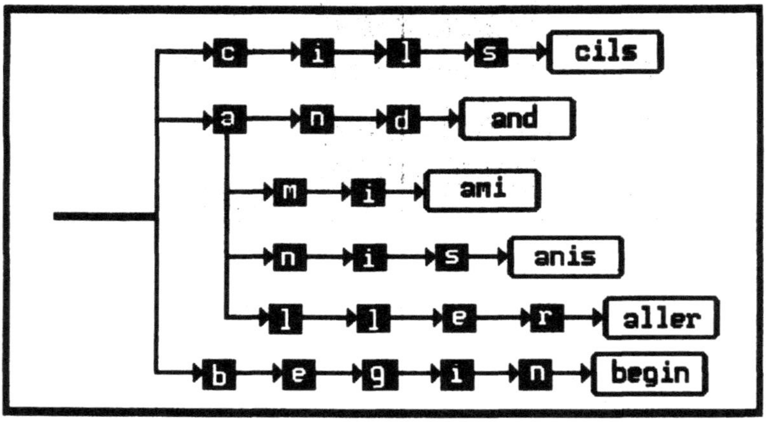

Figure 4. Example or a reduced dictionary involving the following words : "cils", "and", "ami", "anis", "aller" and "begin".

3.2.2 Letter database formation

The second part of the training step is the most difficult and must answer such questions as: How can we best modelize handwriting to obtain an efficient recognition process? Is it possible to define invariant primitives in handwritings (invariant according to the same scriptor or to different scriptors)? How is a machine able to take into account modifications in handwritings? Is it realistic to attempt to develop an omni-scriptor recognition system?

The first step to form the letter database consists in defining a prototype letter database which includes the 26 letters of the alphabet. There are two different ways to acquire these letters. On one hand, one can use a "standard letter representation" (i.e. the representations taught in grade school), idealized via a 4-directional coding; the coded letters can be introduced in the database via the keyboard of the system (Fig. 5). On the other hand, a scriptor can define his personal prototype letter database. In this case, he writes the 26 letters of the alphabet on a graphic digitizing tablet. Each letter is digitized, acquired, filtered, coded to obtain its symbolic representation named primitive P1. Then it is loaded in the prototype letter database according to a tree structure. Nodes of the tree represent directional codes. A leaf represents a letter which consists of all the encountered codes, from the root to this leaf.

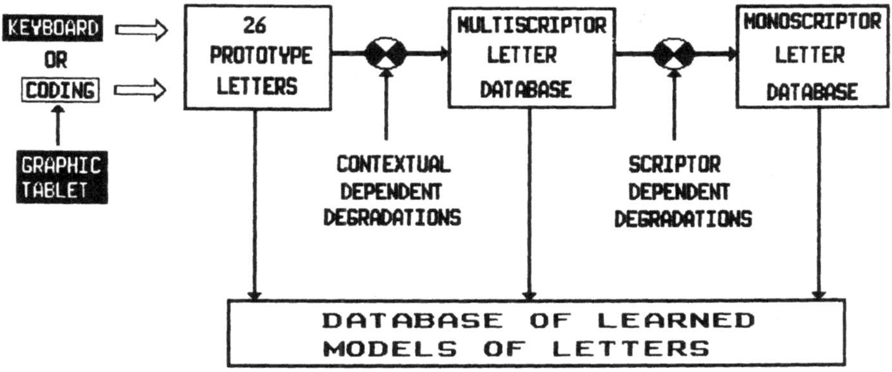

Figure 5. Letter Database Formation.

Figure 5 shows, the modifications of the prototype ideal letter model database allowing the generation of a multi or omni-scriptor letter database, taking into account context-dependent degradations, and how a mono-scriptor-dependent letter database can be obtained by taking into account scriptor-dependent degradations. In both cases, degradations are represented using rewriting rules acting on the codes which represent letters in the database. The set of defined rules constitutes, according to us, a degeneration model deduced from experience.

We distinguish between the context-dependent degradations and the scriptor-dependent degradations.

The context-dependent degradations result from the connection of two letters when the beginning of the second letter is not directly in the continuation of the terminal part of the first letter (for instance, an "r" following a "b"). Such degradations are taken into account by suppressing or adding codes at the beginning or at the end of the symbolic representations of prototype letters which can be affected by this kind of degeneration. Examples of such modifications of coding are shown in Figure 6.

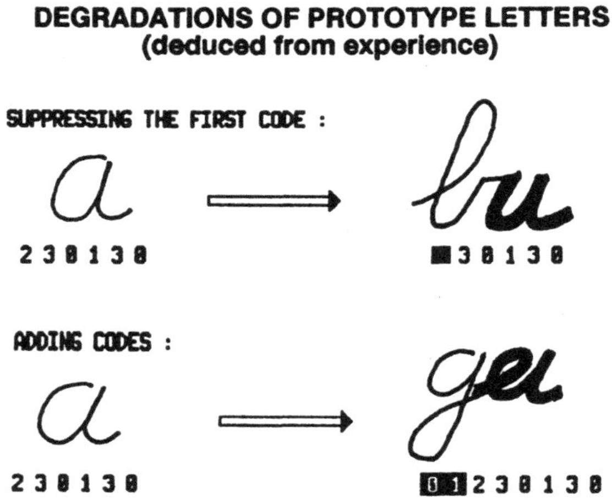

Figure 6. Context Dependent Degradations.

The scriptor-dependent degradations mainly affect the vertical cusps. When they occur, loops and humps may be created or suppressed. Figure 7 shows how the rewriting rules which take into account such degradations are applied. Each time a cusp is detected in the initial coding of a letter belonging to the multi or omni-scriptor database, either code 2 or code 0 is added in its symbolic representation to create a new template in the mono-scriptor database, i.e. a new degraded version of its initial ideal representation is created. To allow more particular degradations, which are not taken into account by this general degeneration model, and to increase the training flexibility of the system, a special function is available, which can be activated when part of a drawing cannot be recognized. In this case, the teacher introduces a new symbolic representation of the letter in the mono-scriptor letter database. This provides the system with some additionnal supervised learning capabilities.

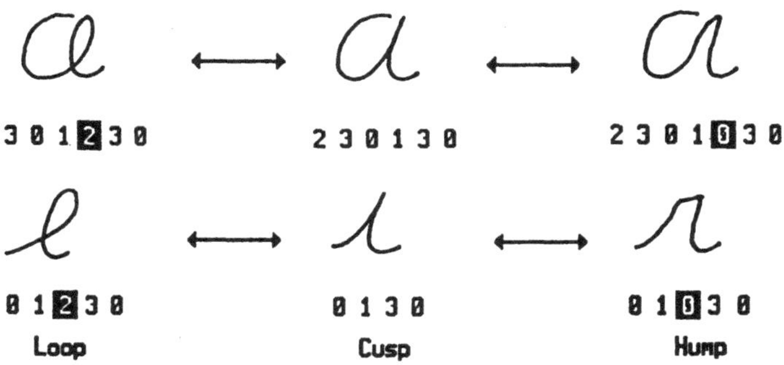

Figure 7. Scriptor Dependent Degradations.

Associated with primitive P1, a second primitive P2 completes the description of a letter. It represents a-priori knowledge of the global shape of the letter. Values assigned to P2 are equal to:

— 0 when a letter has no vertical line as in an "e",

— 1 when a letter has one ascender as in an "h",

— 2 when a letter has one descender as in an "j"

— 3 when a letter has one ascender and one descender as in an "ƒ".

Primitive P2 allows two cursive handwritten letters which have the same topological aspect but not the same size to be distinguished (for instance, an "e" and an "l").

Figure 8 displays part of a mono-scriptor letter database. Letters are classified using a tree structure. Nodes of the tree represent directional codes relative to primitives P1. Each leaf represents a letter and its associated primitive P2.

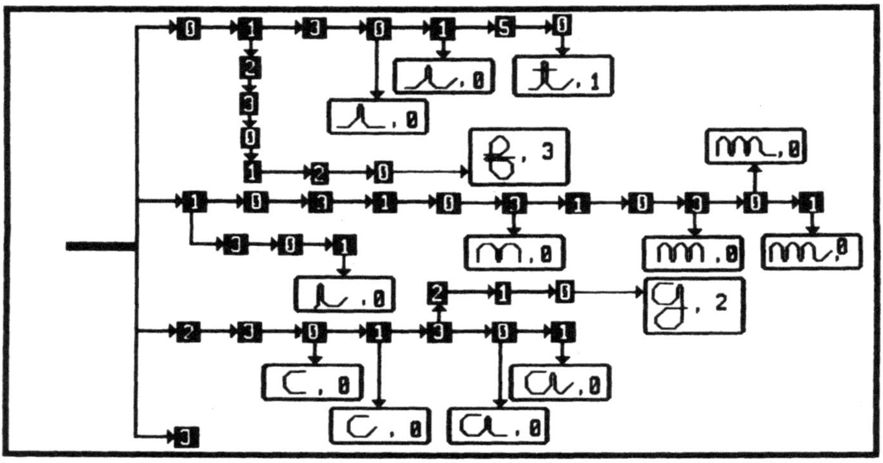

Figure 8. Instance of a reduced letter database involving "a", "c", "f", "g", "i","m", "n", and "t".

3.3 Recognition step

3.3.1 General principle

To recognize an input word, letters are detected using a prediction-verification strategy. From an input sequence of codes a predicted letter is obtained by searching a path in the corresponding tree-structured database. This predicted letter becomes verified if it allows either to initiate or to pursue a path in the dictionary otherwise it is rejected. Finally, the defined paths in the dictionary lead to several candidate words and a global recognition criterion is applied to obtain a ranked list of candidate words.

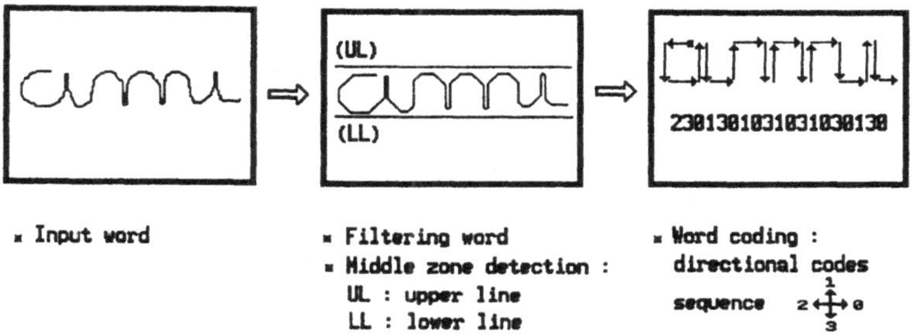

ANALYSIS

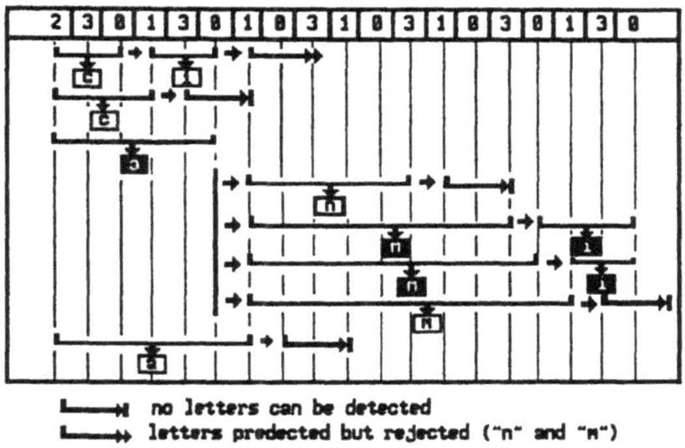

RESULTS

created list	type of the list
ci	eliminated string (list closed by rejected letters "n" and "m")
c	eliminated string (after "c" no letter can be detected)
an	eliminated string (after "n" no letter can be detected)
ami	candidate word
a	eliminated string (after "a" no letter can be detected)

Figure 9. Simplified description of the system recognizing the word "ami" by prediction (using the database defined in Figure 8) or verification (using the dictionary defined in Figure 4).

Figure 9 illustrates the general principle of recognition. It demonstrates how the system recognizes the French word "ami" and distinguishes the different stages of the recognition process which are more accurately described below.

3.3.2 Drawing acquisition and coding

The input drawing is processed to define its symbolic representation which includes both primitive P'1 which is a sequence of four directional and pen-up or pen-down codes, and primitive P'2 which is a sequence of labelled extrema as previously seen.

To define P'2 the extrema points of the input drawing are extracted. This allows to determine the medial zone of the drawing. Finally, we obtain a list of extrema which are labelled according to their position in the drawing (upper zone, middle zone or lower zone).

3.3.3 Prediction on letters

Prediction is carried out by scanning primitives P'1 and P'2 along the writing direction.

Each scanned code of primitive P'1 is used to attempt to define a path in the letter database. Each time a leaf in this tree-structured database is reached, a letter is predicted and a possible segmentation point is positioned in the input code sequence. This acts, in some sense, as a dynamic segmentation process rather than as a-priori static segmentation.

3.3.4 Verification on letters

As seen before, this predicted letter is described in the database not only by a directional code sequence but also by an a-priori knowledge on its spatial extent coded by primitive P2.

During the verification step, primitive P'2 of the input data is also scanned to determine whether or not it matches the spatial extent of the predicted letter.

If the matching is not perfect, the prediction is not rejected but affected with a likelihood weight used during the global decision step.

When a letter is predicted, a node corresponding with this letter is searched for in the dictionary. If one is found the letter is verified, otherwise it is rejected.

Each verified letter allows the placement of a segmentation point in the input sequence. From this point, the prediction-verification process proceeds by attempting to identify the next letter of the input word.

3.3.5 Decision on words

A word is recognized when the prediction-verification process leads to the definition of a complete path in the dictionary. During this first part of the recognition process, all the possible segmentation points are retained and stored. When the verification phase leads to a rejected letter or to a letter corresponding to a leaf in the dictionary, the process is backtracked to attempt to predict other letters from the last retained segmentation point (Fig. 9). In other words, due to its tree-structure, the dictionary guides the exploratory process of the letter tree database in depth first.

All the solutions are generated, therefore several candidate words can be selected and a final decision can be made on these words.

3.3.6 Global decision on words

A quality factor (QF) which has not any absolute meaning is used to rank the list of candidate words. To date, the first element of the list is regarded as the recognized word. When the system is adapted to sentence recognition, the set of ranked lists of candidates will constitute the input data of the syntactic or semantic analysers which will make the final decisions.

The quality factor is an arithmetic criterion deduced from three parameters.

The first is the global masking weight (GMW) of a candidate word. It is proportional to the number of directional codes of primitives P'1, used to define the path of a candidate word in the dictionary. The more input data (directional codes) the system considers, the better recognition is.

$$\mathrm{GMW} = \sum_{i=1}^{i=N} \mathrm{MW}(i)$$

The second is a degeneration weight (DW) which increases with the importance of the degradation of each template letter involved in the candidate word recognition. It allows the definition of the global degeneration weight (DGW) of a word as:

$$GDW = \sum_{i=1}^{i=N} DW(i)$$

where N is number of letters of the candidate word.

The third is a likelihood weight (LW) which increases when primitive P'2 of the candidate word mismatches the primitives P2 of the template letters involved in the input word recognition. It allows the definition of the global likelihood weight (GLW) of a word as:

$$GLW = \sum_{i=1}^{i=N} LW(i)$$

where N is number of letters of the candidate word.

Based on these three parameters, the quality factor of a candidate word is defined as:

$$QF = GMW - GDW - GLW$$

4. Results

The recognition system includes an IBM AT compatible micro-computer, a digitizing tablet (Summagraphics Bit Pad One) and a specific computer program written in Pascal. Five scriptors participated in the evaluation procedure : a system specialist (the teacher) and four scriptors taken at random. The dictionary included 110 candidate words, from which 50 were selected as test words according to such properties as their similarity or their length (e.g. do, to, end, and, for, file, blockwrite, blockread).

The trials of the four scriptors were performed in two stages: before and after the system was trained to their handwriting. During each stage, the input word was written five times.

The following average recognition rates were obtained:

	Success	Confusion	Reject
Before training step	40%	54%	6%
After training step	94%	6%	0%

There are two main reasons for the failures:

1. a bad detection of the middle zone of the handwriting,

2. unpredictible variations of handwriting such as a very important reduction in size.

To avoid the first class of mistakes, we must improve the method that we use to detect the middle zone of the input word.

Other problems can be reduced if we ask the scriptor to improve the general quality of his handwriting. As a matter of fact, people do not write in the same way when they take notes for themselves as when they are writing a letter to someone. Conscious that he is in a communication process, the scriptor must pay attention and follow some elementary rules such as maintaining a sufficient size of the drawing of the letters to avoid undersampling problems.

The average duration of the recognition process is about 5 seconds, and the recognition reproducibility is very good.

It should be noted that confusion and rejection rates are related to an important parameter of the recognition system: a threshold applied to the criterion used to accept words in the list of candidate words. If this threshold is low the confusion rates are high and reject rates are low. The opposite can be observed when the threshold is high. In our results, the low confusion rate is due to the low value of the threshold we used to conduct our trials.

In a more general fashion, the recognition system can be improved by introducing syntactic and semantic knowledges on the sentence from which the word to recognize is extracted. As a matter of fact, when the recognition fails, the right word is always found in the list of candidate words, but with a lower value of the QF factor.

5. Conclusion

The present on-line cursive handwriting recognition system is quite efficient when applied to a reduced vocabulary. It is well adapted to certain realistic problems. Up to now it does not seem possible to consider a great handwriting variability or an extended vocabulary. Nevertheless, this system can solve a lot of handwriting recognition problems which often refer to limited vocabularies, e.g. post address recognition, handwritten computer program recognition or cheque amounts verification.

The recognition system has important adaptation properties. First, it is adaptable to one or several scriptors. Second, it can include various kinds of letter databases and dictionaries in order to be specialized for various applications. Third, the letter database and the dictionary can be modified (by adding or suppressing a letter or a word respectively) in a very simple and efficient manner. Finally, as it is made of rules the degeneration model used to enrich the letterdatabase and to give "omni-scriptor" performances to the recognition system, can easily be extended with new rules.

Currently, the major inconvenience of this system is the small size of its dictionary. It reduces the number of verified letters during the recognition step, thereby decreasing the number of candidate words. Thus, on one hand ambiguities are avoided, and on the other hand time consumption during the process remains reasonable.

This system should be considered as a module of a more general recognition system which will include syntactical and semantical modules so as to easily work on larger vocabularies and be efficiently adapted to specialized application domains.

References

Amin, A., "Machine Recognition of Handwritten Arabics Words by the IRAC II System", *Proc. 6th Conf. on Pattern Recognition,* Munich, 1982, pp. 34-36.

Belaïd, A., Haton, J. P., "La reconnaissance de l'écriture", *La Recherche,* **170,** 1985, pp. 1188-1197.

Berthod, M., "Expérimentation sur l'échantillonnage de tracés manuscrits en temps réel.", *1er Congrès AFCET-IRIA, Reconnaissance de Formes et Traitement des Images,* 1978, pp. 560-567.

Berthod, M., Jancen, P., "Le prétraitement des tracés manuscrits sur une tablette graphique.", *2ème Congrès AFCET-IRIA, Reconnaissances des Formes et Intelligence Artificielle,* 1979, pp. 195-209.

Bozinovic, R. M., Srihari, S. N., "Off-Line Cursive Script Word Recognition", *IEEE Trans. on PAMI,* **11-1,** 1989, pp. 68-83.

Burr, D. J., "A Normalizing Transform for Cursive Script Recognition." *6th IJCPR,* Munich, 1982, pp. 1027-1030.

Chen, C., Lorette, G., Gaudaire, M., "Hierarchical Primitives Matching - A Structural Method for Cursive Script Recognition.", *4th SCIA,* 1985, pp. 59-65.

Earnest, L. D., "Machine Reading of Cursive Script.", *Proc. IFIP Congress 1962,* 1963, pp. 462-466.

Eden, M., "Handwriting Generation and Recognition.", in *Recognizing Patterns.*, P. Kohlers and M. Eden, Eds., MIT Press 1968, pp. 138-154.

Ehrich, R. W., Koehler, K. J., "Experiment in the Contextuel Recognition of Cursive Script.", *I.E.E.E. Trans. Comput,* **C 24,** 1975, pp. 182-194.

Fox, A. S., Tappert, C. C., "On-Line External Word Segmentation for Handwriting Recognition.", *Proc. of the Third International Symposium on Handwriting Computer Application,* 1987, pp. 53-55.

Freeman, H., "Computer Processing of Line Drawing Images.", *Computing Surveys,* **6-1**, 1974, pp. 57-97.

Ford D., Higgins, C. "Fast Dictionary Search Techniques", *4th I.G.S. Conference*, 1989, p. 25. (Also in this volume)

Frishopf, L. S., Harmon, L. D., "Machine Reading of Cursive Script", *4th London Symposium on Information Theory,* 1961, pp. 300-316.

Hollerbach, J. M., "A simple model for handwriting", *Proc. 5th Conf. on Artificial Inteligence,* 1977, p. 881.

Lingren, N., "Cursive Script Recognition", *IEEE Spectrum,* **2-5,** 1965, pp. 104-116.

Lorette, G.,*"Méthode de traitement automatique de points et de lignes en reconnaissance des formes - Application à l'analyse de scènes, à la reconnaissance de signatures et de mots manuscrits."*, Université Paris Val-de-Marne, 1984, 330 pp.

Mermelstein, P., Eden, M., "Experiment on Computer Recognition of Connected Handwritten Words.", *Information and Control,* **7,** 1964, pp. 255-270.

Miller, G. M. "Real Time Classification of Handwritten Script Words." - *Proc. IFIP Congress,* 1972, pp. 218-223.

Mori, S., Yammamoto, K., Yasuda, M., "Research on machine recognition of handprinted characters", *IEEE Trans. Pattern Analysis and Machine Intelligence*, **6,** 1984, pp. 386-405.

Plamondon, R., Lorette, G., "25 Years of Automatic Personal Identification Through Handwriting.", *Abstract 4th IGS Conference*, Trondheim, 1989, p. 54. (See Section 1 of this volume)

Plamondon, R., Maarse, F. J., "An Evaluation of Motor Models of Handwriting", *IEEE Transactions on Systems, Man, and Cybernetics,* **19-5,** 1989, pp. 1060-1072.

Suen, C. Y., Berthod, M., Mori, S., "Automatic Recognition of Handprinted Characters - The State of the Art.", *Proc. IEEE*, **68**, 1980, pp. 469-487.

Srihari, S. N., Hull, J. J., Choudhari, R., "Integrating Diverse Knowledge Sources in Text Recognition", *ACM Trans. on D.I.S*, **1-1,** 1983, pp. 68-87.

Tappert, C. C., "Adaptative On-Line Handwriting Recognition.", *Proc. 7th ICPR,* Montreal, 1984, pp. 1004-1007.

Tappert, C. C., Suen, C. Y., Wakahara, T., "On-Line Handwriting Recognition - A Survey", *Proc. 9th ICPR*, Rome, 1988, pp. 1123-1131.

Teulings, H. L., Schomaker, L. R. B., Gerristen, J., Drexler, H., Albers, M., "An On-Line Handwriting-Recognition System Based on Unreliable Modules", *Abstract 4th IGS Conference,* 1989, p. 33.

Teulings, H. L., Thomassen, A. J. W. M., Van Galen, G. P., "Invariants in Handwriting : the Information Contained in a Motor Program", in *Graphonomics : Contemporary Research in Handwriting*, Kao, Van Galen, Hoosain Eds, Elsevier Science Publ., 1986, pp. 305-315.

Thomassen, A. J. W. M., Maarse, F. J, Tibosch, H. J. C. M., "Spontaneous Versus Habitual Stroke Sequence in Generating Characters", *Proc. 3rd Int. Symp. on Handwriting and Computer Applications*, Montreal, 1987, pp. 25-27.

Van Der Gon, J. J., Thuring, J.Ph., "The Guiding of Human Writing Movements.", *Kybernetik*, **2,** 1965, pp. 145-148.

Vredenbreght, J., Koster, W. G., " Analysis and synthesis of handwriting", Philips *Technical rev.*, 1971, **32**, pp. 73-78.

Computer Processing of Handwriting
Eds. R. Plamondon & C. G. Leedham

ON DETERMINING DISTINCTIVE FEATURES OF HANDPRINTED CHARACTERS USING STATISTICAL TECHNIQUES

SHOZO KONDO
Department of Communications Engineering, Faculty of Engineering. Tokai Univerisity, Hiratsuka-shi, 259-12 JAPAN

The goal of this study is not to design a particular computer algorithm. We have attempted first to create a theory of characters and a methodology for designing recognition algorithms. In this paper, distinctive features of a troublesome pair of Roman cursive lower case letters are investigated using the principal component analysis. From the investigation it is shown that distinctive features of the pair of letters depend upon authors, and we cannot find common distinctive features for all authors.

1. Introduction

The state of the art in character recognition has advanced from the use of primitive schemes for recognizing machine-printed numerics to the application of sophisticated techniques for recognizing wide variety of handprinted characters such as Kanji, Katakana and Hiragana, Suen et al. (1980); Mori et al. (1984); Tappert et al. (1988). Owing to the great variation in shapes and styles of characters handwritten by different people, however, the recognition accuracy of the various OCR systems on free writing is far from perfect, Shinghal and Suen (1982).

The most important problem in the handprinted character recognition theory is that of detecting features which are hardly affected by handwriting distortions. Blesser et al. (1974) have argued that the failures of previous feature detection techniques lie in the ad hoc nature by which the features were chosen, and in order to build a machine which will recognize characters as accurately as humans, specific knowledge about human classification of characters must first be obtained.

The stroke structure of handprinted characters is one of the features which is relatively invariant under writing distortions. By the stroke structure of a character, we mean a set of elementary functional segments, called strokes, which form the character. A number of biological handwriting models for investigating the stroke structure of cursive handwriting have been published to date, Eden (1962); Plamondon and Maarse (1989); Yasahura (1971). These models are based on the study of human motor processes. In these theories, a stroke is defined as a sequence of stimuli, which results in the production of a fundamental unit of handwriting movement Plamondon and Maarse (1989). Therefore, the stroke structure of a character corresponds to a motor program and is independent of phenomenological shape of the character, Yasahura (1971).

In the field of speech analysis, sufficient definition of phonemes of speech has never been obtained from the study of human speech organs, Trubetskoy (1958). From the above fact, it seems to be seen that we cannot obtain a sufficient definition of stroke from the study of human motor process of writing.

Essential principles of information transmission, using physical entities such as speech sound and writing, are identity and difference, that is, identity of those entities which are not the same, and difference of those entities which are similar to each other, Saussure (1949); Mounin (1968). For example, a linguistic system is a series of differences of sounds combined with a series of differences of ideas, Saussure (1949). From this viewpoint, researchers in the field of speech analysis were able to obtain a definition of a phoneme as a bundle of distinctive features by using opposition of similar speech sounds, Trubetskoy (1958); Jakobson et al (1951);.

Following this theory of speech analysis, techniques for investigating distinctive features of handprinted characters, which are called functional attributes, have been published to date, Blesser et al (1974); Shillman et al (1976); Suen and Shillman (1977). These techniques are based on the study of ambiguous Roman upper case letters. The goal of these studies is not to design a particular computer algorithm. Researchers in this field have attempted first to create a theory of characters and a methodology for designing recognition algorithms.

In this paper, statistical techniques for investigating distinctive features of isolated handprinted Roman lower case letters, are proposed. Especially distinctive features of the lower case letters will be determined, based on the study of ambiguous characters. Since the stroke structures of Roman upper class letter are given without any analysis, it is relatively easy to find their distinctive features; but it is difficult to find distinctive features of cursive lower case letters because their stroke structures are not given clearly. Therefore, to find distinctive features of the lower case letters, we use some statistical techniques. The main results of our investigation are as follows.

1. Distinctive features of isolated handprinted characters of some Roman lower case letters depend upon authors and therefore, in those cases, we cannot find common distinctive features independent of authors.
2. If we want to make a machine capable of recognizing handprinted characters at low error rate of about 4% exhibited by humans, then the machine must have the ability to learn distinctive features for each author from his writings.

2. Determining Distinctive Features of Handprinted Characters

2.1. Statistical Techniques

In this paper we use on-line isolated handprinted characters as data for our investigation.

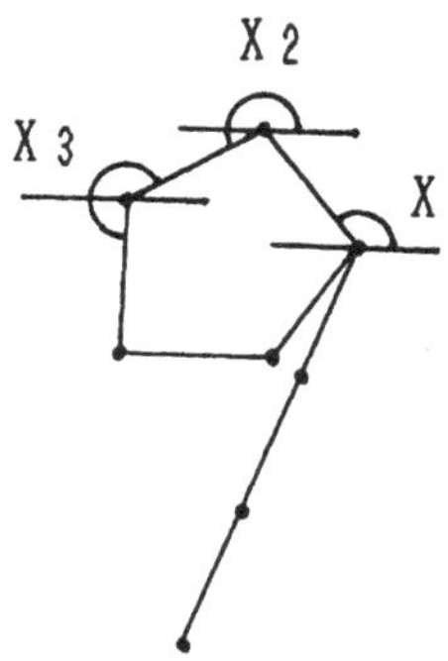

Figure 1. Representation of an on-line handprinted character by a sequence of n short equal-length segments.

A character handwritten on a tablet digitizer by an author is converted to a sequence of 2-dimensional points. The sequence is spatially filtered and represented by a series of n short equal-length line segments as shown in Fig. 1. The series can be specified by an n-dimensional vector whose components are the inclination angles of the line segments. The vector will be referred to as a character vector.

Let

$$\{x^1, x^2, \ldots, x^N\}$$
$$x^k = (x_1{}^k, x_2{}^k, \ldots, x_n{}^k)$$

be a set of N character vectors given for the following statistical analysis, called the Principal Component Analysis (PCA for short), Oja (1983). This set will be referred to as a basic data set for the PCA. The covariance matrix from the basic data set is given by

$$E[(x^k - \hat{x})^T (x^k - \hat{x})] = N^{-1} \{(x^1 - \hat{x})^T (x^1 - \hat{x}) + \ldots + (x^N - \hat{x})^T (x^N - \hat{x})\}$$

where $\hat{x}$ is the average character vector in the basic data set.

Let

$$c_1 \geq c_2 \geq \ldots \geq c_n \geq 0$$

be the eigenvalues of the covariance matrix and

$$\{g_1, g_2, \ldots, g_n\}$$

be an orthonormal basis made from the eigenvectors corresponding to the eigenvalues.

Let x(m) be the m-order truncated Fourier series of a character vector x by the orthonormal basis, Oja (1983).

$$x(m) = \hat{x} + (x - \hat{x}, g_1)\, g_1 + (x - \hat{x}, g_2)\, g_2 + \ldots + (x - \hat{x}, g_n)\, g_n \qquad (1)$$

Then the mean error of this approximation by the truncated series, e(m), is given as follows.

$$e(m) = E\,[\,||x - x(m)||^2\,]$$
$$= c_{m+1} + c_{m+2} + \ldots + c_n \qquad (2)$$

where $||x - x(m)||$ is the norm of x - x(m).

If the first and second eigenvalues c_1, c_2 are relatively large compared with the others, then the 2-order truncated Fourier series

$$x(2) = \hat{x} + k_1 g_1 + k_2 g_2 \tag{3}$$

$$k_i = (x - \hat{x}, g_i) \qquad (i = 1, 2)$$

gives a good approximation of the original character vector x and does not lose physical attributes which humans use to classify it. In this case, any character vector can be specified by a 2-dimensional coordinate point (k_1, k_2) where k_1, k_2 are the coefficients in the 2-order series (Eq. 3).

It will be referred to as a 2-dimensional coordinate representation of the character vector. Conversely, a character can be generated from a 2-dimensional coordinate point using the series (Eq. 3).

2.2 Determining Distinctive Features of a Troublesome Pair of Letters

Here we investigate distinctive features of isolated cursive lower case letters.

Shillman et al. (1976) developed a number of techniques for investigating distinctive features, which are called functional attributes, used by humans in recognizing isolated Roman upper case letters. Since the stroke structure of the upper case letters are given without any analysis, it is relatively easy to find distinctive features of upper case letters. However, it is difficult to find distinctive features of cursive lower case letters because their stroke structures are not given clearly. Thus we investigate distinctive features of the cursive lower case letters using the statistical techniques described above.

Trubetzkoy (1958) has argued that the concept of distinction premises the concept of opposition, and distinctive features of speech sounds can be obtained only by phonetic opposition of the speech sounds. It follows from this viewpoint that distinctive features of handprinted characters can be obtained only by opposition of the characters. The concept of opposition premises the concept of similarity. Thus distinctive features can be obtained by the opposition of similar character shapes. In this paper, distinctive features of a troublesome pair of letters, b and f, are mainly investigated.

We take as a basic data set for the PCA a set of 40 isolated handprinted characters of the letters, b and f, written by two authors, A and B, as shown in Fig. 2. Using the first and second principal components with respect to the basic data set, the 2-dimensional coordinate representations of the character vectors in the basic data set are shown in Fig. 3. In this PCA we have

$$(c_1 + c_2) / (c_1 + c_2 + \ldots + c_n) = 0.78$$

which can be taken as a measure of average efficiency of approximation by the series (3). Therefore, the 2-order Fourier series (3) gives relatively good approximations of original characters in the basic data set.

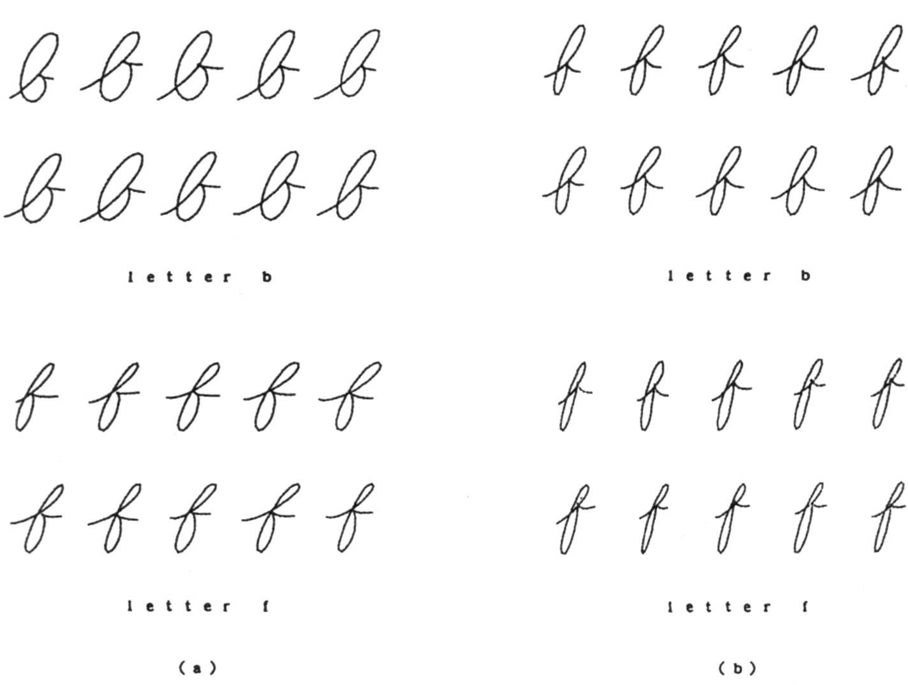

Figure 2. A basic data set of handprinted characters.
(a) written by author A.
(b) written by author B.

As seen in Fig. 3, the clusters for b and f overlap because f's of the author A are similar to b's of the author B. Thus the two clusters cannot be separated using any mathematical technique. Two 2-dimensional coordinate representations of characters for the authors A and B, are separately shown in Fig. 4(a) and 4(b), respectively. The clusters of the letters, b and f, of each author, are easily discriminated. It can be concluded from this fact, that distinctive features of the letters, b and f, for the authors, are different from each other, and we cannot find common distinctive features of b and f for the mixed data.

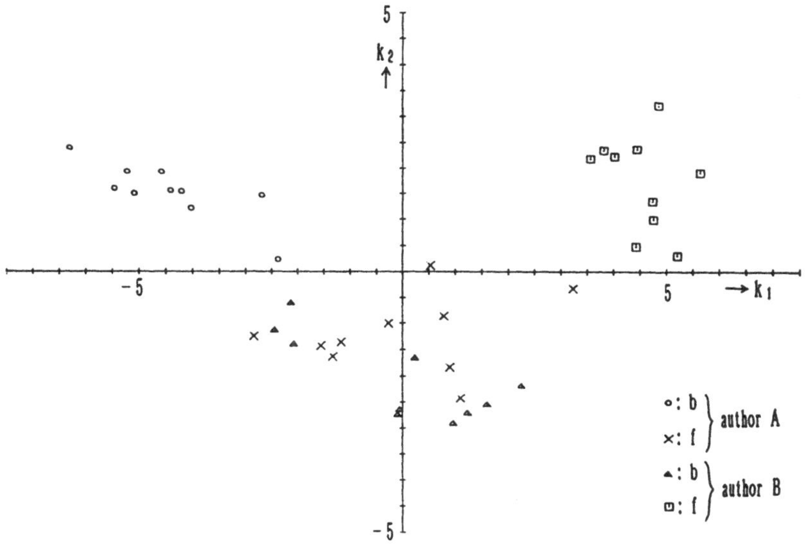

Figure 3. The 2-dimensional coordinate representations of the characters in the basic data set.

In order to find distinctive features of b and f for the authors A and B, we generate two sequences of ambiguous characters from b to f along the trajectories TA, TB in Fig. 4(a), (b) for the authors A, B, respectively, as shown in

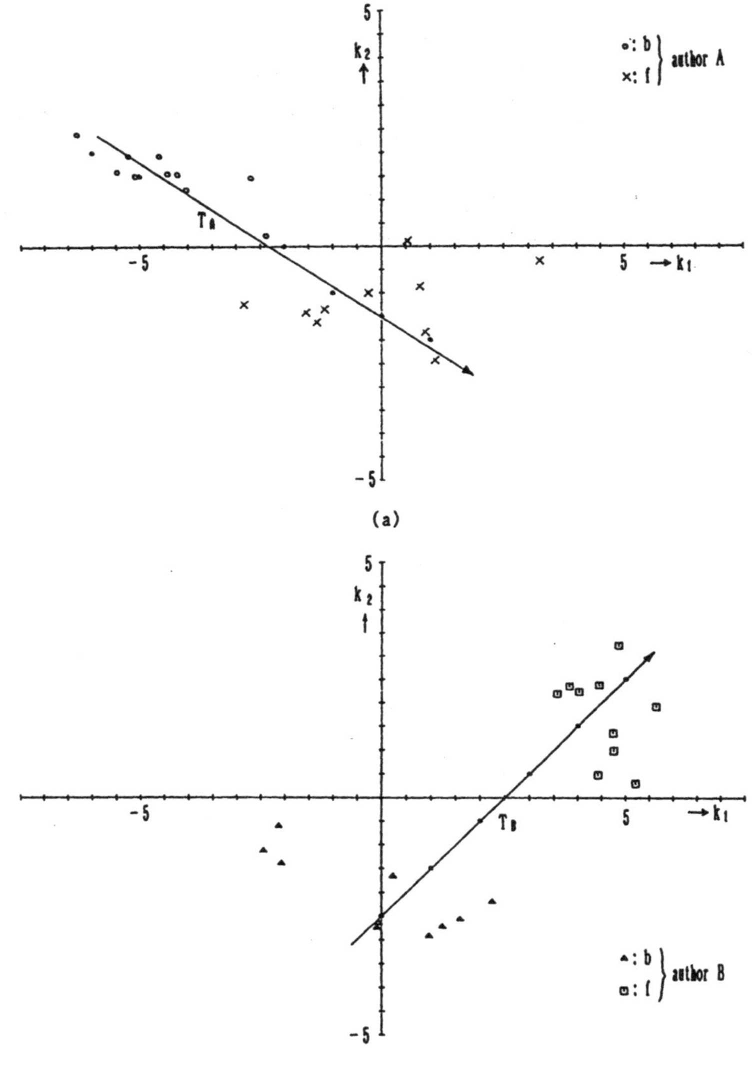

Figure 4. The two-dimensional coordinate representations of characters.
(a) written by author A
(b) written by author B.

Fig. 5(a), (b). As easily seen from the sequences, distinctive features of b and f for the author A are the position of the starting point of writing, the lengths of the first upstroke, and the final wing of writing, while distinctive features for the author B are the thickness and length of the lower loop of writing. Thus the distinctive features of the two authors are different from each other. We can discriminate between characters of b and f written by each author using each set of distinctive features, but can not discriminate between f's written by the author A and b's written by the author B.

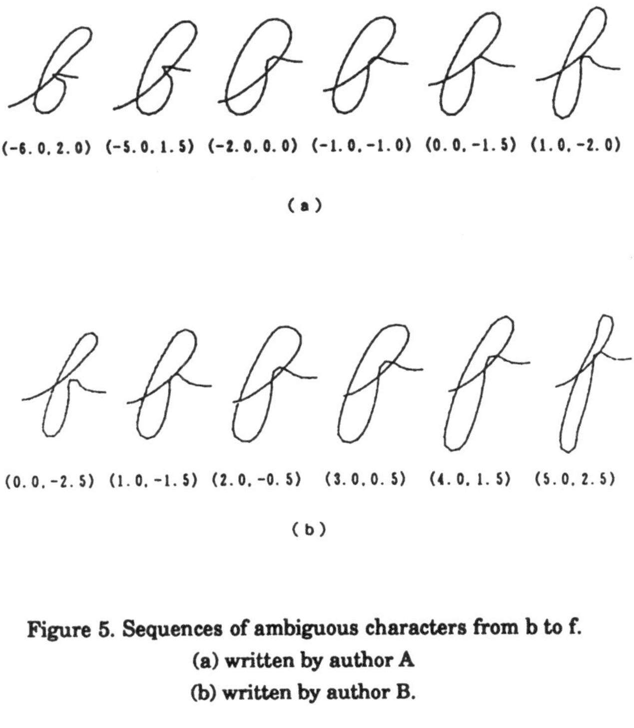

Figure 5. Sequences of ambiguous characters from b to f.
(a) written by author A
(b) written by author B.

In order to examine the above fact, the two PCAs are attempted with respect to the two basic data sets of characters of b and f written by the two authors. The 2-dimensional coordinate representations of the characters in the basic data set of each author are shown in Fig.6 (a), (b). As seen from these figures, two clusters of b and f for each author lie along the vertical axis,

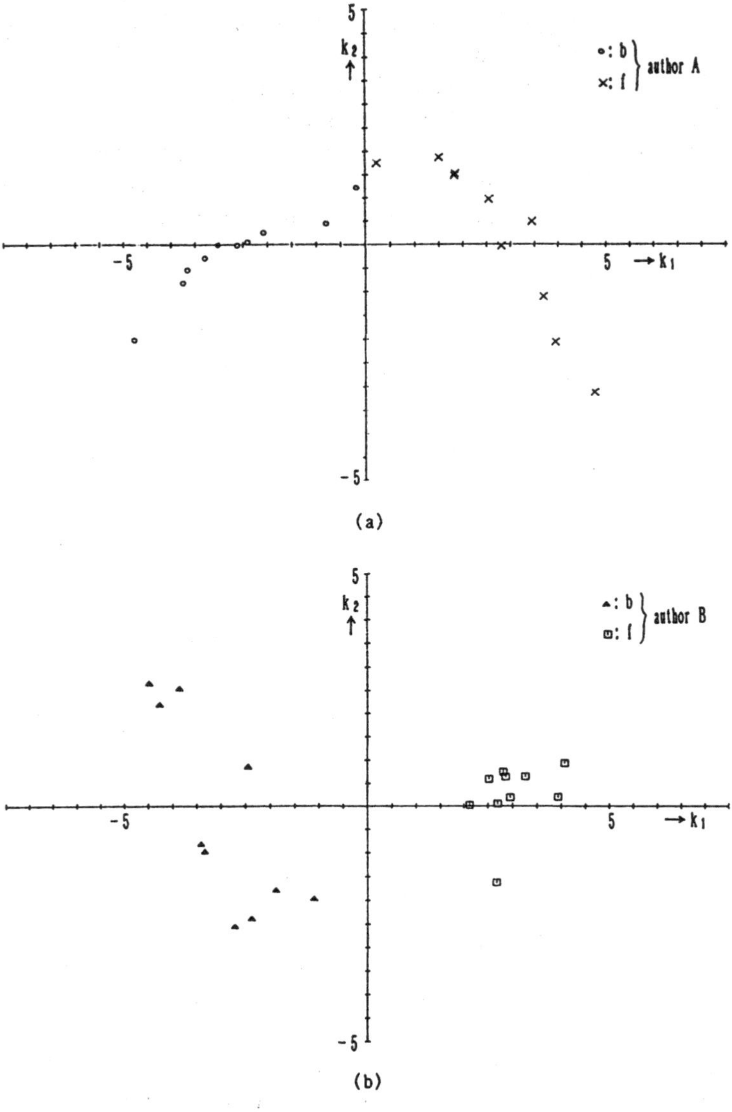

Figure 6. The two-dimensional coordinate representations of characters.
(a) written by author A
(b) written by author B.

which corresponds to the first principle component for each basic data set. Therefore, the first principal component can represent distinctive features of b and f for each author. Fig. 7 shows the first two principal components for the authors. From Fig. 7, differentiation of the distinctive features of the authors can be seen.

From the above investigation, it can be concluded that the distinctive features of b and f for the authors A and B are different from each other, and we cannot find common distinctive features for both authors.

A similar investigation can be done for every troublesome pair of lower case letters.

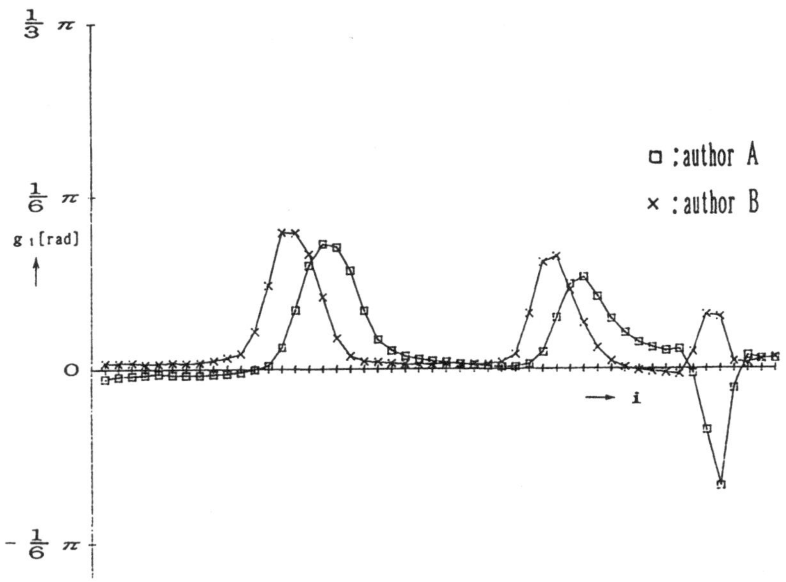

Figure 7. The first principal component of the authors A and B.

3. Conclusions

In this paper, we investigated distinctive features of a troublesome pair of roman lower case letters using the principal component analysis. From the investigation, it can be seen that distinctive features of b and f depend upon the author, and we can not find common distinctive features for all authors. Therefore, if we want to make a machine capable of recognizing hand-printed characters at low error rate of about 4% exhibited by humans, then the machine must have the ability to learn distinctive features for each author from his writings.

References

Blesser, B., Shillman, R., Kuklinski, T., Cox, C., Eden, M., Ventura, J., "A Theoretical Approach for Character Recognition Based on Phenomenological Attributes", *International Journal of Man-Machine Studies*, **6**, 1974, pp. 701-714.

Eden, M., "Handwriting and Pattern Recognition", *IRE Trans. Inform. Theory*, **IT-8**, 1962, pp. 160-166.

Jakobson, R., Fant, C. G. M., Halle, M., *Preliminaries to Speech Analysis, The Distinctive Features and their Correlates*, MIT Press, 1951.

Mounin, G., *Saussure ou le structuraliste sans le savoir*, Editions Seghers, 1968.

Mori, S., Yamamoto, K., Yasuda, M., "Research on Machine Recognition of Handprinted Characters", *IEEE Trans. Pattern Anal. Machine Intell.*, **PAMI-6**, **4**, 1984, pp. 386-405.

Oja, E., *Subspace Methods of Pattern Recognition*, John Wiley and Sons, 1983.

Plamondon, R., Maarse, F. J., "An Evaluation of Motor Models of Handwriting", *IEEE Trans. Systems, Man, and Cybernetics*, **19**, **5**, 1989, pp. 1060-1072.

Saussure, F., *Cours de linguuistique generale*, Charles Bally et Albert Schehaye, 1949.

Shillman, R. H. J., Kuklinski, T., Blesser, B. A., "Psychological Techniques for Investigating the Distinctive Features of Letters", *International Journal of Man-Machine Studies*, **8**, 1976, pp. 195-205.

Shinghal, R., Suen, C. Y., "A Method for Selecting Constrained Hand-Printed Character Shapes for Machine Recognition", *IEEE Trans. Pattern Anal. Machine Intell.*, **PAMI-4, 1**, 1982, pp. 74-76.

Suen, C. Y., Berthod, M., Mori, S., "Automatic Recognition of Handprinted Character-The state of the Art", *Proc. IEEE*, **68**, 1980, pp. 469-487.

Suen, C. Y., Shillman, R. J., "Low Error Rate Optical Character Recognition of Unconstrained Handprinted Letters Based on Model of Human Perception", *IEEE Trans. Systems, Man, and Cybernetics*, **7**, 1977, pp. 491-495.

Tappert, C. C., Suen, C. Y., Wakahara, T., "On-line Handwriting Recognition-A Survey", *Proc. 9th Int. Conf. Patt. Rec.*, Rome, 1988, pp. 1123-1132.

Trubetzkoy, N., *Grundzuge der Phonologie*, Vanderhoeck and Ruprecht, Goettingen, 1958.

Yasuhara, M., "Some Remarks on the Automatic Recognition of Handwritten Characters", (in Japanese) *Journal of Information Processing Society of Japan*, **12, 2**, 1971, pp. 88-92.

Computer Processing of Handwriting
Eds. R. Plamondon & C. G. Leedham

A GESTURE BASED TEXT AND DIAGRAM EDITOR

Lisa. K. WELBOURN and R. J. WHITROW
Department of Computing, Nottingham Polytechnic, Burton Street, Nottingham, United Kingdom

This paper describes an editor which has been designed to mimic the usual pen and paper type of editing. Hand-drawn gestures are used to specify the editing task. The use of gestures as an interface has become more important with the advent of electronic paper. The user specifies the editing task and its range by drawing a gesture on a tablet. The "ink" of the pen appears on a screen, allowing the user to see what has been drawn. The recognition of the gesture or gestures is on-line in its nature and the results of the edit are displayed immediately. The editor has been designed to work on cursive and ASCII text together with diagrams. The output from the editor is directed to one of two separate recognizers. The cursive script is recognized and turned into ASCII characters, and the diagrams are recognized as straight lines, curves and circles etc. The choice of gestures and their recognition is described and discussion of the user acceptability given.

1. Introduction

This paper describes work undertaken as part of Esprit project 295, "The Paper Interface". The project aimed to construct an interface to an Office System that was based around "paper". Documents that already exist on paper could be entered into the system via a scanner and some static recognition. The documents could consist of hand-written script, hand-drawn diagrams, typescript, photographs or a mixture of all them. Once into the system, these documents could be edited and added to using a stylus on electronic paper or a graphics tablet, and some dynamic recognition. Finally, the results could be output to a colour ink-jet printer. The part of the work presented here deals with the editing of the documents. From past work, on the subject, the words "gesture based editing" have evolved to describe the technique of using hand-drawn symbols to indicate the editing tasks, hence the title of the paper "A Gesture Based Text and Diagram Editor".

2. Gesture Based Editors - The State of the Art

Vannevar Bush was probably the first person to imagine using devices other than keyboards to communicate with computers. In his paper "As We May Think" (Bush, 1945), outlined his MEMEX machine: "A device which types when talked to, and uses a stylus to add marginal notes to a document...." The publication of the paper stimulated thought and other ideas followed.

Licklider and Clark (1962) foresaw the need for: "An electronic input-output surface for the display and communication of correlated symbolic and pictorial information" and "combined speech recognition, hand printed character recognition and light pen editing".

Sutherland (1963) is the earliest reference found which starts to head towards these goals. The system is called SketchPad and can be described as follows: "A SketchPad user sketches directly on to a computer display with a light pen. The light pen is used to both position the parts of the drawing and to point to change them".

Coleman (1969) took this a stage further to actually provide the pen driven editing. Whilst Sutherland's system used buttons to specify the editing command, Coleman's made use of proof reading symbols drawn on a tablet. However, as soon as any additional text was required the keyboard was enabled.

The early 1970's saw little reported work, however, but a number of systems then appeared in the middle-to-late 1970's. They used varying input and output technologies and worked on either text or diagrams. The Line Drawing Editor designed by Jarvis (1976) works specifically on schematic diagrams representing logic circuits. The specific category of diagram leads to a small library of symbols that need to be recognized. These symbols are constructed of lines of four orientations with angles of 0, 90 and ± 45 degrees. However, allowing the recognition of lines of only these orientations means that only a finite number of symbols will be available.

Buxton et al (1979) allowed the drawing of a line based symbol to be interpreted as a musical note. The shape of the symbol indicates the length of the note, e.g. minim, quaver, etc., whilst its starting position indicates the sound of the note, e.g. E, F, G, A, etc.

Suenaga and Nagura (1980 & 1983) used a FAX machine as the input/output technology for their editor. Their papers describe two editors, FLEX-T for text and FLEX-G for graphics. In both editors, the original document is fed into the machine first. Next the user must produce a sheet of markings. This is made by placing a thin sheet of paper over the original, and marking each line or block of text with a symbol to indicate its type. For instance, making an 'A' mark over a line indicates that it should be a circular arc. After recognizing the marks and their position on the paper, the lines and text can be amended and the results output to the FAX.

Doster and Oed (1984) took a slightly different approach to the others. They designed a symbol driven front-end for a conventional text editor.

More recently another couple of groups have reported their interest in the subject of the use of gestures for editing. IBM (1986) have published some results of pen and paper experiments to find out if people can use gestures for editing and if they are consistent in those that they use. Within the last year, Wang have announced a new product, Wang Freestyle. This system allows exactly what Bush envisaged - "the use of a stylus to add notes to a document". No recognition occurs in the system, the pages displayed on the screen are simply stored in bit map form. The user can use a stylus on a tablet to make notes on that page, or a telephone to record a message that can be heard when that page is displayed.

The input technologies are now beginning to catch up with the ideas. Electronic paper, is a thin flat display that can be held in the hand, and on which a stylus can be used directly. It is becoming cheaper and more readily available. Even so, there are still problems - many have parallax problems caused by the thick layers of glass covering the 'paper' and a surface which is difficult to write on. Many people think that as the technology improves new interest will be shown in cursive script recognition and gesture based editing techniques.

3. The Interface

The editor has been developed on a Sun workstation using a Datapath graphics terminal for the display and a Summagraphics Bit Pad Two digitizing tablet for the input. For various reasons, including the high cost, and the problems with parallax and the writing surfaces, it was impossible to use an Electronic Paper. However, it is hoped that sometime in the near future, the system will be moved on to such a system, and allow further testing for users reactions.

The editor is used in conjunction with two recognizers, also developed as part of this Esprit project. At present, one of the recognizers recognizes the script, the other the diagrams. However, they both use the same techniques to perform the recognition, and it is hoped that they will soon be combined. With this in mind, the editor has been designed to work on both script and diagrams, but in a moded editing environment. One mode for script, another for diagrams. The same system can be used to create and edit all types of documents containing either script, diagrams, or a mixture of both. More details about the recognizers can be found in the Esprit reports for project 295, and papers by Wright (1988) and Wells (1989).

All input to the editor is made by the stylus moving on the surface of the tablet. This combination is used like a pen on paper, however the user cannot see what is being written at the site of the pen instead he must look at the screen. The screen displays the document being edited, together with a small cross hair cursor that tracks the movement of the stylus. The user is therefore able to see the position of the pen in relation to the text and diagrams of the document. As a gesture is made, the 'ink' from the pen appears on the screen showing the user what he has drawn.

The tablet can be set, via various dip switches, to report co-ordinates in different formats, different resolutions, and using many operating characteristics.

The tablet has a 3D active volume, that is the region in which the stylus must remain to continue reporting different positions. Two dimensions extend over the surface of the tablet, a square of 11". The third dimension of the volume is established by a height above the surface of the tablet. This is termed the proximity - the maximum distance above the active area that the stylus can be held to report a position. For the Bit Pad Two, it is approximately 0.5".

One reporting mode of the tablet is Stream Mode. In this mode, the position of the pen is continuously reported whilst the tip is in proximity. When the pen moves out of proximity the same position is reported again and again until the pen returns to proximity. It was found that this mode was best in a writing situation, a writer will often lift his pen just enough to stop full pressure on the paper, but not enough to stop the ink marking the paper. The stylus is different to a pen, in that it can either be in contact with the tablet or not. This can cause problems, as valuable word information can be missed if reporting stops immediately the pen leaves the surface. Stream mode overcomes this, by reporting constantly whether or not the pen is directly in contact with the surface. Rarely will a writer lift his pen more than 0.5" of the page without meaning it.

Another of the most important characteristics that needs to be set for writing is the report rate, the number of reports that the tablet passes to the host computer per minute. The report rate is dependent on the report format and the baud rate of the connection.

When writing on the tablet and examining the output on the screen, it was found that a high report rate was necessary. With lower rates the reported script became very misshaped and incomprehensible even to a human eye. Figure 1 shows the trace on the screen made with a slow report rate, in comparison to the trace made by the stylus on the tablet.

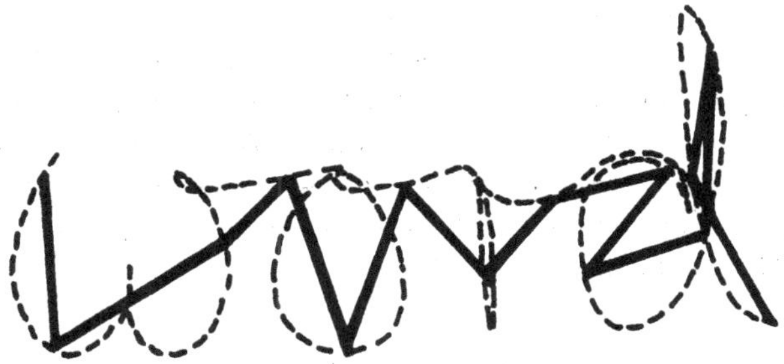

Figure 1. Comparison of traces on tablet and screen.

Once received by the host computer the reports can be decoded to give the string of co-ordinate pairs describing the movement of the stylus. As mentioned earlier, whilst set in stream mode, reports are made continuously. This has the disadvantage that there is no information to indicate whether or not the stylus is touching the tablet or not. A combination of the distance between, and an estimate of the time between, co-ordinates was eventually used to separate the string of co-ordinates into individual strokes. The choice of time and distance was merely a question of trial and error whilst trying to write clearly on the screen. They were set at 10 screen pixels and 2 seconds.

4. The Editing Symbols

4.1 The Choice

Before making any choices or firm decisions about the types of gestures to be used, a small experiment was conducted. 38 subjects were given a copy of a document and asked to carry out a number of editing operations. The people involved included those who were not from any computing background. Some of them did not proof-read or correct text or anyone but themselves to read. Others had secretary's who carried out the re-typing of documents, so they were more used to making symbols for others to understand. Finally there were the secretary's themselves who understood the problems associated with reading, understanding and interpreting other peoples symbols. An analysis of the results showed that the subjects were consistent in their use of the 'insert' symbol. 24 of the 38 used the same symbol to indicate the insertion of a character, word or phrase. The least inter-subject agreement was found with the 'new paragraph' symbol. Of the 38 symbols collected, 26 were different, and none had more than three occurrences. Further details of the results of the experiment can be found in Welbourn (1988 & 1989). The symbols chosen for implementation are shown in Figure 2. They were chosen after studying the results of the experiment, and attempting to draw the symbols using the stylus on the tablet.

4.2 Symbol Recognition in the Text Editor

The recognition of gestures and symbols has been designed to be as relaxed as possible. This allows for the variablility of people's symbols, their position within the text etc. The symbols have been categorized by the number of gestures they contain, and the types of these gestures (Figure 3). As long as the symbol drawn contains the correct number of strokes, it requires only one distinguishing feature to identify it from the others in the group. Gestures are recognized as belonging to one of 5 categories: horizontal and straight, vertical and straight, an arrow head, a dot or anything else. In practice, however, on the tablet it is virtually impossible to draw a perfect symbol, or gesture. Therefore the tests for the gesture types have to be as relaxed as possible, whilst still managing not to confuse any.

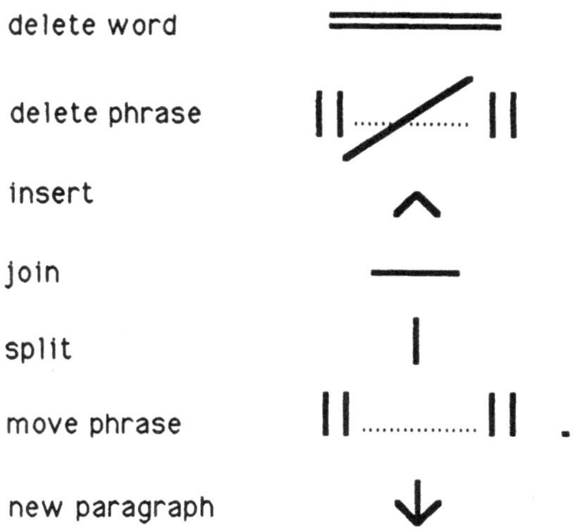

Figure 2. The symbols chosen.

No. of Strokes	Editing Symbol
1	join split
2	delete word insert one stroke new paragraph
3	insert two strokes
4	insert three strokes
5	delete phrase move phrase insert four strokes
6	insert five+ strokes

Figure 3. The set of recognized editing strokes.

For instance, the tests for horizontal and vertical gestures simply looks at the two end points of the gesture. The angle, taken to the horizontal, of a line joining these two points is calculated. If this angle is less than 30 degrees (Figure 4), a test for horizontal straightness is carried out. Alternatively, if the angle is greater than 60 degrees, a test for vertical straightness is carried out. Gestures with angles falling between 30 and 60 degrees are categorized as "anything else". The tests for straightness consider segments of the gesture. A segment is taken to be five consecutive co-ordinates (Figure 5). In the horizontal test, if all the segment gradients are less than or equal to 45 degrees then the gesture is said to be horizontal and straight. For a gesture to be considered to be straight and vertical, all the segment gradients must have angles greater than or equal to 45 degrees.

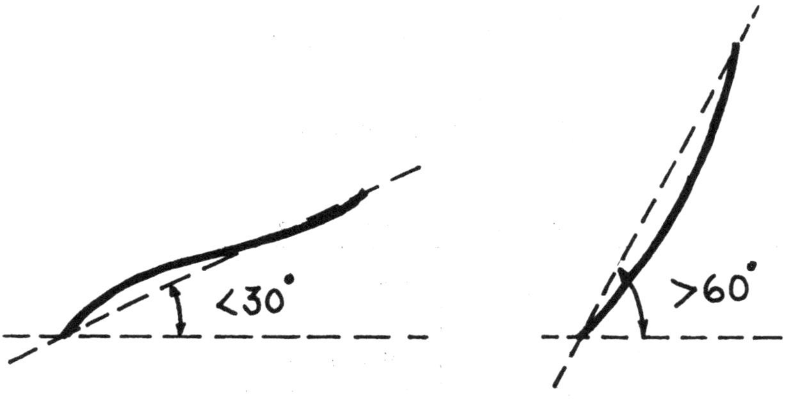

Figure 4. Recognition of horizontal and vertical strokes.

At present the set of symbols is limited to those shown previously, but it was felt that a user should be free to use almost any symbol for any editing task. Certainly the others that proved popular in the experiment should be available. The way in which the symbols are recognized has been designed to be extendable. Extra checks about gesture position can easily be included to allow one symbol to be distinguished from another.

Considering the one stroke symbols. At present, to distinguish between a split and a join simply requires the type of gesture to be evaluated. However, in the experiment to choose the symbols, a vertical line proved a popular choice for a "delete character" symbol, and a single horizontal line was also

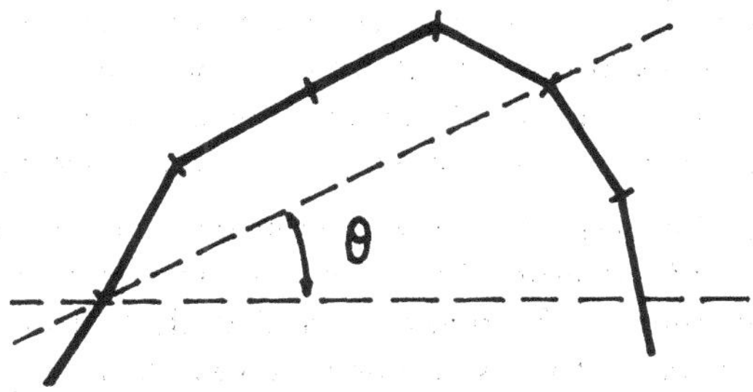

Figure 5. Segment gradients.

used to mean delete. To distinguish between these pairs, split/delete character, join/delete word, some contextual information is necessary. The position of the symbol in relation to the surrounding characters and words gives the information. If a vertical gesture is found to be covering a character then it can be considered as a delete, however, if it falls between two characters then it should be dealt with as a split. In the case of the horizontal gestures, the positions of the end points of the gesture in relation to the words can be used. For a join, the stroke ends will be closer to the end of one word, and close to the start of the next, whilst for a delete the stroke should cover a larger part of the words (Figure 6).

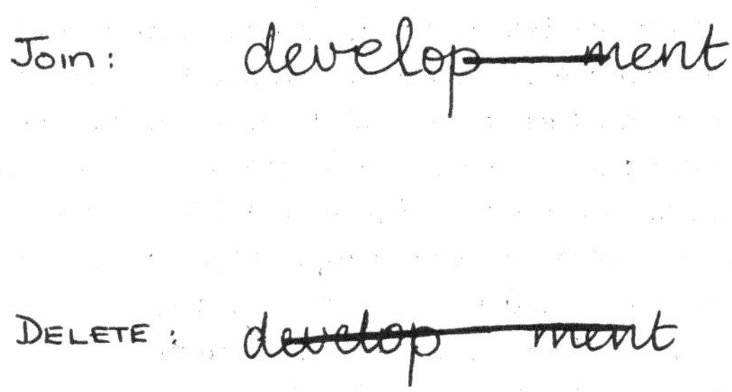

Figure 6. Comparing straight lines for Join and Delete.

4.3 Symbol Recognition in the Diagram Editor

As mentioned earlier, the editor has been work in a moded fashion. This was found to be necessary as the recognizers were separate and could not make any decisions about the type of data they were being passed. For example, the graphics recognizer would decide that a capital letter 'L' was really two straight lines, one vertical and one horizontal, joined to form a corner.

It was also found necessary to have two modes within the diagram editor, one for drawing and the other for editing. A user may want to include two small horizontal lines in the diagram, and if the editor was to recognize them as a delete symbol, they would be removed along with any other stroke they covered. Such a case arises when drawing a capacitor in a circuit diagram.

The modes are chosen by tapping the stylus in an area at the top of the editor screen. These areas, shown in Figures 7 and 8, act as toggle type buttons, allowing a user to switch between modes quickly and easily.

The symbols recognized and acted upon in the diagram editor, are at present limited to just delete and move. The delete symbol is similar to that used in the text editor. A move, however, is acheived by touching the stylus down on a stroke, and drawing a line to the new position. This cannot be considered as a satisfactory method as only one diagram line can be moved at a time, and it may be necessary to move shapes made of a number of lines.

5. The Data Structure

The document being created or edited is stored in a data structure whilst the editor is in use (Figure 9). It was developed to increase the speed of the editor and facilitate the recognition of the position of the editing symbols. Along with its components, each structure stores the upper and lower boundary values in each direction (Figure 10). The x and y co-ordinates that make up any editing symbol can be compared with these values to identify the structures being operated upon. The data structure has been designed along the lines of an Office Document Architecture (ODA) document. An ODA document is made of composite objects which in turn comprise of basic objects. Only one type of basic object may be contained in each composite object. Basic objects can be for instance, alpha-numeric characters, or graphic

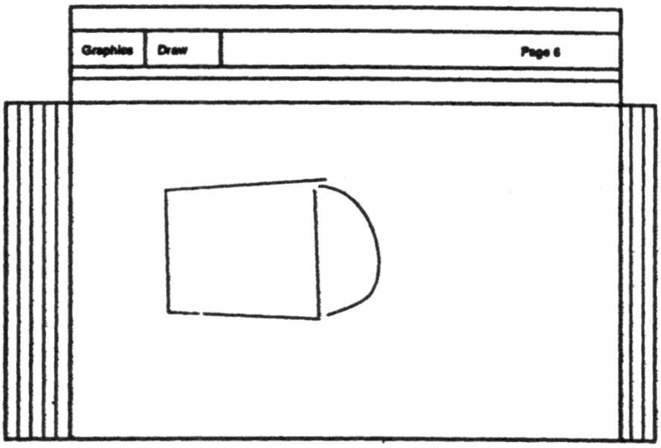

Figure 7. The Diagram Editor.

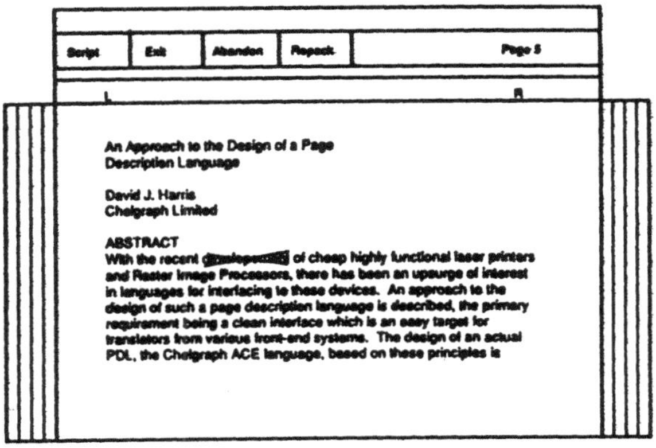

Figure 8. The Text Editor.

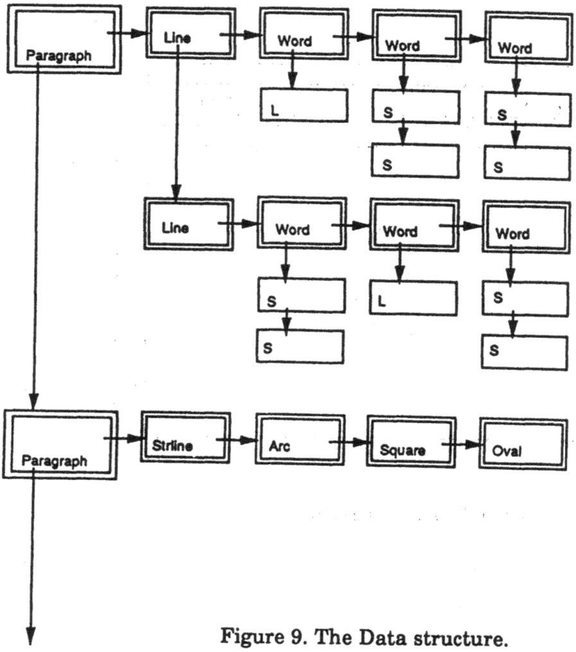

Figure 9. The Data structure.

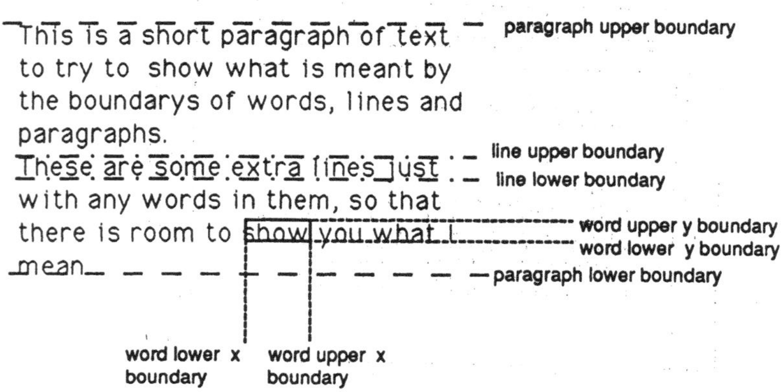

Figure 10. The structure boundaries.

characters. Our structure is based around a paragraph. A paragraph can be made up of words or diagrams. The pages of the data structure have also been designed to comply with the ODA definitions. In ODA a page is considerd to be a rectangular area that corresponds to a unit of presentation medium. Our page is therefore the area of the document that can be displayed on the screen at anyone time. The page structure is super-imposed onto the paragraph structure, and can be altered independently of it. This is an important feature since, whilst editing electronically, be it with the editor described here, or with a more conventional wordprocessor, the contents of a page are liable to change frequently. As words are deleted for instance, the gaps are closed, lines repacked between margins, and consequently there may be room for extra words at the bottom of the page.

6. Discussion

The text editor was used by a number of people before the diagram editor was incorporated. Further testing has not been carried out since then, because of the difficulties experienced by the users at that time. Users took time in learning to use the system, they felt it was because they had to become accustomed to the fact that they wrote on the tablet and saw the results appear on the screen, whereas with pen and paper the results appear at the site of the writing. Once they had become familiar with the system, most managed to edit a piece of text successfully. There were problems the delicate positioning of a split symbol, but again it was felt that if the user could write at the same site as the output appeared, this would not be a problem.

It is hoped that these problems will be overcome once the system is implemented on electronic paper. More testing of user reactions will be carried out at that time.

References

Bush, V., (1945) "As we may think", *The Atlantic Monthly* **176**, July 1945, pp. 101-108.

Buxton, W., Sneiderman, R., Reeves, W., Patel, S. and Baecker, R.,"The evolution of the SSSP score editing tools", *Computer Music Journal*, **3**, No. 4, 1979, pp. 14-25 & 60.

Buxton, W., Fiume, E., Hill, R., Lee, A. and Woo, C., "Continuous hand gesture driven input", *Proc. Graphics Interface.*, 1983, pp.191-195.

Coleman, M.L., "Text editing on a graphic display device using handdrawn proof reader's symbols", *Proc. 2nd Univ. Illinois Conf. on Comp. Graphics.* In "Pertinent Concepts in Computer Graphics" Faiman & Nievergelt, (Eds.), 1969.

Doster, W., and Oed, R., "Word processing with on-line script recognition", *IEEE Micro*, Oct. 1984, pp. 36-43.

Jarvis, J.F., "The line drawing editor, An experiment with computer vision", *Computer Graphics (ACM)*, **10**, No. 2, Summer 1976, pp.288-292.

Konneker, L.K., "A graphical interaction technique which uses gestures", *Proc. IEEE 1st Int. Conf. Office Automation*, Dec 1984, pp.51-55.

Licklider, J., Clark, W., "On-line machine computer communication", *AFIPS Conf. Proc.* 21, pp. 113-128.

Nagura, M., and Suenaga, Y., "A facsimile based graphics editing system by auxilliary mark recognition", *IEEE Trans. Patt. Analysis & Mach. Intell.*, **PAMI-5**, No. 4, July 1983, pp. 433-441.

Suenaga, Y., and Nagura, M., "A facsimile based manuscript layout and editing system by auxiliary mark recognition", *Proc. 5th Int. Conf. Patt. Recog.*, Dec 1980, pp. 856-858.

Sutherland, I.E., "*Sketchpad, A man-machine graphical communication system*" PhD Thesis, Massachusetts Institute of Technology. Publs: Garland 1980.

Welbourn, L.K., "A gesture based text editor", *Proc. HCI 1988*, pp.363-371.

Wells, C.J., Evett, L., Whitby, P., Whitrow, R., "The use of orthographic information for script recognition", *4th IGS Conf*, Trondheim, Norway, July 1989. (Also in this volume)

Wright, P.T., "*Algorithms for the recognition of handwriting in real-time*", Ph.D. Thesis, Oct. 1988, Nottingham Polytechnic, UK.

Wells, C.J., "*The use of higher level knowledge to aid the machine recognition of handwriting*", M.Phil. Thesis (in preparation), Nottingham Polytechnic, UK.

Welbourn, L.K., "*Using gestures for editing documents*", Ph.D. Thesis (in preparation), Nottingham Polytechnic, UK.

Computer Processing of Handwriting
Eds. R. Plamondon & C. G. Leedham

AUTOMATIC RECOGNITION AND TRANSCRIPTION OF PITMAN'S HANDWRITTEN SHORTHAND

GRAHAM LEEDHAM
Department of Electronic Systems Engineering, University of Essex, Wivenhoe Park, Colchester, Essex CO4 3SQ, England

This paper describes the current state of research into the automatic recognition and transcription of Pitman's handwritten shorthand.

The paper starts by describing the Pitman shorthand convention and justifying its use as a means of fast textual input to computer in term of its speed, popularity and close similarity to a machinography (machine compatible script). The variability of the real shorthand script is discussed.

The real-time acquisition and preprocessing of shorthand is considered and the design of a novel instrumented pen for monitoring writing pressure is presented. An approach to the recognition of Pitman's shorthand which splits the script into two classes (shortforms and vocalized outlines) is proposed and investigated. A template matching technique is described which has been successfully implemented for the recognition of shortforms. An algorithm is described for the recognition of vocalized outlines.

Finally a method of transcribing the recognized pattern primitives of vocalized outlines into correctly spelt English is presented.

1. Introduction and Overview

A machine capable of producing a simultaneous verbatim written transcript of speech spoken at the normal speaking rate of 180 words per minute would have many applications in commercial and professional environments. These include rapid text entry into office word processor systems and verbatim reporting of meetings and court proceedings. In addition such a machine could be used as an aid for the deaf by providing a real-time transcript of lectures and meetings and would also facilitate real-time subtitling of television programmes.

1.1 Methods of Simultaneous Transcription

The obvious method of achieving simultaneous transcription of speech is by using a speech recognizer. Unfortunately, the difficulties involved in the recognition of unconstrained, unlimited vocabulary connected speech are daunting and will not be solved for many years (Holmes, 1987). Therefore, alternative methods which require a human intermediary to carry out the computationally difficult tasks of isolating the speakers voice from other background sounds, separating words in the speech, ignoring the redundant speech utterances and simultaneously encoding the essential speech components of each word via some other medium such as keyboard or handwriting.

Figure 1 shows a speed *versus* popularity comparison of various methods of recording textual information. The only keyboards which can be used to accurately record speech at more than 180 words per minute are the phonetic chord keyboards designed for machine shorthand. The use of the Palantype and Stenotype keyboards for verbatim recording applications has been extensively investigated (Delaplain, 1980; Downton, 1982) and Computer Aided Transcription (CAT) systems using these keyboards systems are currently used in court and other reporting.

Handwriting is a more natural form of human communication than that offered by keyboards and could be an attractive alternative method of textual input to a computer if the associated pattern recognition problems could be overcome.

There are a number of handwritten shorthand and speedwriting systems which have been developed for verbatim reporting and dictation purposes. Although the maximum speed at which they can be used is generally less than normal spoken English they are typically 20%-30% faster than the QWERTY keyboard.

The simultaneous recognition and transcription of handwritten shorthand would form a popular and fast input method to computers if the associated technical problems could be overcome. Work at the University of Southampton (Brooks, 1985) initially considered the alternatives of Pitman and Gregg shorthand. These shorthand systems are used extensively: Pitman's shorthand is mostly used in the United Kingdom and Commonwealth countries whilst Gregg's shorthand is more popular in the

USA. The Pitman shorthand notation was selected for consideration because of the large number of existing users and, as will be explained in section 1.3, it has a number of features which suggest it is particularly suitable for machine recognition purposes.

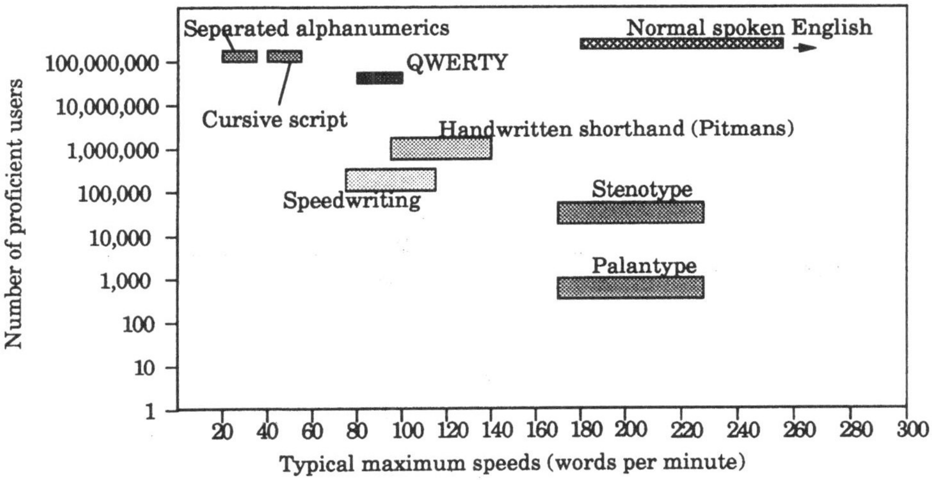

Figure 1. Speed versus popularity comparison of recording textual information (English language only).

1.2 Description of Pitmans Shorthand

Pitman's shorthand is a notation which was developed to allow English, spoken at between 120 and 180 words per minute to be recorded using a pen and paper. It is most widely used in offices in the UK but is also taught in 74 other countries. Based upon examination results obtained from the Pitman Examination Institute it is estimated that worldwide there are over 500,000 practising Pitman shorthand writers who are capable of recording at more than 100 words per minute. It is also estimated that there are several million people who have received some training in its use. There are two forms of the Pitman shorthand system; New Era and Pitman 2000. Both are based on the same basic principles but the New Era notation has several conventions which make it slightly faster but more difficult to learn than the Pitman 2000 notation. In the initial work (Brooks, 1985; Brooks and Newell, 1981 and 1985) the New Era notation was used but in the later work (Leedham et al., 1984; Leedham, 1985; Leedham and Downton, 1984, 1986 and 1987; Qiao and Leedham, 1989) the Pitman 2000 notation has been used. The following basic description of Pitman shorthand is common to both notations.

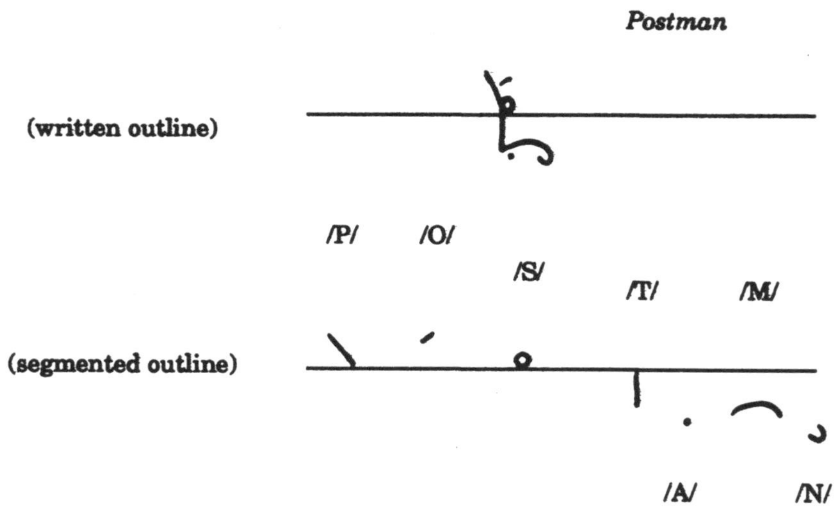

Figure 2. Illustration of the construction of a Pitman shorthand outline.

Pitman's shorthand records speech phonetically rather than orthographically as in longhand. The system divides the English language into 40 phonemes comprising 24 consonants, 12 vowels and 4 dipthongs; a simple written symbol is assigned to each sound, Pitman, 1975. Each word is represented by a separately written outline consisting of a combination of these phonetic symbols. The consonants of the word are recorded first as a connected string of strokes, hooks, loops and circles. Any vowels and diphthongs occurring in the word are written as small marks around this consonant kernel in one of three positions relative to an associated consonant stroke. The consonant kernel is also written in one of three positions relative to the writing line (above, on or through it) in order to indicate the first vowel sound. Stroke thickness is used to distinguish between voiced and unvoiced sounds. Figure 2 shows a typical shorthand outline and its construction from the basic units of strokes, hooks and circles.

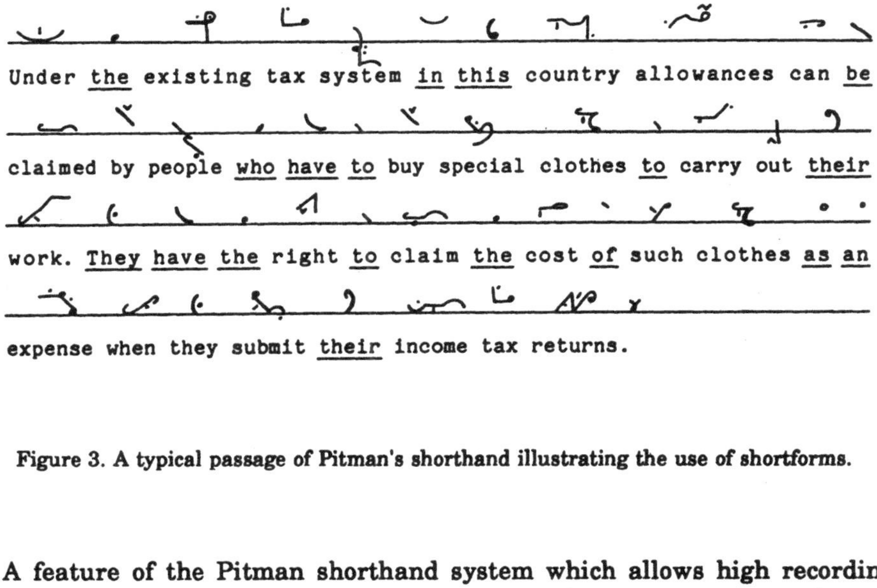

Figure 3. A typical passage of Pitman's shorthand illustrating the use of shortforms.

A feature of the Pitman shorthand system which allows high recording speeds to be achieved is the use of shortforms. These are simple single pen-strokes which, when they occur as separate outlines, represent commonly occurring words or phrases. Shortforms can be easily distinguished from vocalized outlines by the absence of any separate vowel symbols in their formation. Shortforms account for as much as 50% of a typical shorthand passage as is illustrated by the extract of Pitman's shorthand shown in Figure 3.

1.3 Pitman Shorthand as a Machinography

Two factors affect the selection of a suitable written shorthand notation for computer-aided transcription; these are 1) a theoretical consideration of how suitable a specific shorthand notation might be for machine transcription and 2) practical consideration of the popularity and effectiveness of that shorthand system. The second factor was considered in the previous section but in order to have some basis upon which to judge the first of these factors, it is necessary to determine the criteria which governs the degree of 'machine compatibility' of any given script.

One approach is to consider how the script in question differs from the concept of a machinography; that is, a script ideally suited for machine recognition. The concept of a machinography was first introduced by Wellisch (1978) but for the purposes of this discussion a machinography is defined as *"a script specifically designed or tailored for the purpose of man-machine communication"*.

Although most natural scripts are by no means ideally suited to machine recognition, it would nevertheless be possible to devise a script with this objective in mind. Brooks (1985) proposed the following guidelines for a machinography:-

- The script should represent an efficient, error-free, and reasonably natural method of interaction between men and machines.
- For reasons of compatibility, the script should possess the same linguistic structure as the local dominant script.
- The script should consist of a small number of simple pattern primitives, which are both easily written and recognized. Further, the ease of writing and recognition of each symbol should be proportional to the frequency of use of that symbol.
- Ideally, the symbols of the alphabet should correspond directly to the individual pattern primitives constituting the alphabet of the structural recognition grammar.

- The script should possess structure on numerous levels, and there should be good correspondence between the linguistic and physical significance of primitives, so that a small physical change represents only a minor linguistic change.
- The script must also possess a simple well-defined physical and linguistic syntax, which may be described in terms of a straightforward structural recognition grammar.
- Finally, for practical reasons, the physical boundaries between adjacent pattern primitives should be marked in some obvious way.

The Pitman shorthand system exhibits many of these guidelines and is potentially well suited to automatic recognition and transcription because the fundamental components of the system (hooks, loops, strokes etc.) can be used as the pattern primitives of a syntactic recognition process and in many cases represent a single phoneme,

- each outline represents a single word or phrase,
- consonants and vowels are easily separated,
- the script is fully defined in outline stroke formation sequences and
- most boundaries between phonetic features are clearly indicated by closure, angularity or high curvature.

1.4 Variability and Readability of Pitmans Shorthand

With any handwritten data there is bound to be variation in symbol formation from writer to writer and even from the same writer from time to time. There are many reasons for this, including the writing implement, the writing surface, fatigue, age and education (Dooijes, 1979; Kuklinski, 1984). The writing of Pitman shorthand is no exception. In Figure 4 three distinctive styles of Pitman shorthand are shown: "text-book" shorthand and the scripts produced by the student and the professional shorthand writer.

The "text book" shorthand (Fig. 4(a)) conforms to all the Pitman rules. The length of consonant strokes are consistent throughout the passage as are the size of circles, hooks and loops. Light and heavy strokes are easily distinguished and the vowels and dipthongs are correctly placed relative to the consonant strokes. The machine recognition of Pitman shorthand would be

greatly simplified if this quality of shorthand were consistently supplied to the recognition system. Unfortunately, this is not the case.

a) Text book shorthand

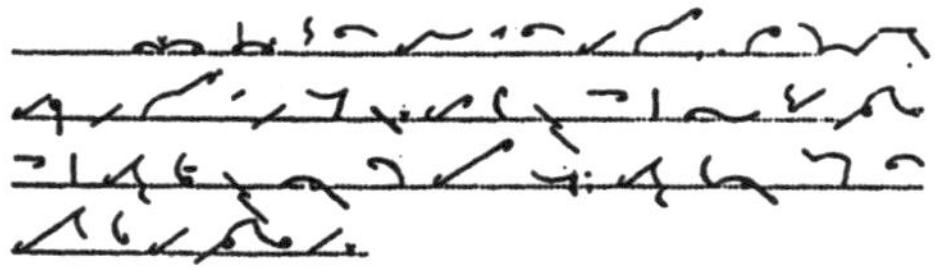

Sometimes it seems that the more we know and the more we learn the less remarkable we consider our learning and our knowledge to be. When other people can do something that we ourselves cannot do we think that those people must be very wonderful indeed; we think that they must be of much more worth than we ourselves are.

b) Student examination script

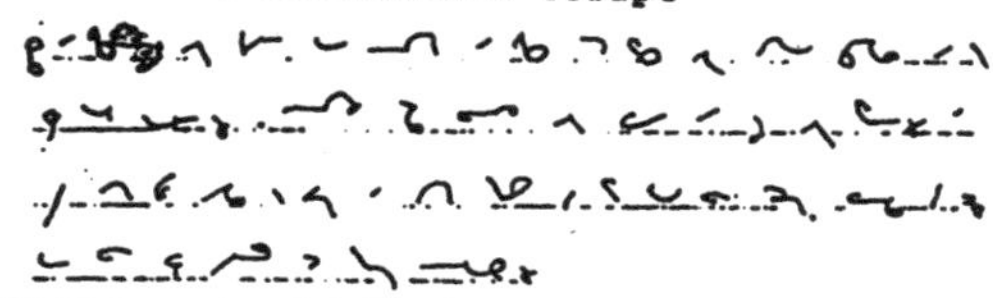

Suits and dresses should be dark in colour and dresses or blouses should have long sleeves and be high in the neck. The collars of these garments should be white and shoes should be black. A judge might well refuse to hear a lady barrister who appeared in court wearing clothes which were not in accordance with the regulations of the Bar Council.

c) Professional recorder script

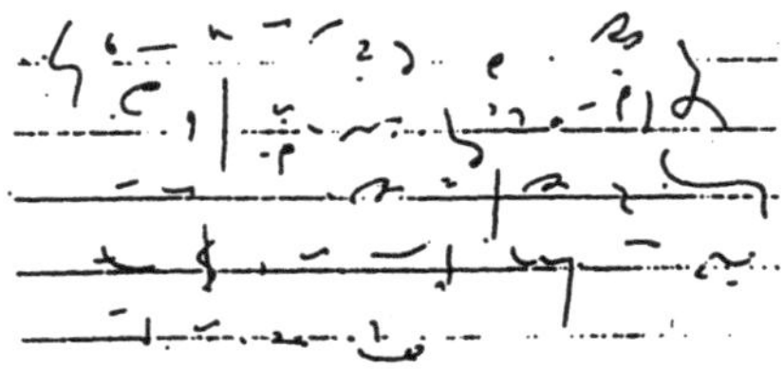

The change that has come about in the last two years is that the Secretary of States predecessor.....limit the power of the trade unions.......want to carry out a vendetta against the unions.............

Figure 4. Three examples of Pitman's handwritten shorthand.

The next style of shorthand (Fig. 4(b)) is that produced by the examination candidate. This shorthand is produced by relative newcomers to the Pitman system, and under examination conditions they will be attempting to reproduce all that they have been taught. The examination shorthand shown is that taken from an examination candidate during a 100 words per minute dictation exercise. From a study of a number of other student scripts taken from the same examination session, it was noticed that some of the scripts were very small (encouraged by Pitmans) which would cause problems in recognition due to the larger relative sizes of hooks (and loops) causing some confusion between hooks and strokes. In the particular example shown, there is no obvious distinction between thick and thin strokes and many intended straight strokes show a marked tendency to be curved. It would appear that often this is due to the writer anticipating a following curved feature such as a circle or hook. There is a natural tendency to produce curvy writing rather than angular features. The distinction between halved and normal length consonant strokes is small in many cases (partly caused by the generally small script) and some stroke inclinations to the writing line are considerably removed from the ideal.

In the case of the professional shorthand reporter (Fig. 4(c)) a very distinctive personal style has developed. Very little restraint is imposed on the style of professional writers as it is unusual for anyone other than the original writer to attempt to transcribe their shorthand notes. Therefore, in many cases, especially when recording at high speed, only the briefest of shorthand outlines are recorded, enough to jog the reporter's memory sufficiently to recall what was said. Generally the transcription of these notes is only possible by making extensive use of context and sometimes memory.

These examples illustrate the recognition problems which an automatic transcription system would encounter when subjected to unconstrained shorthand from existing writers. Most existing shorthand writers produce shorthand scripts which, in terms of their variability and readability, can be placed somewhere between the student script and the professional script. It might reasonably be expected, however, that the use of a transcription system providing continuous feedback to the writer, together with the positive incentive to minimize errors instilled by an automatic transcription system, would help to maintain the quality of writing at or above the standard of the examination candidate.

Experiments and measurements have show that even with the benefit of contextual clues, humans are unable to read back perfectly a shorthand record produced by another writer (Leedham, 1985). Performance accuracy ranges from about 95% correct with 'textbook' shorthand and students' examination scripts down to less than 50% correct with professional verbatim writers' scripts. It is clear from these experiments that the techniques used by verbatim reporters to increase their recording speed above normal dictation speeds of 120 words per minute produce a personalized script which is only decipherable by the original writer making use of context and memory.

At normal dictation speeds a full and complete shorthand transcript of the speech can be produced and hence machine transcription is feasible. In this case, the major limitation on transcription performance is imposed by variability in the way shorthand strokes are written. Clearly, this will depend on the writing style of the individual and the ergonomics of the writing equipment. Experiments indicate a typical error rate of about 15% in the length of consonant strokes, and about 5% in stroke angle (Leedham, 1985; Leedham et al., 1984 and 1986). This level of performance is certainly not adequate for any practical purpose except, perhaps, as an aid for the deaf, since the cumulative effect of consonant strokes length and angle errors in multistroke vocalized outlines would give an average error rate of 20% in single-stroke outlines, 33% in two-stroke outlines, 47% in three-stroke outlines etc., unless higher level syntactic and semantic knowledge can be used to increase stroke redundancy.

2. Acquisition and Recognition

2.1 Data Collection

Most shorthand writers prefer to use a ring bound notepad of roughly A5 size (210mm x 149mm) which is lined at approximately 8mm intervals. The short line length minimizes the lateral hand movement whilst writing and the ring binding allows rapid page turning: both are necessary to maintain a high recording speed. Shorthand writers mostly use either a pencil or ball-point pen: a pencil is preferred since it produces a visible distinction between thick and thin strokes which encourages the writer to distinguish voiced sounds.

In order to acquire Pitman shorthand in real-time a digitising writing tablet is required to monitor the pen position and since commercial writing tablets only monitor pen movement, some means must be provided to monitor line thickness. Line thickness is closely associated with writing pressure and consequently an instrumented pen which detects the pressure exerted by the pen tip on the paper can be used to obtain this information. The writing surface must be as natural and easy to use as a notebook and pen. A pressure sensitive writing tablet surface of about A5 size and sensitive enough to detect variation in handwriting pressure is prone to accidental activation from many extraneous objects and thus requires a handrest. Alternatively a small writing area of one line height and width could be used with a step paper feed each time a new line is required. This solution may be practical but requires further development and trials to assess its potential and user acceptability.

On-line acquisition of shorthand data allows not only the shape and pressure of the outline to be recorded but also dynamic information about the writing sequence and pen velocity. This dynamic information is invaluable in the automatic recognition process since knowledge of the direction in which a particular stroke was written or the rotational direction of a circle, hook or loop avoids many possible confusions in the outline classification.

An early study (Brooks, 1985; Brooks and Newell, 1981 and 1985) successfully investigated the feasibility of both recognizing and transcribing Pitman handwritten shorthand. This study used a Datapad writing tablet to record pen movement and a ballpoint pen fitted with a click-action switch to record writing pressure. During the feasibility study it was concluded that a writing tablet with a static resolution of better than 0.2mm in both axes was required to represent accurately the smaller features of the Pitman notation. The Datapad tablet operates by a pen causing a point contact between two parallel resistive sheets separated by a small air gap. The tablet surface does not require a specialised pen and can be activated by hand contact alone. Therefore a handrest has to be provided to ensure that this does not happen. A modified and improved version of this tablet, the Micropad, was used in later experiments and samples the pen position approximately 220 times per second and a new sample is only sent it is differs from the previous sample. This feature of the tablet enables some indication of the pen velocity at any point on the outline to be recorded since the time duration between transmitted samples will, on average, be greater when the pen is moving with low velocity as occurs at sudden changes in pen direction.

At the present time an ideal tablet digitiser for the collection of normal handwriting does not exist. All currently available tablet digitisers impose some constraints on the writer making it difficult for them to write in their normal manner.

Evaluation of the initial feasibility study showed that the original click action ballpoint pen was neither ergonomically satisfactory not did it provide accurate pressure information of shorthand strokes. A novel instrumented stylus was therefore developed for use with the Micropad tablet. The stylus design was based on a clutch pencil and a small displacement transducer. The clutch of the pencil was sprung relative to the pen body over the 0.5mm working range of the displacement transducer as shown in Figure 5(a). The principle of operation of the displacement transducer is shown in Figure 5(b). The transformer primary winding is driven by an alternating voltage which induces varying voltages in the two secondary windings depending upon the position of the movable core. The secondary windings are wound and connected such that the output voltage varies proportionally to the core position. This pen design has the following advantages:

- The displacement transducer is inexpensive in comparison to other load transducers (strain gauges and piezoresistive cells) and there is no need for overload protection. The majority of people who write cursive script (including shorthand) typically use a writing pressure of up to 200g. However, pressure peaks of several kilograms are frequently encountered at dots and full stops.
- The small displacement of the transducer provides a cushioned feel as if writing on a wad of paper (as in the case of a ring bound notebook). This is preferred by most writers.
- With variation in writing pressure soft graphite (2B) provides some visible feedback between thick and thin strokes.
- The increased friction between graphite and paper helps to reduce 'skate' on the writing tablet surface and consequently improves the quality of the digitised shorthand.

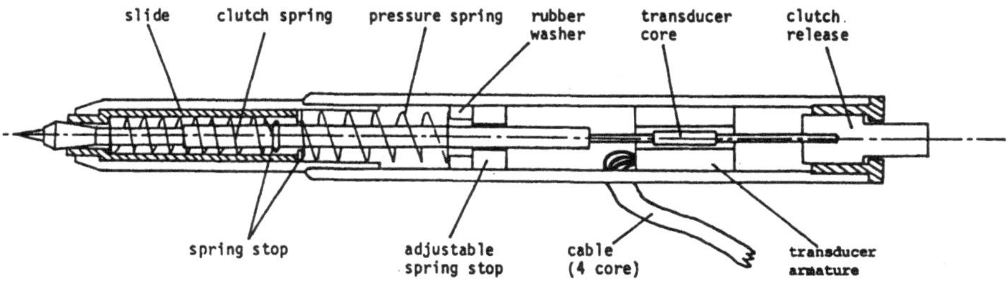

(a) Constructional details of the clutch pencil and transducer.

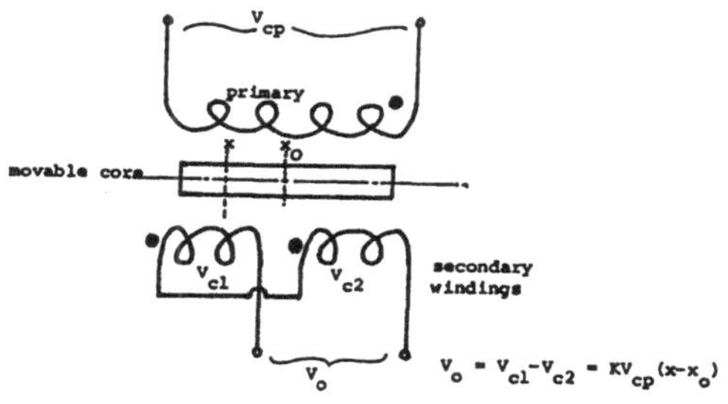

(b) The LVDT principle.

Figure 5. An instrumented pencil for monitoring writing pressure.

2.2 Preprocessing the Raw Data

From a detailed study of shorthand collected using the Micropad writing tablet a number of imperfections were observed in the digitised data. Recognition of shorthand outlines is considerably more successful if these imperfections are removed before attempting the recognition process. Some of these imperfections are specific to the chosen writing tablet but most will be encountered on any writing tablet.

One problem is the occurrence of accidently penlifts during the formation of each ouline. These are found mainly at the start and end of shorthand oulines and at thick to thin stroke transitions in the consonant kernel, and occur when the writing pressure is less than the minimum required to activate the tablet. It is not possible to correct for lost data at the beginning and end of shorthand outlines but accidental penlifts which occur between stroke endpoints can often be identified. Accidental penlifts were detected by measuring the absolute distance betwen the pen-up and pen-down point and the time duration of the penlift at all occurrences of a pen-up/pen-down in each shorthand outline. Practical trials using the Micropad and instrumented pencil showed that if the absolute distance between the pen-up and pen-down point was less than 1mm and the time duration of the penlift was less than 150ms then the penlift was not intended and should be deleted.

At points of low pen velocity the sampling frequency of the writing tablet interacts with the quantisation levels of the tablet surface to cause the samples to oscillate between adjacent coordinates. In addition there is a small amount of noise present in the digitised outline caused partly by the thickness of the pen tip providing an imprecise point contact and partly by imperfections in the tablet surface. The outline must be smoothed to reduce the effect of this noise. Repetitive oscillatory samples, which occur when the pen is moving slowly, were detected by searching for two identical samples which have only one other sample between them. When this occurrence was detected the central sample and the second repetitive sample were deleted. The inter-coordinate time for each of the deleted samples was added to the following sample to ensure the pen velocity information was maintained. The pen pressure information of the deleted sample was averaged with the following sample value of pen pressure. Typically this technique removed approximately 10% of all samples in the shorthand outlines.

2.3 Recognition

The feasibility study at Southampton University (Brooks, 1985; Brooks and Newell, 1981 and 1985) investigated the whole process of automatic recognition and transcription of Pitman's shorthand as the basis of an aid for the deaf. The resulting prototype system was based on a Z80 microprocessor system and designed to operate in real time. Pitman shorthand was written on a Datapad writing tablet using a ballpoint pen fitted with a click-action switch to monitor writing pressure. The tablet had a static resolution of better than 0.2mm in each axis and sampled the pen position approximately 220 times per second. The system transcribed shorthand into pseudo-orthographic text as follows:

1. Each consonant outline was segmented and classified into primitives of hooks, loops, circles and strokes. The segmentation process commenced at the beginning of each outline and parsed the outline on a feature by feature basis to the end of the outline.
2. A set of production rules converted the string of primitives into a phonetic string.
3. A set of spelling rules converted the phonetic string into a pseudo-orthographic representation.

The performance of this prototype system was not sufficient for it to be used in any practical application but it demonstrated the feasibility of the technique.

Evaluation of the prototype system showed that two main factors limited its performance. Firstly, the ergonomics of the data acquisition devices (writing tablet and instrumented pen) required a writing technique which was disliked by many writers and consequently affected the quality of the shorthand outlines. Secondly, the segmentation and classification of primitives in the consonant part of each outline was not sufficiently accurate. In particular, the feature by feature segmentation process produced the most probable feature at each stage but did not necessarily produce the most likely overall interpretation of the outline. As a result of this the prototype system frequently recognized 'impossible' outlines, that is, ones which implied a phonetic structure which does not occur in the English language.

2.3.1 Modified strategy for shorthand recognition

As previously explained between 30% and 50% of the outlines in typical shorthand are shortforms representing the most frequently used words and phrases in the English language (e.g. *the*, *of*, *and*, *to_the* etc.). The number of shortforms in common use is small (less than 100) and each is formed from a single penstroke. Although shortforms themselves convey very little of the meaning in a sentence they do provide useful landmarks which enable the reader to perceive its syntactic structure. This improves the readability of the transcript by reducing the time spent trying to decipher a poorly transcribed word which conveys little meaning. The frequency of the shortforms and their simplified structure compared to vocalized outlines suggests that they should be treated as characters and recognized as individual units rather than be segmented and recognized from their topological features.

Pitman outlines are fully defined in terms of the order and direction in which strokes must be written and since shortforms consist of only one pen-stroke they are always formed in the same way. Real-time collection of shorthand outlines using a writing tablet maintains the sequence and direction of pen strokes and consequently allows shortforms to be efficiently and fully represented by a vector sequence. Therefore only local shape variation will occur between handwritten shortforms. Rather than attempt to classify the shortform by extracting structural features a technique which has been found particularly effective at matching warped patterns was selected Ney, 1982; Burr, 1980; Ikeda et al., 1981). This technique is based upon dynamic programming principles and is guaranteed to find the optimum match between two patterns as well as being computationally efficient (Leedham and Downton, 1984 and 1987).

The remaining vocalized outlines can be recognized using a syntactic approach based upon the same primitives selected in the feasibility study. To improve recognition accuracy however, a more sophisticated means of segmenting and classifying the primitives must be used. Structural knowledge obtained from an analysis of the shorthand outlines for the most frequently used one thousand words in the English language (Kucera and Francis, 1967) is available and can be incorporated into the isolation and classification of pattern primitives. The classification of primitives needs to interact with this knowledge source to obtain the optimum valid path through the outline which is both *consistent with the knowledge source* and *composed of high match primitives*. This approach, which has been successfully applied to speech recognition (Lowerre and Reddy, 1980) creates a recognition tree for each vocalized outline assigning a new level to each pattern primitive. A similarity measure is assigned at each node based on the primitive's shape match and the probability of the primitive occurring in that position of the outline. A path probability is also assigned to each node based on the current and previous similarity measures. All possible routes through the shorthand outline are considered and only discarded if the path probability falls below a predefined value. This process continues to the end of the outline where the path with the maximum probability is the optimum route which satisfies the above two criteria. The significant advantage of this approach is the avoidance of 'impossible' outlines which were so evident in the feasibility study. In conjunction with the shortform recognition this approach should improve the performance to a standard whereby the pseudo-phonetic output is decipherable and usable as an aid for the deaf. Alternatively, this interaction between feature classification and knowledge may be achieved using a neural network.

2.4 Shortform Recognition

The selection of which shortforms should be included in the set of allowable shortforms was based on an analysis of the 1000 most commonly occurring words in the English language. Since the 1000 most used words represent the largest subject independent dictionary (Schwartz, 1963) and the shorthand transcription system is intended for general use, it was decided to use these words to obtain a subject independent shortform dictionary. It would be possible to substitute or add other personal or subject dependent shortforms to the dictionary if required.

The Pitman notation defines 90 of the most frequently used 1000 words as shortforms. These 90 words account for over 37% of normal English. The 1000 most frequently used words account for approximately 70% of normal English. Of these 90 shortforms 63 were selected for inclusion in the shortform dictionary. These 63 shortforms account for approximately 36% of normal English. The 27 shortforms which were disregarded in this analysis were all at the low frequency end of the 1000 word list and were not considered particularly useful as few shorthand writers use them.

The total shortform dictionary comprised this set of 63 words and an additional 34 commonly occurring phrases derived from these words. From a study of the resulting 97 outlines it transpired that 66 different shapes were used. Of these shapes 46 were uniquely associated with one shortform, 14 were associated with two shortforms, 4 were associated with three shortforms and 2 were associated with four shortforms. Resolution of the shapes associated with more than one shortform is made by considering the outline's position relative to the writing line (one of three possible positions) and the thickness of the written line (a binary decision). Seven of these shapes cannot be uniquely associated with one shortform without taking into account the context of the sentence. This level of processing was not attempted in this investigation as the resulting errors, whilst not grammatically correct are not sufficient to make the text unreadable. Typical examples of words which have identical Pitman shortforms are 'A' and 'An', 'To' and 'Too' and 'Is' and 'His'.

2.4.1 Classification

One shortform is a single dot which could not be satisfactorily classified using the template match described below. Instead, a check was made to see if the whole outline could be contained within a 1mm square. If it could the outline was considered to be a dot and classified separately.

Input outlines which would not fit within the 1mm square were classified from a set of 65 trained reference templates. Firstly, the number of possible alternatives to be considered was reduced. This was achieved by classifying the input pattern into one of three classes depending upon its position relative to the writing line: above, on or through it. These were referred to as class A, class B and class C respectively. This reduced the problem to one of selecting 1 shape from 25 for class A, 1 from 42 for class B and 1 from 13 for class C. The shape classification within each group was achieved using the following dynamic programming template match.

Two patterns $A = \{a_1, a_2, a_3, a_4 \ldots a_i \ldots a_I\}$ the input pattern and $B = \{b_1, b_2, b_3, b_4 \ldots b_j \ldots b_J\}$ the reference pattern are represented by a string of I and J vectors with a_i and b_j as individual vectors in the input and reference patterns respectively.

The optimum template match TM between these two patterns can be calculated using the formula:

$$TM = s(I,J) \,/\, (I+J-1)$$

where $s(I,J)$ is given by the recurrence relation

$$s(I,J) = f(i,j) + \min\,[\, s(i-1,j)\,,\, s(i-1,j-1)\,,\, s(i,j-1)\,]\,.$$

In this relationship the overall distance $f(i,j)$ is given by

$$f(i,j) = d(i,j) + p(i,j)$$

and $d(i,j)$ is the Lee metric (Berlkamp, 1968) between two vectors a_i and b_j.

The penalty function $p(i,j)$ is given by

$$p(i,j) = |\, (i-j)\ DIV\ w \,|$$

where DIV is the integer divisor and w defines the penalty strength.

The recurrence relation is calculated with the following initial conditions

$$s(1,1) = f(1,1)$$

$$s(1,j) = f(1,j) + s(1,j\text{-}1)$$

$$s(i,1) = f(i,1) + s(i\text{-}1,1)$$

$s(I,J)$ represents the optimum path through the $f(i,j)$ matrix from $f(1,1)$ to $f(I,J)$ and is subject to local and global constraints. The effect of the local constraint (the minimization term) is to restrict the incremental step path through the matrix and the penalty function $p(i,j)$ is a global constraint which weights the minimum path through the $f(i,j)$ matrix to favour a straight line diagonal through a square matrix. Experimentation with the set of shortform patterns has shown w=4 to be a suitable weighting. This discourages the template match from considering matches where excessive warping is required. The penalty function also incorporates the relative sizes of the two shapes in the calculation. Size is an important factor in Pitman shorthand since a number of features only differ in this respect.

The numeric value of $s(I,J)$ is a measure of the differences between the input and reference pattern. If $s(I,J)=0$ the reference and input patterns are identical in size and shape. The larger the value of $s(I,J)$ (and consequently *TM*) the more dissimilar are the two shapes under scrutiny.

2.4.2 Performance

Four sets of 66 different shortform shapes were collected from one writer and one of the sets of data was used as the reference templates and experiments were performed with smoothed and filtered data. The results of this analysis are shown by the solid line in Figure 6. This template matching technique recognized 82.5% of the outlines correctly when applied to the preprocessed and vectorised data.

Although these results were encouraging the remaining misrecognitions can change the meaning of some passages to such an extent that the shortforms would have been better left unrecognized. The correct recognition of shortforms is particularly important if they are to be used as landmarks in the interpretation of poorly transcribed vocalized outlines. An incorrectly recognized and transcribed shortform will cause the reader to have some preconceived interpretation of the words surrounding the shortform and thus make it difficult to interpret poorly transcribed vocalized outlines preceding or following the shortform.

The outlines which were misrecognized most frequently are shown in Figure 7. The misrecognitions were not surprising because of the very similar shapes involved. Extracting simple structural features such as straightness of downward stroke, enclosed area of the circle and amount of crossing overshoot from the shapes in Fig. 7(a) would be sufficient to resolve most confusions between the five shortform shapes. In addition measuring such features as direction of the intial pen movement for the shapes in Fig. 7(b), angularity in the centre of the stroke (Figs. 7(c) and 7(e)) and straightness of the upward stroke in Fig. 7(d) will improve the recognition performance to that shown by the dotted line in Figure 6. That is, the correct recognition rate would improve to 90.8%.

The remaining similarity between some shortform shapes coupled with the variation in shape of shortforms due to the natural variation in each persons handwriting indicates that for the group of shortforms chosen it is unlikely that 100% correct recognition of all outlines can be made without taking into account the context of the script. This was concluded from the observation that people could not classify accurately a number of the shortforms when they were viewed in isolation.

2.5 Recognition of Vocalized Outlines

The on-line recognition of vocalized outlines is more difficult than the recognition of shortforms because it is necessary to isolate each consonant primitive within the outline before it can be classified. As with normal cursive handwriting, accurate positioning of boundaries between the main features (letters for cursive handwriting and phonetic features for Pitman's shorthand) is often dependent on previously recognizing the overall outline and backtracking to define the feature boundaries.

Figure 8 illustrates the approach under investigation for the segmentation and classification of the consonant part of a vocalized Pitman outline into its phonetic features. This involves three stages. In the first stage, after initial processing has been performed, loops and circles are located by searching for closure or near closure. In the second stage possible segmentation positions are placed at points where a sudden change in pen direction is observed. In the third stage regions of high curvature combined with writing pauses are located and a segmentation position is determined. Each segment is examined to determine whether it is long enough to be a primitive (Qiao and Leedham, 1989).

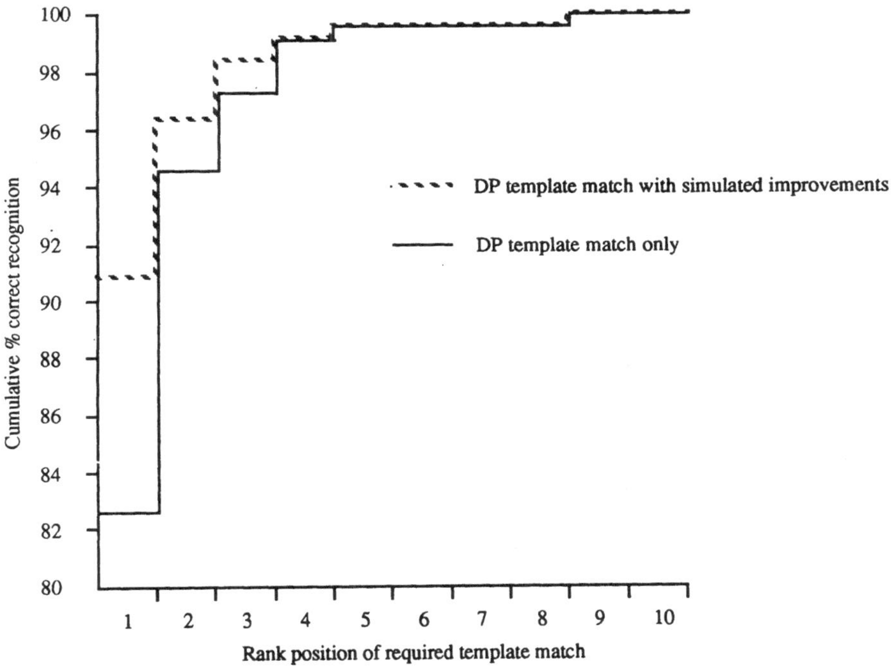

Figure 6. Cumulative recognition performance for shortform template matching.

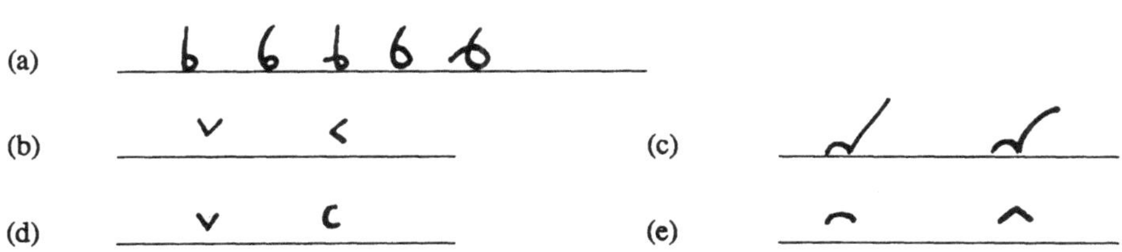

Figure 7. Groups of similar shortform shapes frequently confused by the template match.

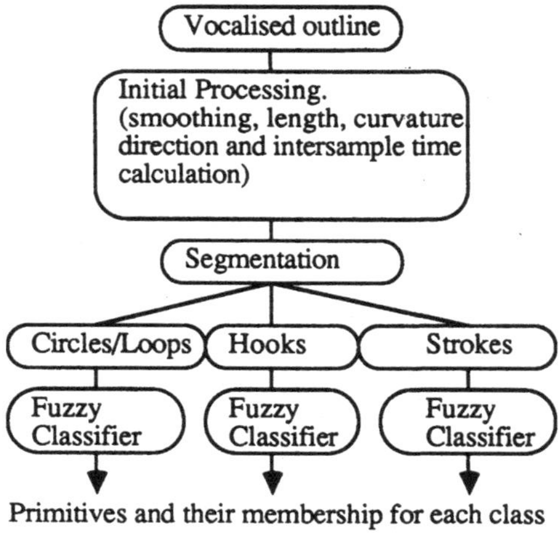

Figure 8. Initial segmentation and classification of features in the consonant kernel of vocalized outlines.

Once the segmentation has been completed the features are classified using a fuzzy classifier which classifies each feature into categories in terms of the degree of membership. Finally, in this system, decision making is achieved using the maximum likelihood principle.

This work is still in the process of being investigated and implementation and evaluation is at the preliminary stages and so results are not presented here. In future work an overall interpretation of the outline will be searched for which comprises a legal combination of phonetic features which are classified with high certainty. This interpretation will be accomplished with the aid of a knowledge source of all legal pair combinations of phonetic features and a dictionary of the most commonly used 1000 shorthand outlines.

2.6 Discussion

It has been accepted that because the Pitman notation has very little redundancy and that the variation in shorthand from writer to writer is such that many peoples shorthand is illegible by anyone other than the original writer (and even the original writer after a period of time). It is therefore necessary to restrict the shorthand to a 'readable' quality as, for example, is produced by the newly trained shorthand writer.

Because of the low redundancy of the Pitman notation the misclassification of a single feature in the consonant outline is likely to lead to a phonetic interpretation and written transcription of the outline which bears very little relationship to the intended word. Therefore a very high segmentation and recognition performance is required. In the current scheme an error in the segmentation of an outline cannot be recovered from. Loops and circles are relatively easily and accurately located. The smooth boundaries between strokes and the boundary between a hook and a stroke are the most difficult to locate accurately. In the future a more flexible segmentation algorithm must be developed which allows initial segmentation errors to be modified when poor or unlikely primitive classification occurs.

Further work will focus on using minimum path searching and neural networks for the overall interaction with a knowledge source of all legal pair combination of phonetic features and a dictionary of the most commonly used 1000 shorthand outlines. This algorithm will search for a legal combination of phonetic features which are classified with high certainty.

3. Transcription

3.1 The Production of Pitman Phonemes

Once the consonant kernel of a vocalized shorthand outline has been segmented into its basic features of hooks, loops, circles and strokes it must be converted into a phonetic representation. Brooks, 1985 investigated this aspect of the problem extensively and his work in this area is summarised below. To illustrate how these principles apply to the transcription of real shorthand outlines, consider the three very simple outlines shown in Figure 9. Each outline consists of two pattern primitives, but their meaning varies significantly.

The outline in Figure 9(a) consists of an initial small hook primitive attached to a horizontal straight stroke primitive of unit length. These primitives represent the two phonemes /L/ and /K/ respectively, with the linguistic order reversed to form the consonant cluster /K/,/L/ as in the word "claim".

The outline in Figure 9(b) consists of a similar initial small hook primitive attached to an upward straight stroke of unit length. These primitives represent the single phoneme /W/ as in the word "win".

The third outline, as in Figure 9(c), consists of a final small hook attached to a horizontal straight stroke of unit length. This outline represents one of the following pairs of phonemes /K/,/F/ or /K/,/V/; the Pitman shorthand notation does not specify the meaning of the final hook primitive exactly in this instance.

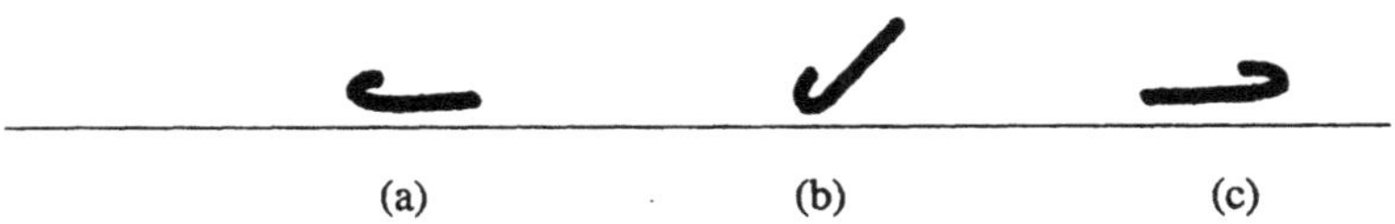

Figure 9. Three simple outlines with similar primitives but different meaning.

Thus, it will be appreciated that the relationship between the classified sequence of primitives and the implied sequence of phonemes is a complex one, governed by the rules of the shorthand system. When formalized this system amounts to a language with a defined grammar.

The production rules serve not only to recognize the implied phonemes, but also as a 'knowledge-base' to determine the position of the associated vowel and auxillary phonemes assigned to each stroke by virtue of its other attributes.

3.2 The Pitman Shorthand Description Language

Brooks, 1985 derived production rules for the shorthand system by reading the Pitman shorthand manuals and systematically encoding the text into a set of formalized rules. He called this terminology the Pitman Shorthand Description Language (PSDL). The production rules of this language form the basis of a simple non-recursive grammar in a structural pattern recognition strategy; the grammar may be non-recursive because of the general similarity between the selected pattern primitives and the shorthand characters used to denote individual phonemes.

For the purpose of application of the production rules, the primitives of this grammar were sub-classified into two groups, termed 'primary' and 'secondary' primitives. Although both groups affect the linguistic meaning of the shorthand outline, only those primitives within the primary group have any inherent linguistic value. In contrast, primitives within the secondary group possess linguistic value only by virtue of their relationship with primary primitives, and furthermore, the meaning attributed changes when they were written in a different context. In this scheme, strokes are the only primary primitives; loops, hooks and vowels are all secondary primitives. For this reason, the production rules are both expressed and applied in terms of strokes and their adjacent primitives.

"Western"

Classified Primitives in Sequential order of writing

1	Hook (small counterclockwise)
2	Stroke (straight, standard length, sector 1)
3	Loop (small, counterclockwise)
4	Stroke (straight, standard length, sector 6)
5	Stroke (straight, standard length, sector 1)
6	Hook (small, clockwise)

Figure 10. Example of pattern primitives in the consonant kernel of an outline.

3.3 Ordering the Pattern Primitives

The component primitives of an outline are neither written nor classified in linguistic order. Thus, before the phonemes implied by the outline can be recognized and subsequently transliterated, the various primitives comprising the consonant kernel of the outline must be sorted.

Consider the shorthand outline for the word "western" shown in Figure 10. After the recognition processes, the consonant component of this outline will be described in terms of a sequence of primitives listed in order of classification. This list must be restored to the correct linguistic order before application of the production rules. This may be done by assigning each classified primitive a value, based upon the address of a particular vector within the sequence of vectors describing the primitive in the linear recognition buffer. This value has been termed the 'effective start address' as, for stroke primitives at least, it is actually equivalent to the starting address of the vector data describing the primitive. For both loop and hook primitives, the effective start address is taken as the address of the feature maxima in the linear recognition buffer. Before application of the production rules then, the string of classified primitives is sorted into order of increasing effective start address which will result in true linguistic order.

3.4 Application of the Production Rules

As an example of the use of the production rules consider the outline shown in Figure 10. The production rules are applied only with respect to the stroke primitives of the consonant kernel, and their adjacent primitives. Therefore, the first stroke primitive in Figure 10 normally represents the phoneme /R/. In this case, there is no auxillary phoneme because the stroke is of unit length. There is no preceding vowel assigned to this stroke, but there is a following vowel.

The first of the production rules relevant to this stroke concerns the effect of the preceding hook primitive. This rule may be read:

> "IF the stroke primitive is preceded by a normal, anti-clockwise, small hook primitive, THEN this combination denotes the phoneme /W/. Any preceding vowel assigned to the stroke should be written before the phoneme /W/."

The normal meaning of this stroke primitive (i.e. /R/) is completely altered by the presence of the proceding hook primitive.

The second rule concerns the effect of the following loop primitive. This rule is slightly more complicated and may be read as:

> "IF the stroke is followed by a normal, anti-clockwise, small circular loop primitive, AND IF this loop is NOT followed by a straight thin stroke primitive written in sector 1 (North East), THEN the loop appends /S/ to the stroke auxillary phoneme /ap/. Any following vowel assigned to the stroke should be written before both of these phonemes."

The contribution of the first stroke (and adjacent primitives) is then complete, and the second stroke may be examined. It is important that individual loop and hook primitives only feature once in the application of the production rules for a given outline. For example, a loop primitive occuring between two stroke primitives can only feature in the production rules for one (if any) of the two strokes.

The second stroke in the outline normally denotes the primary phoneme /T/ and once again, there is no auxillary phoneme as the stroke is of unit length. There are no vowel primitives assigned to this stroke. Since there are effectively no adjacent secondary primitives, rules more generalized rules are used to complete the phonetic meaning of the stroke primitive in the event that no other more specific rules apply. For example, the third rule may be read:

> "IF no more specific rule is matched dealing with primitives immediately before the current stroke, THEN append any preceding vowel assigned to the stroke prior to the primary phoneme."

This process continues for every stroke primitive within the consonant kernel of the shorthand outline, until all of the phonemes implied by the outline have been recognized. The remaining phonetic string is passed to the linguistic processing algorithms to be transliterated into orthographic English.

3.5 The Transliteration of Vocalized Outlines

Transliteration is required to convert the quasi-phonetic source script (the shorthand notation) into an orthographic target script. A set of spelling rules need to be derived governing how best to represent a phoneme graphemically in any given context. Unfortunately, however, the English language has very few established spelling rules.

Work by Hanna, 1966 allows an estimate to be made of the possible range of graphemes. On average, there are approximately 3.1 possible graphemes for every consonant phoneme and 5.1 possible graphemes per vowel phoneme.

By way of example, consider development of the set of rules governing transliteration of the Pitman long /A/ vowel. Table 1 illustrates the wide range of possible graphemes for this phoneme. In order to derive such spelling rules, a large number of examples of words containing a particular phoneme in a given context are required, so that patterns may be observed.

The set of spelling rules for the /A/ phoneme are shown in Table 2. Brooks isolated a similar set of rules for every other phoneme (and distinguishing orthographic feature) in the Pitman alphabet.

Consider the first of the rules in Table 2, governing use of the grapheme "<ay>". This rule would be read:

> "IF the /A/ phoneme is preceded by any phonetic consonant AND followed by a word boundary, THEN the /A/ sound is best represented by the grapheme <ay>."

This rule would thus be satisfied, for example, by the sequence of phonemes occuring in the words "pay", "may" and "say". Similarly, the second rule would be read:

> "IF and /A/ phoneme is followed by the phonetic consonant /N/, AND this in turn is followed by any inflexion or word boundary, THEN the /A/ sound is best represented by the grapheme <ai>."

This rule would thus be satisfied, for example, by the sequence of phonemes occurring in the words "pain", "rain" and "training".

In addition to these groups of rules which govern the transliteration of specific phonemes, the overall transliteration scheme needs to be sensitive to a number of the more 'general' rules of English spelling such as the silent 'e' at the end of a word.

By the time that transliteration is called for, the recognition procedure will have assembled the phonemes corresponding to the written outline in the correct linguistic order. For each phoneme in this source script, the transliteration algorithm attempts to match one of the context rules like those shown in Table 2, with the actual phonetic context that exists in the shorthand outline. If a specific context rule is matched, then the grapheme suggested by that rule is added to the target script. If no specific context rule is matched, then the most common grapheme for that particular phoneme is added to the target script. When appropriate, the transcription algorithm also checks to determine whether any of the more general spelling rules are applicable, and, if so, takes the necessary action. This process continues sequentially for each of the phonemes in the source script until the whole outline has been transliterated.

4 Concluding Remarks

This chapter has given an overview of the progress which has been made into the automatic recognition and transcription of Pitman's handwritten shorthand. This is work which has been carried out since 1978 first at the University of Southampton and latterly at the University of Essex.

This overview has included a justification of why the work is of interest and of commercial relevence and described some of the problems associated with the automatic recognition of the script and why the Pitman shorthand notation is potentially well suited to automatic recognition.

An approach to the automatic recognition and transcription of Pitmans handwritten shorthand has been proposed which isolates shortforms (between 30% and 50% of normal shorthand) from the script and recognizes them using a dynamic programming template match technique. The remaining vocalized outlines would be recognized using a more complex procedure which segments each outline into its constituent pattern primitives before attempting to classify them and interpret the whole outline on the basis of the most likely overall interpretation. The criteria for the most likely overall interpretation is based upon pattern primitive shape matches and interaction with a knowledge source of Pitman's shorthand.

From the results obtained it is concluded that the template matching technique proposed for shortforms is able to distinguish shortform shapes with a reasonably high level of accuracy. Although the 90.8% recognition accuracy obtained is not sufficient for a commercial automatic transcription system it is encouraging and could be improved with further research; there are a number of variations in the formation of the recurrence relation and its constraints which may produce better recognition results for the set of shortform shapes. The main advantage of this template matching technique is its general purpose nature and ability to compare any two shapes which are represented as a string of vectors. Pitman shorthand writers frequently create their own personalised set of shortforms to cope with words and phrases which are specific to their particular employment. The template match described here would allow users to enter their own personalised shortforms, and their transcriptions, into the transcription system and thus 'tune' it to their own personal requirements.

Grapheme	Estimated %	Cumulative %	Examples
a	43.19	43.19	mAbel / mAry
a..e	34.03	77.22	lAte / cAre
ai	10.29	87.51	rAIn / fAIr
ay	5.30	92.81	pAy
e..e	1.62	94.43	fEtE / thErE
ea	1.09	95.52	EAch / tEAr
ai..e	0.85	96.37	rAIse / question-nAIrE
e	0.81	97.18	cafE / sombrEro /
ei	0.76	97.94	vEIn / thEIr
eigh	0.72	98.66	slEIGH
ey	0.60	99.26	thEY / EYrie
et	0.36	99.62	bouquET
aigh	0.16	99.78	strAIGHt
ei..e	0.08	99.86	sEIne
au..e	0.04	99.90	gAUge
ay..e	0.04	99.94	EYE

Table 1. The phoneme-grapheme correspondence for the Pitman /A/ vowel (after Brooks, 1985).

RULE	Phonetic Context	Grapheme
1	(consonant), A, (word boundary)	ay
2	A, N, (inflexion or boundary)	ai
3	A, D>2, (word boundary)	a..e
4	A, D<3, (word boundary)	ai
5	A, (consonant; but not N or D), (no vowel)	a..e
6	else	a

Table 2. The transliteration rules for the Pitman /A/ vowel (after Brooks, 1985).

All the figures for recognition performance quoted in this paper were obtained 'off-line'. That is, there was no real-time feedback to the user of the recognition and transcription performance. If the algorithm were implemented in a real-time recognition and transcription system the effect of the feedback would undoubtedly improve the recognition performance to some extent. Indeed, it is likely that the need to incorporate context in the classification of a number of shortform shapes would reduce because of the 'training' effect of the machine on the user: the user would tend to form the frequently confused shortforms slightly differently in order to minimize the chance of confusion. Further research and field trials are needed to verify this.

Finally, it is concluded that the approach to the recognition and transcription of Pitman's shorthand described here will, with further research work, enable a readable, simultaneous transcription of speech to be produced at speeds of up to 140 words per minute.

There is still a considerable amount of work to be accomplished before a prototype recognition and transcription with acceptable performance can be demonstrated. It is still necessary to:

- Further consider the constraints which must be imposed on the user and the system to enable automatic recognition to be performed. Currently these restrictions are:

  — The writer must write with near text book quality shorthand or, more realistically, must produce the style and quality observed in most examination scripts.

  — Only a limited number of shortforms can be used (maximum of about 100) and the system must be trained by the operator with examples of these shortforms.

  — Shortforms must be written with one penstroke and other outlines must have at least one vowel included and the consonant part of the outline written completely.

  — The writer must make a clear distinction between thick and thin penstrokes in the form of light and heavy writing pressure using an instrumented pen.

  — The shorthand must conform to the rules of Pitman shorthand (for consistency with the knowledge source).

- Other constraints need to be considered. For example, difficulties are encountered in the recognition process when there is a smooth transition from one consonant stroke to the next as occurs when the /P/ stroke precedes the /N/ stroke as in the outline for the word 'company'. If any simple change in the notation could be made to avoid these features the automatic recognition would be greatly simplified.

- The template matching algorithm for the shortforms needs further refinement.

- The recognition algorithm for the more complex outlines needs to be investigated further, implemented and evaluated. The generation and incorporation of a suitable knowledge source into the recognition process is a key issue in the successful outcome of this work. It is these outlines which contain most of the information in a shorthand passage and must be classified with a very high degree of accuracy if the transcription is to be intelligible. Only a small error in the recognition stage can produce a totally incomprehensible transcription, or worse, may change the meaning of the sentence.

- The data acquisition system needs further improvement. A better method of digitising the shorthand and obtaining line thickness information is needed.
- Further work is needed to convert the recognized phonetic structure of the written shorthand into correctly spelt English. Much of the existing work carried out by Brooks, 1985 with Pitman's handwritten shorthand and Palantype machine shorthand can be used in this task but there are still outstanding problems to be overcome.
- The man-machine interface requirements of a shorthand transcription system need to be evaluated. It is not clear how the transcription performance should be conveyed to the operator; how they should correct misrecognitions or how the system should be configured.

Pitman's shorthand has over the last few years been losing popularity in favour of speedwriting systems because it no longer has a prominent place in the modern electronic office. If automatic handwritten shorthand transcription does not enter the information technology race soon and find a place alongside the word processor and other multi-media systems which will soon enter the modern office the number of proficient shorthand users will have fallen to a level where it would prove very difficult to regain a foothold in the market.

References

Berlkamp E.R., *Algebraic Coding Theory*, McGraw Hill International Book Co., Ch. 8 pp. 204-205, 1968.

Brooks C.P., *Computer Transcription of Written Shorthand for the Deaf*, PhD Thesis, Faculty of Engineering, University of Southampton, 1985.

Brooks C.P. and Newell A.F., Simultaneous Transcription of Pitman's New Era Shorthand, *Int. Conf. on Microprocessors in Automation and Communications*, pp.171-179, London, 27-29 Jan. 1981.

Brooks C.P., Newell A.F., Computer Transcription of Handwritten Shorthand as an Aid for the Deaf - A feasibility Study, *International Journal of Man-Machine Studies*, Vol. 23, No. 1, pp. 45-60, 1985.

Burr, D.J. Designing a Handwriting Reader. *Proc. 5th Int. Conf. Pattern Recognition*, pp.715-722, Miami 1980.

Delaplain R., *Computer Aided Transcription - State of the Art*, Report of the CAT Analysis Project at the National Centre for State Courts, 300 Newport Ave., Williamsbury, Virginia, 23185, 1980.

Dooijes E.H., A description of handwriting dynamics, *Proc. 9th IMACS Conference*, pp.727-731, 1979.

Downton A.C., *Simultaneous Transcription of Machine Shorthand for the Deaf*, PhD Thesis, Faculty of Engineering, University of Southampton, 1982.

Hanna P.R., Hanna J.S., Hodges R.E., Rudorf Jnr, *Phoneme-Grapheme Correspondences as cues to spelling improvement*, US Government Printing Office, 1966.

Ikeda, et.al. On-line recognition of hand-written characters utilising positional and stroke vector sequences, *Pattern Recognition* 14, pp.191-206, 1981.

Kandel A., *Fuzzy Techniques in Pattern Recognition*, John Wiley & Sons Inc., 1982.

Kucera H., Francis W.N., *Computational Analysis of Present Day American English*. Brown University Press, 1967.

Kuklinski T.T., Components of handprint style variability, *Proc. 7th Intr. Conf. on Pattern Recognition*, pp.2.924-2.926, 1984.

Leedham C.G., *Computer Acquisition and Recognition of Pitman's Handwritten Shorthand*, PhD Thesis, Faculty of Engineering, University of Southampton, 1985.

Leedham C.G., Downton A.C., On-line recognition of shortforms in Pitman's handwritten shorthand, *Proceedings of the 7th Int. Conference on Pattern Recognition*, pp. 2.1058-2.1060, Montreal, 30th July-4th August 1984.

Leedham C.G., Downton A.C., Automatic Recognition and Transcription of Pitman's Handwritten Shorthand - an approach to shortforms, *Pattern Recognition*, Vol. 20, No. 3, pp. 341-348, 1987.

Leedham C.G., Downton A.C., Brooks C.P., Newell A.F., On-line acquisition of Pitman's handwritten shorthand as a means of rapid data entry, *Proceedings of the 1st Int. Conference on Human-Computer Interaction*, pp. 2.86-2.91, London, September 1984.

Leedham C.G., Downton A.C., On-line Recognition of Pitman's Handwritten Shorthand - an evaluation of potential, *Int. Journal of Man-Machine Studies*, Vol. 24, pp. 375-393, 1986.

Lowerre, B., Reddy, R., The Harpy Speech Understanding System, in *Trends in Speech Recognition*. Wayne A. Lea (ed). pp.340-460, Prentice Hall, 1980.

Masamichi Shimura, An Approach to Pattern Recognition and Associative Memories using Fuzzy Logic, in *'Fuzzy Sets and Their Applications to Cognitive and Decision Processes'*, Edited by L.A. Zadeh, K.S. Fu, K. Tanaka and M. Shimura, Academic Press, 1974.

Ney H., Dynamic Programming as a Technique for Pattern Recognition. *Proc. 6th Int. Conf. Pattern Recognition*, pp.1119-1125, Munich 1982.

Pitman 2000 Shorthand - First course. Pitman Publishing Ltd., 1975.

Qiao Y., Leedham C.G., Segmentation and Classification of Consonant Features in Vocalized Pitman Shorthand Outlines, *Proc. International Conference on Image Processing*, Singapore, September 1989.

Schwartz, E.S., A dictionary for minimum redundancy encoding. *Journal of the Association for Computing Machinery*, Vol.10, No.3, pp.413-419, 1963.

Wellisch H.H., *The Conversion of Scripts - Its Nature, History and Utilisation*, John Wiley and Sons, 1978.

PART 3

CONTEXT AND KNOWLEDGE

Overview

Reports and comparative studies on handwriting recognition have, in the last few years, clearly demonstrated that good performance is ultimately linked not only to the capacity of the system to extract and compare good feature sets but also to the integration of context and knowledge in the different processing stages. The last section of this book contains five chapters dealing with some aspects of this difficult problem.

The first chapter, by Wells, Evett, Whitby and Whitrow, considers methods of using orthographic knowledge to reduce uncertainty between a number of letter string candidates produced by a recognition system while the chapter by Ford and Higgins demonstrates a practical dictionary look-up technique, applicable to many forms of automatic script recognition.

The chapter by Parisse, Rosenthal, Imadache, Andreewsky and Cochu examines the relationship between reading and the automatic recognition of handwriting and develops arguments in favour of a task-oriented approach to analyse these problems. The fourth chapter, by Faure and Wang, deals with the recognition of handwritten mathematical expressions based on a modular segmentation scheme relying on data-driven or knowledge-driven strategies.

Finally, the last chapter of this book, by Zeng, Zhang, Inoue, Sanada and Tezuka focusses on and applies a flexible model of brush-writing to study and analyse the generation of handwritten Chinese characters.

Computer Processing of Handwriting
Eds. R. Plamondon & C. G. Leedham

THE USE OF ORTHOGRAPHIC INFORMATION FOR SCRIPT RECOGNITION

C. J. Wells, L. J. Evett, P. E. Whitby and R. J. Whitrow
Nottingham Polytechnic, Nottingham, England

A pattern recognition stage of a machine script recognition system produces a number of letter string candidates for any word processed by it. The present paper considers methods of using orthographic information to reduce this uncertainty. Letter string candidates may be rejected if they contain letter sequences which are not allowable in English, or are not real English words. The best way of deciding which candidates to reject is by checking them against a list of acceptable English words. For a practical system, a reasonably large vocabulary is necessary. Such a large lexicon must be searchable in real time. The present paper describes methods of representing and searching large lists of words so that the performance of the recognition system is improved significantly. It examines methods of optimizing search times and memory requirements, and discusses the use of higher level knowledge.

1. Introduction

Two natural modes of communicating with computers are via spoken and handwritten input. For recognition of both script and speech the information physically present in the signal is not sufficient for unambiguous identification of words. Human readers/listeners use their knowledge of language and the world to guide their processing. The present paper describes methods for using knowledge about orthography to aid the machine recognition of handwritten script.

A basic pattern recognition system codes handwritten input and identifies the most likely characters that it represents. This system may produce a number of alternative characters for each character processed by it. Higher-level knowledge can improve recognition performance by helping to choose between the alternative characters. There are a number of types of knowledge which can be applied to this problem. These are of word structure, of syntax and of semantics. More general context may also be applied. Figure 1 illustrates the main types of knowledge which can be used to aid recognition. Knowledge from these sources may also interact in the recognition process.

The alternative characters from the pattern recognition stage can be combined to produce a number (sometimes very large) of alternative letter strings. Many of these are not acceptable since they do not constitute acceptable English words. The first stage of higher level processing uses knowledge about words to reduce the number of candidate strings.

2. Orthographic Constraints

A number of methods have been used to exploit the constraints of letter patterns, or orthography, to improve recognition. Goshtaby and Ehrich (1988) describe several methods of implementing orthographic constraints. One method treats written language as a Markov process (e.g., Farag, 1979; Raviv, 1967). This method basically takes the output of the recognizer and finds the letter string with the maximum probability, using the transitional probabilities of the letters and the probabilities of the letters being correct from the recognizer. Another method uses n-grams to filter out unacceptable letter string candidates from the recognizer (e.g., Higgins and Whitrow, 1984). N-grams are letter strings of size n letters. These are collected from text and used either to judge whether or not letter string candidates are legal in English and/or their probabilities of occurrence.

These methods have similar problems. One is that, since these methods consider parts of words, the information they contain is averaged over positions and so may not reflect accurately letter occurrences for specific positions. Their accuracy will also depend upon the size and representativeness of the subset of language from which they have been collected. Methods using probabilities have a built in error rate. For example, Sayre (1973) describes two cases where the use of probabilities leads to mistakes. If the pattern recognizer produces "a" and "o" with equal probability, the input word "far" would never be correctly recognized, since the bigram "fo" is approximately three times as common in English as the bigram "fa". "po" and "pa" are of equal probability so that these methods would not be able to choose between "pod" and "pad" for example. A final problem is that these methods do not guarantee that the output will consist of real words, only of legal letter sequences.

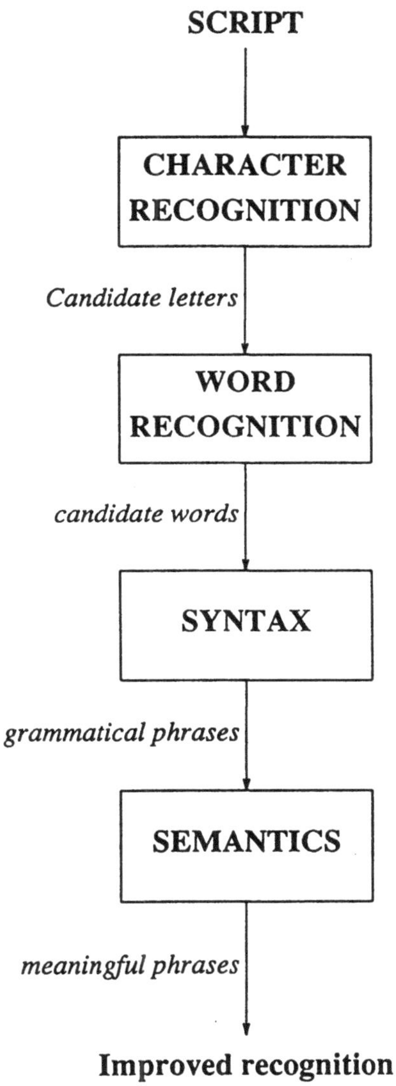

Figure 1. Stages of word recognition.

Given these problems, it is clear that checking candidates against a list of acceptable words is preferable. Words represent both occurrence and position information and will guarantee the output is of acceptable words. The selection of words on the list will still require care, but checking candidates against a list of words will reduce that list of candidates more than checking them against a list of grams from either a similar length list or a longer length list (see Table 1).

List	No. of items:	% * reduction
4-grams from short word list (11,795)	13166	94.5
4-grams from long word list (71279)	35412	36.2
4 letter words from short word list (21,211)	1323	98.5

*n.b. 100 candidate strings, 10 allowable = 90% reduction.
100 candidate strings, 1 allowable = 99% reduction.

Table 1. Percentage reduction in number of candidates by checking them against 4-grams and words.

This comparison was made using binary tree structures which are not the most computationally efficient structures for representing either grams or words. The main problem with using a word look-up method is that in order for the system to be able to process a realistic vocabulary the number of words used needs to be large (of the order of 50,000 words, depending upon the application). This can lead to large memory requirements and slow search times, although reductions in memory costs means this is becoming less of a problem. The next section considers methods of representing and searching a large vocabulary, and how this stage of processing may interact with higher level processes.

3. Methods for Word Look-Up

3.1 Vocabulary Size

The size of the lexicon for a practical recognition system is a matter for debate. It has been estimated that 8,000 morphemes, representing a much larger number of words, will cover most text (Crystal, 1989). (Morphemes are the root forms of words from which other forms are derived (e.g., "dogs", "dogged" and "doggy" are derived from "dog"), so that a number of morphemes represents a much larger number of words). A general purpose recognition system might be required to recognize more words than this, since handwritten input from many different people about many different topics might be more varied in content than printed text. Domain dependent systems might need a smaller, more specialised vocabulary; for example, in a doctor's surgery, or an estate agent's office, a smaller but application-specific vocabulary might be used. The full Oxford English Dictionary contains about 300,000 headwords. The content of the lexicon needs to reflect the contents of any likely to-be-processed text. The dictionary should not be too short, so that some words are rejected because they do not occur in the dictionary. It should not be too long, so that it accepts words that are not commonly in use and unlikely to be present in any text sample.

The authors have found that about 15,000 words is adequate for most purposes, although this fails occasionally. For the purposes of testing the present system, a lexicon of about 60,000 words, which is inclusive for most text and enables a good test of time and memory properties has also been used. The shorter word list was taken from the MRC word list (Wilson, 1987), excluding words of rare usage, derivations, and of word frequency (Kucera and Francis, 1967) less than one per million. The longer word list represents the headwords of a machine-readable version of the Oxford

Advanced Learners' Dictionary (Hornby, 1988), which gives good coverage of English text (Sampson, 1989).

3.2 Data Structures for Word-List Representation

For real-time search, a data structure is required that can represent a large vocabulary in memory. Such a structure would not require continuous alteration, so long as it is possible to change without too much difficulty. The main properties required are memory efficiency and speed of search. From these criteria it is clear that either tree structures or hash tables are suitable. A hash table would not be a good choice, since there would be difficulty in avoiding collisions if the vocabulary were changed, and it would not enable analysis of or access to subword letter structure. For these reasons, a tree structure was preferred.

Multi-way trees may be searched faster than binary trees and so multi-way trees were investigated. For this particular application a 26-way tree, with a route for each letter at every node, is an effective structure to take full advantage of the alphabet. Unsuccessful searches are faster in a multi-way tree compared to a binary tree (Knuth, 1973), which is ideal for this purpose as the majority of strings being searched for will not be in the word list. (For example, an input word "pack" produced 1204 possible candidate strings from combining candidate characters from a pattern recognition stage. Only three were allowable, namely "pack", "pact" and "pant"). Searching is simple, efficient and fast in the 26-way tree, as is its construction. However, the memory overheads are too large for practical purposes; a word list of 14,769 items requires nearly 5 Megabytes. An example of how the 26-way tree might represent a limited lexicon is given in Figure 2.

Each node contains one letter and 26 pointers. To make the searching algorithm more efficient, all 26 routes from every node in the tree must be allowed for, and on average most of these are wasted. Using two experimental word lists of 14,769 items and 79,065 items, the mean number of routes used is 1/26, and just less than 1/26, respectively. Each node in the tree also has a flag which is set if that letter is an end-of-word.

Using this tree, a search is successful if each of the candidate letters in the string can be followed to the next one, and if the last letter in the string is flagged for end-of-word in the tree. If at any stage the required pointer to the next letter does not exist, the search is unsuccessful, and that particular candidate string is therefore not allowable. No further searches are made for strings beginning with these letter combinations.

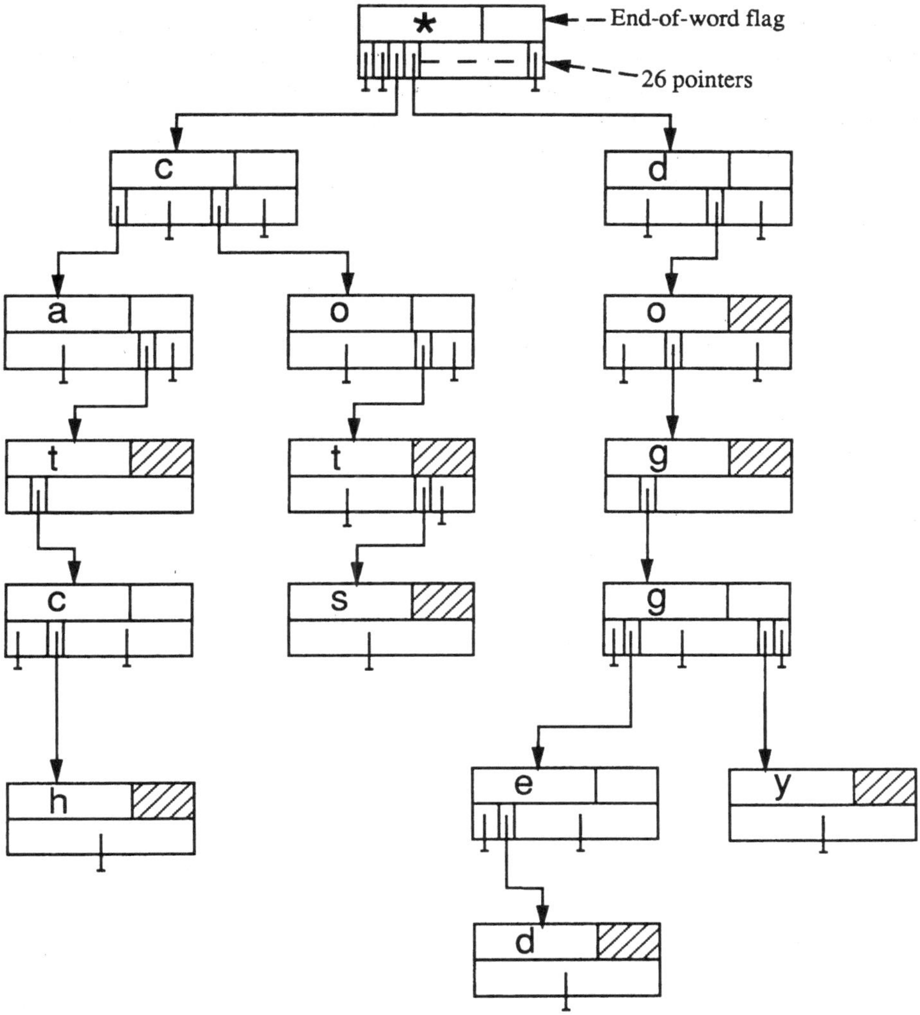

Figure 2. A representation of a 26-way tree for the lexicon: cat, catch, cot, do, dog, dogged, doggy.

3.3 Memory Reduction

An alternative method was devised which includes the advantages of the 26-way tree method, but reduces memory considerably. For example, the 14,769 word list is reduced to just over 0.5 Mb. Instead of having 26 pointers to indicate the next letters in words, the new method uses a 32-bit integer. Twenty-six bits represent the possible routes from any node, being set (ie: = 1) if a letter is allowable, and not (ie: = 0) if a letter is not allowable. The end-of-word flag is the most significant bit, leaving 5 bits unused in which to encode various attributes of the words; for example, whether it is a proper noun, an acronym, or the start or part of a commonly used compound or phrase. Figure 3 illustrates the construction of this reduced memory tree with the same limited vocabulary as for Figure 2. Figure 4 illustrates a node of the tree in more detail.

Each node also has one pointer which, if used, points to a variable length array of pointers to other nodes (ie: 32 bits + 1 pointer). When searching for a particular string, it is immediately apparent at any node whether the required route from that node exists or not, by checking the relevant bit of the 26 flags. If and only if it does, that route is followed. For example, if the search is for the string "dog", 'd' is the second bit of the 26 to be set, so follow the second pointer of the 2-element array to reach the 'd' node on the level below, and so on until the entire string is found, at the 'g', where the word flag is set.

3.4 Comparison of Alternative Tree Structures

Sussenguth (1963) describes a mixed-method data structure involving an existing tree. There is an optimum point, below which it is more economical to represent the remaining parts of the words in a linked list than in the standard tree nodes. This structure was investigated and implemented for both the 26-way tree and the reduced memory tree. This method gave good memory reduction for the 26-way tree, but greatly increased complexity for construction and search for both types of tree.

The following results were obtained for the cursively written test sentence *"mary had a little lamb its fleece was white as snow and everywhere that mary went the lamb was sure to go"* to compare the 2 alternative tree struc-

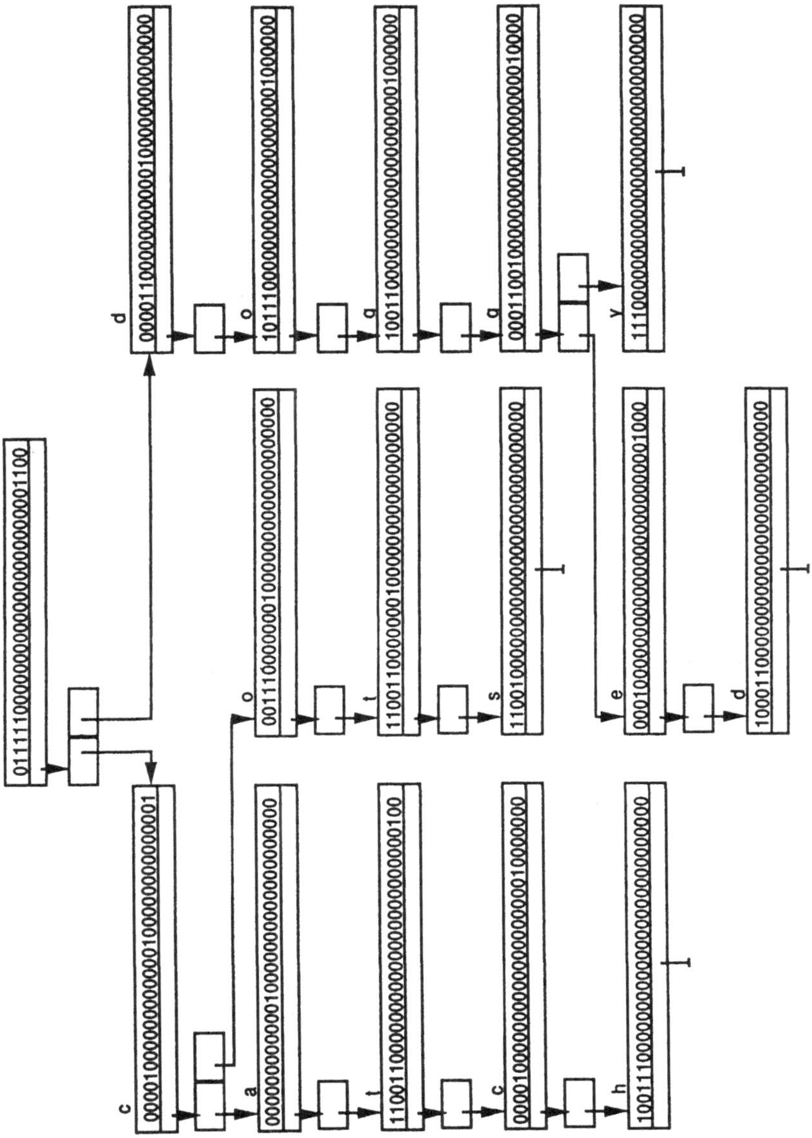

Figure 3. A reduced memory tree for the lexicon: cat, catch, cot, cots, dog, dogged, doggy.

tures. Table 2 gives the results of this comparison. The trees were implemented in C and tested on a SUN 3/160 file server with 12 Megabytes of main memory.

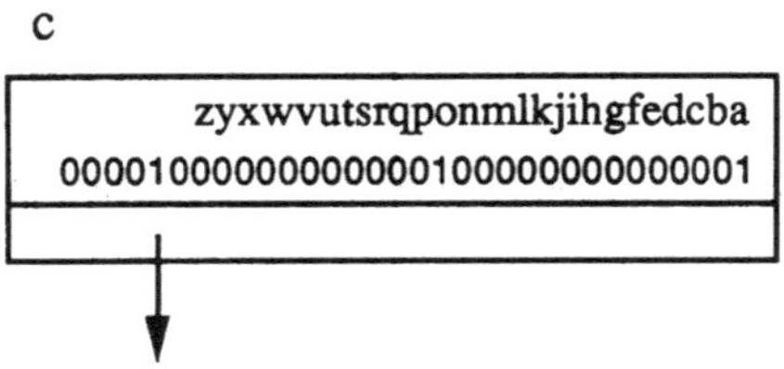

Figure 4. Detail of a node of the reduced memory tree.

word list length:	tree type:	memory used (Mb)	search time* (mins:secs)
14769 words	1. 26-way	4.9	0:06
	2. reduced mem.	0.5	0:12
60144 words	3. 26-way	15.1	0:33
	4. reduced mem.	1.6	0:16

*n.b. These times should be considered only as relative measures.

Table 2. Memory requirements and processing times for the different types of trees.

Although the 26-way tree is extremely wasteful of memory, with a short word list the search times are faster. The reduced memory tree has extremely reasonable memory requirements, although longer search times for the short word list compared with the 26-way tree. The most interesting result is that it gives quicker search times for a long word list. It may be that the longer search times for the 26-way tree with the long list are due to disc accesses and transfers during swapping. This is not the whole answer, since the relative increase in search time is greater for the 26-way tree than for the reduced memory tree (27 *vs* 3 secs) when the linked-list method is used for both types of tree with the long word list; neither structure approaches the size of memory.

Experiments have shown that the shorter (14,769 item) word list contains insufficient items for practical purposes. That is, the correct words (input script) were not found in the list, so search times for longer lists should be considered more important. There are disadvantages in using too long a list in that more of the candidate strings will be allowable, and depending on the Machine-Readable Dictionary (MRD) that the list is taken from, can contain rare words that most native speakers would not recognize as "real" words. It is considered better to have a longer set of allowable candidate strings which includes the correct word, rather than a short set without the correct item.

Given all these factors, this reduced memory tree was chosen as the most appropriate for the situation, being the optimum for memory size and search times for the longer word list. Further improvements could be made by using a different method of end compression. Common endings of words (for example -ed, -ing) which are at present duplicated in the tree for words like walked, walking, shouted, shouting, can be grouped together. This would not affect the speed of look-up, but would again decrease the memory requirements.

Using tree structures has consequences for other aspects of the recognition process. For example if the recognizer cannot produce any character at one letter position within a word, we can use the tree of words to suggest what this missing character might be (eg: in the case "c a * e" the * could be an 'f', 'k', 'r' or 's' etc.). The path through the tree is followed as normal, ie: through the 'c' and 'a' nodes, then followed down all routes from the 'a' node to see if a path exists which ends in 'e', where the end-of-word flag is set. This method can be used even if the first letter position is unknown, or if there is more than one letter unknown within a word, whatever the word length.

3.5 Ordering the Words

Once the allowable words have been selected from the candidate word strings, some way of ordering them is necessary so that the most likely can be displayed as output from the recognition and look-up system, with the other candidates available to be displayed if necessary.

The confidences that the recognizer provides for each candidate letter can be used for this purpose. These may be combined to give a likelihood of a string being correct. However, these confidences only apply at a particular letter position for the set of possible alternative characters for that position, and are therefore not strictly comparable across positions. They are also not dependent on letters preceding or following. A more reliable measure is the rank order of candidate letters within a letter position, based on the confidences. These can be combined to give an overall measure of likelihood for the allowable words. There are various ways of combining the ranks across letter positions. It was found that choosing the allowable string with the lowest score for the mean of the ranks across letter positions gave the best results.

4. Recognition Results

Two example passages have been written by a small number of subjects, and the script analysed by the current recognition system. The recognition system uses Freeman encoding and matching to the database for the pattern recognition stage (Wright, 1989). This produces a number of candidate characters for each character position, with alternative segmentations for some characters (e.g., a "d" may be recognized as a "d" or as a "c" and an "l"). The character candidates are combined to produce candidate character strings, which are then looked up in the tree. The first passage (of 57 words) was written cursively by 6 people, and the second (of 69 words) by 4 people. The database in the pattern recognizer against which their writing was matched was not obtained from these writers. If the first four positions of the ordered list of allowable candidate strings are considered, the correct words are present between 82.4% and 94% of the time, depending on the particular writing being analysed. If the first 10 positions are considered, recognition can reach 100%. These are encouraging results for a writer-independent system, given that other stages of analysis which should further improve the recognition rate, are at present being developed.

It is difficult to compare recognition performance with different recognition systems. This is because systems vary in terms of the set of acceptable symbols, restrictions on input, the size of the acceptable vocabulary, the number of writers permitted, the amount of training received and whether the system was trained by the writers from whom the results were obtained, and the recognition score reported (which may be of per cent correct strokes, characters, or words). A system which is similar in some respects to the present one is that of Srihari and Bozinovic (1987). They used multiple features for pattern recognition with matching to a database, and partial word look-up in a trie structure. They report word recognition scores of 64% correct among the top 4 using a large vocabulary (7,800 words) for look-up, and with the test material having been written by the writer from whom the database was obtained. The same system performed at 65% correct word recognition among the top 4 with a small vocabulary (710 words) and a test writer who had only contributed a small proportion of writing to the database, which was mainly constructed from the writing of 2 other writers. Other authors report much higher word recognition rates (e.g., Ehrich and Koehler, 1975), but these are from writer-dependent limited vocabulary systems.

Detailed analysis of another passage written by only one person is given below. An example of the input is given in Figure 5.

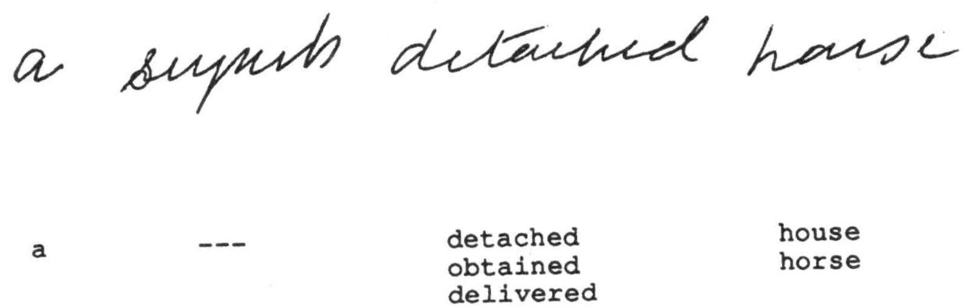

Figure 5. Example of handwritten input and the output after word look-up.

The actual passage is given first, followed by the recognized version. Note that cursive writing is the most difficult case, since segmentation errors mean that the number of possible candidates can reach explosive proportions. The numbers in brackets are the rank of the correct word in its list and the length of the list of allowable words; the word given is the top ranked word. For example, "house(1/2)" means that "house" was ranked first out of 2 possible words found in the dictionary search; "flu(3/10)" means that "flu" was the top ranked word out of 10, with the correct word "the" ranked third. Occasionally the correct word was not recognized at all and this is noted. In some of these cases no acceptable word was found and this is indicated by a dash. The reason for this in all cases was that the recognizer did not give the correct letter as one of the candidates in at least one of the letter positions in the input word.

The actual passage was:

> "a superb detached house situated in a most sought after residential area on the carters estate within easy walking distance of western grove with its multiple shopping facilities. the cliff top with its access to sandy bathing beaches is also within easy walking distance. the area is well served with excellent local sporting facilities including numerous local golf courses, tennis courts and local sports centres which are all within easy reach. the area is also well served with excellent local schools for children of all ages. the property is in need of modernisation and redecoration but offers good-sized accommodation with three reception rooms, four good bedrooms and large secluded well stocked rear garden."

The recognition results were:

> "a(1/1) -(0) detached(1/3) house(1/2) situated(1/3) in(1/5) a(1/1) most(1/10) sought(1/1) after(1/10) residential(1/1) area(1/4) on(1/4) flu(3/10) carters(1/7) estate(1/2) within(1/5) easy(1/8) walking(1/4) distance(1/1) of(1/4) western(1/2) grove(1/1) with(1/8) its(1/2) multiple(1/1)-(0) families(2/9). the(1/10) cliff(1/4) top(1/4) with(1/10) its(1/3) access(1/10) to(1/1) sandy(1/4) framing(2/4) beaches(1/10) is(1/1) dine(2/4) whim(2/4) easy(1/5) walking(1/3) distance(1/3). the(1/8) area(1/6) is(1/1) coal(9;correct word not recognized) sewed(4/5) mill(1;correct word not recognized) excellent(1/1) local(1/1) spoiling(1;correct word not recognized) facilities(1/8) including(1/2) numerous(1/2) local(1/10) golf(1/1) courses(1/1) ten-

nis(1/10) comb(1;correct word not recognized) duel(4/6) local(1/1) sports(1/2) cams(2/10) which(1/7) arc(3/8) cry(4/5) whim(2/7) easy(1/4) reach(1/10). he(2/10) and(2/10) us(2/2) also(1/1) wool(2/10) saved(3/7) with(1/10) excellent(1/2) local(1/2) schools(1/1) for(1/4) children(1/3) of(1/8) all(1/6) ages(1/7). the(1/7) property(1/2) is(1/1) in(1/5) weed(2/4) off(2/6) modernisation(1/1) anal(2/7) redecoration(1/1) but(1/5) offers(1/3) good(3/4) sized(1/1) accommodation(1/1) with(1/5) three(1/3) reception(1/1) rooms(1/2) four(1/10) grid(2/4) bedrooms(1/1) and(1/4) large(1/1) secluded(1/1) vial(4/6) stocked(1/2) wear(9;correct word not recognized) garden(1/1)"

5. Interface to Higher Level Knowledge

Further stages of the script recognition system use various types of linguistic information, in order to reduce the word-level ambiguity following the lexical check (e.g., Keenan and Evett, 1989). For example, information about how words combine to form compounds would choose "sports centres" over "sports cams" in the above data, and syntax would give "which are" in preference to "which arc". Analysis of the syntax and semantics of the input sentences requires information about the particular words involved, some of which can be obtained from a standard dictionary in machine readable form.

The tree structure described above forms the interface to these other levels of analysis. The tree contains an index number for each root word (i.e. at each node where the end-of-word flag is set) which gives the address of the word's full dictionary entry in a machine-readable dictionary stored on disk. This MRD has been modified and coded specifically for this system. Morphological indexing allows derivatives to be grouped together under one headword so they can be accessed on disk with ease. The dictionary contains information about how derivations are generated, grammatical word class, the definitions of words, and appropriate topic or context in which the words are commonly used.

The words in the tree are flagged for other attributes. If words can be members of compounds (such as phrasal verbs, idioms, common phrases), then they are flagged in the tree and another tree representing these compounds is checked. If all the members of a compound occur amongst the alternative candidates in the appropriate adjacent positions, then these candidates are given preference for further processing.

In these ways, the tree structure not only acts as a list of words against which to check the letter string candidates, but also contains information about word properties and enables access to fuller information as and when required. The tree structure acts as a lexical access device.

6. Conclusions

This paper has described ways of using orthographic information in order to reduce the uncertainty in the output from a pattern recognizer for handwriting. It was shown that checking the output against words is more efficient than checking it against sub-word orthographic information. Ways of representing large vocabularies in memory were investigated, and several were described. These show that large vocabularies may be represented so that they may be searched in real time, and have reasonable memory requirements. The tree structure is a suitable place in which to encode certain types of information (such as morphology, compounding, case) and can be used as a place at which to obtain access to fuller information if required. In this way, the tree structure operates as a lexical access device.

Acknowledgement

This work was supported by ESPRIT Project 295: The Paper Interface.

References

Crystal, D., *Encyclopaedia of Language*, CUP, 1989.

Ehrich, R. W., Koehler, K. J., "Experiments in the contextual recognition of cursive script," *IEEE Trans. Computers*, **24**, 1975, pp. 182-194.

Farag, R. F., "Word-level recognition of cursive script," *IEEE Trans. Computers*, **28**, 1979, pp. 172-175.

Goshtaby, A., Ehrich, R. W., "Contextual word recognition using probabilistic relaxation labeling," *Pattern Recognition*, **21**, 1988, pp. 455-462.

Higgins, C. A., Whitrow, R. J., "On-line cursive script recognition," *Proc. Interact 84, 1st IFIP Conf. HCI*, London 1984, pp.140-144.

Hornby, A. S., *Oxford Advanced Learners' Dictionary*, Oxford University Press, Oxford, 1988.

Keenan, F. G., Evett, L. J., "Lexical structure for natural language processing," *Proc. First Int. Workshop on Language Acquisition, IJCAI-89,* 1989.

Knuth, D. E., *Art of Computer Programming Vol. 3: Sorting and Searching,* Addison Wesley, 1973.

Kucera, H., Francis, W. N., *Computational Analysis of Present-Day American English,* Brown University Press, 1967.

Raviv, J., "Decision making in Markov chains applied to the problem of pattern recognition," *IEEE Trans. Inform. Theory,* **IT-3,** 1967, pp. 536-551.

Sampson, G., "How fully does a machine-usable dictionary cover English text?", *Literary and Linguistic Computing,* **4,** 1989, pp. 29-35.

Sayre, K. M., "Machine recognition of handwritten words: A project report", *Pattern Recognition,* **5,** 1973, pp. 213-228.

Srihari, S. N., Bozinovic, R. M., "A multi-level perception approach to reading cursive script", *Artificial Intelligence,* **33,** 1987, pp. 217-255.

Sussenguth, E. H., "Use of tree structures for processing files", *Comm. of the ACM,* **6,** 1963, pp. 272-279.

Wilson, M., "*MRC Psycholinguistic Database: Machine Usable Dictionary, Version 2.00.*", *RAL-87-054,* 1987.

Wright, P. T., "*Algorithms for the Recognition of Handwriting in Real-Time*", Ph.D. thesis, Nottingham Polytechnic, 1989.

Computer Processing of Handwriting
Eds. R. Plamondon & C. G. Leedham

A TREE-BASED DICTIONARY SEARCH TECHNIQUE AND COMPARISON WITH *N*-GRAM LETTER GRAPH REDUCTION

DAVID M FORD and COLIN A HIGGINS
Department of Computer Science, University of Nottingham, Nottingham NG7 2RD, United Kingdom

Segmentation methods are commonly used in automatic handwriting recognition systems. Pre-processed input data is divided into smaller segments, which are then recognized, perhaps by comparing with pre-defined templates, and these recognized pieces recombined to form the required output. Recognition of any one segment is often ambiguous, resulting in a substitution set of possible templates. The segmentation itself may also be ambiguous.

Higher level context can be used to resolve some of this ambiguity by defining a set of valid output which can be expected from the system. This context can be applied at various levels ranging from legal letter sequences, to valid words, to correct sentence semantics. The letter and word context can be applied using a purely computational approach, whereas the sentence level semantics require techniques from the field of Artificial Intelligence.

This paper discusses the application of letter and word context, and in particular demonstrates a dictionary lookup technique which is applicable to many forms of automatic script recognition. A comparison is made between this method and a binary *n*-gram approach.

1. Introduction

In many syntactical and statistical handwriting recognition systems, unique classification is impossible and an ambiguous substitution set may be produced. This ambiguity can often be resolved if contextual information at a higher level can define a limited set of valid output. A dictionary or lexicon of this information can then be constructed, and the list produced by the recognition process reduced and validated. Often very large substitution sets are produced which need to be checked quickly and efficiently against a large dictionary. This is especially the case with cursive script recognition systems.

A number of different techniques have been proposed to make use of contextual information.

Statistical information about the transition probabilities between letters can be used to remove or reduce unlikely letter sequences and produce the required output (e.g. Neuhoff (1975); Riseman and Hanson (1974); Hull and Srihari (1982)). This is often referred to as a *bottom-up* technique. These techniques are usually very efficient, but do not guarantee valid output words.

Another approach assumes that the written word comes from a fixed dictionary and the nearest matching word provides the required output (e.g. Duda and Hart (1968)). This is often referred to as a *top-down* technique. These techniques always produce a valid output word but can be inefficient to implement.

A number of hybrid methods have been suggested to balance out the advantages and disadvantages of the two approaches (e.g. Srihari *et al.* (1983); Shinghal and Toussaint (1979)). Some of these techniques are briefly discussed below.

A technique incorporating the advantages of both top-down and bottom-up approaches is presented here. This appears to be ideally suited to input data which can be represented as a directed letter graph, for example as is produced by cursive script or character recognition systems. Each node of the letter graph contains information concerning the match of a segment to a particular template, together with links to any possible succeeding nodes. The data structures used for the dictionary and graph are directly comparable, leading to a very efficient technique. An algorithm is designed and implemented, and experiments are reported which investigate the characteristics and suitability of this technique. Different internal structures for the dictionary are discussed and compared.

Experiments have been carried out to compare the performance of this algorithm with a method which has been used previously to reduce the letter graph using binary n-grams. Results are presented which show that the new technique is usually much faster than a 4-gram system, and a manifold increase in performance is achieved with very large graphs.

2. Typical Output from Segmentation Based Text Recognition Systems

There are two main approaches to the problem of text recognition. The first involves attempting to recognize a complete word as a whole unit - *whole word recognition*. The second involves segmenting the word into smaller units which are then recognized - *segmentation based recognition*. Only the second of these will be discussed in this paper.

A typical segmentation based text recognition system operates in several distinct stages. Firstly, the raw data, whether it be a scanned image or a time-ordered sequence of coordinates from an on-line digitizing tablet, is preprocessed to smooth, straighten and normalise the script. Secondly, a list of possible segmentation points is produced and thirdly, features are extracted from the segments which allow recognition to take place. The segmentation and subsequent recognition may be achieved at the letter level, or at the subletter level, where more than one segment may be combined to produce a single letter.

The letter-matching routine can fall into one of two main categories:

(a) those which produce a single word as output consisting of the most likely letters which span the written sample;

(b) those which produce a list of possible letters in each segmentation position together with a weighting of their likelihood.

Recognition systems belonging to type (a) make a binary decision at this stage as to where letter segmentation occurs - we will refer to this as a *fixed letter-segmentation*. Those belonging to type (b) may retain all of the possible segmentation points for future use and will be referred to as *ambiguous letter-segmentation* (no decision has been made at this stage as to exactly where the letters occur in the written word).

Example

With a fixed letter-segmentation recognition system, the most likely word is produced as output. For example, if the word *dog* was presented to the system, the letter matching algorithm may make a mistake, for instance, in the second letter position and produce the output *d-a-g*. It is then the task of any contextual post-processor to correct the output and produce the most likely real word.

An ambiguous letter-segmentation system produces a list of possible candidates in each segment position, together with a *certainty weight* to indicate the closeness of the match. If the word *dog* was presented to the system, it may produce two options for the second character, perhaps an '*a*' with high certainty weight of 60 or an '*o*' with a lower certainty weight of 40.

In cursive script recognition systems, individual characters are permitted to run into each other, with the problem that segmentation of the script into letters is ambiguous. For example, if the word *dog* is written cursively, there is the possibility that the word may be *clog*, where the segmentation after the first letter is ambiguous.

One way to represent the output from such a system is as a directed graph (Hayes (1980); Higgins and Whitrow (1985); Peleg (1979)). Figure 1 shows a simplified letter graph that might be produced by a recognition system acting on the input word *dog*. The '@' symbol represents the start of a word and the '#' symbol represents the end of a word. The graph is traversed from left to right, yielding a list of all possible combinations of letters that the original data might represent, i.e. {dcig, dog, dag, clcig, clog, clag}. The certainty weights can be attached to each letter on the graph and combined as the graph is traced to produce a ranking for each word that is produced.

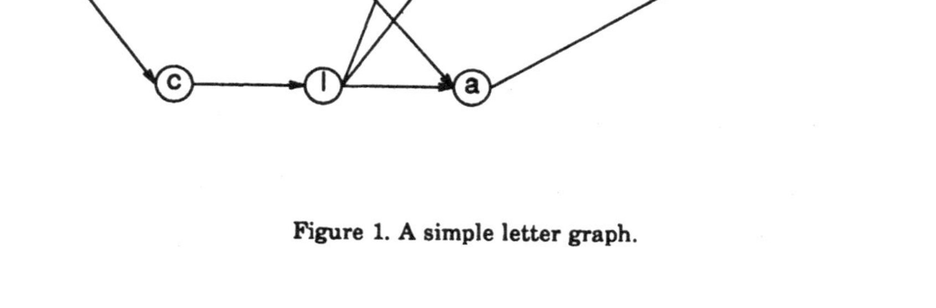

Figure 1. A simple letter graph.

3. Use of Contextual Information from Dictionary Sources

3.1 Fixed Letter-Segmentation

Most of the approaches to using contextual information have been based on constructing a post-processor for the output from fixed letter-segmentation recognition systems, the output from the recognition stage is a single, most likely word. The contextual disambiguation process takes a *word*, consisting of the most likely letters that span the word, as its input and returns the most likely written word as its output. There are several techniques which have been widely reported, the most common are listed below.

3.1.1 N-gram techniques

The probability of any individual n-letter sequence occurring can be calculated by examining large pieces of text. These probabilities can then be used to calculate the most likely written word given the output word from the recognition system, (see Riseman and Ehrich (1971)).

3.1.2 Viterbi algorithm

The Viterbi Algorithm (VA) takes the output word from the recognition system and, using statistical information on the sequence of letters in English and likely errors from the recognition system, calculates the most likely input word. Viterbi (1967) first described the algorithm. Forney (1973) provides a thorough tutorial introduction to the theory behind the algorithm. Neuhoff (1975) described how it could be applied to the problem of text recognition. Various authors have discussed its application, including Riseman and Hanson (1974) and Hull and Srihari (1982) who compared its performance against a binary n-gram approach.

The VA makes use of a *confusion matrix* of *a priori* probabilities observed from the recognition system, together with the transition probabilities between characters. In other words, the probability that a given letter may be mis-recognized as another letter is calculated and stored, together with the probability that it can be preceded or followed by any other character.

A 26 x l node trellis is constructed, where l is the length of the word, linking every letter with every other letter (see Figure 2). On the nodes of the trellis are the confusion probabilities, and on the arcs are the transition probabilities. [Notation: The confusion probability that a written letter X will be mis-recognized as a Y is written P(X I Y). The transition probability that a letter X may be followed by a letter Y is written P(X - Y).] By tracing a path through this trellis, and combining the probabilities on the arcs and nodes included in the path, the probability that the traced word could have been the original input word can be calculated. The algorithm finds the most likely path through this trellis, so yielding the most likely interpretation of the input word.

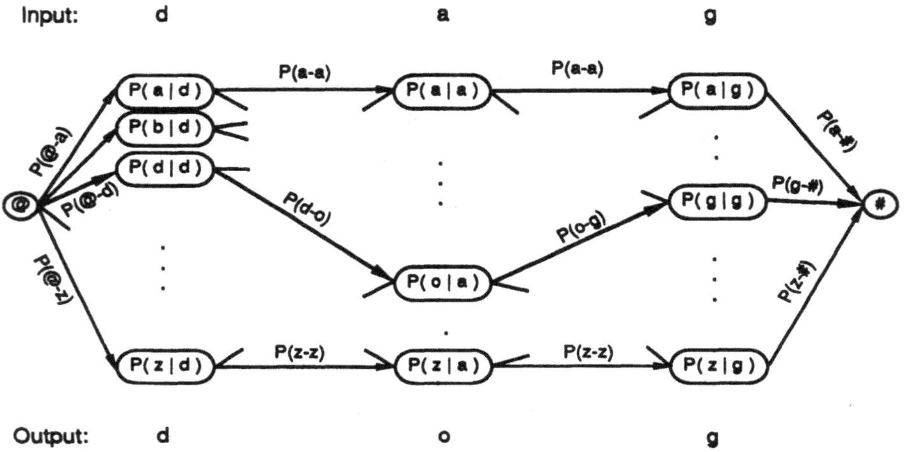

Figure 2. Viterbi Algorithm Trellis.

3.1.3 Dictionary Viterbi algorithm

The VA produces the most likely interpretation of the written word, but does not guarantee that the word exists. The Dictionary Viterbi Algorithm (DVA), developed by Srihari *et al.* (1983), is an improvement on the VA making use of a simultaneous full dictionary search, in place of the transition probabilities, to ensure that only a valid word is produced. (A similar technique was used by the HARPY speech understanding system, using words and sentences in place of letters and words, see Lowerre and Reddy (1980).) The dictionary lookup uses a *trie* structure to store the dictionary. This is described in Section 3.2.2.

3.1.4 Modified Viterbi algorithm

Shinghal and Toussaint (1979a) describe another variant of the VA, called the Modified Viterbi Algorithm (MVA). Here a heuristic depth of search *d* is set by the user so that only the *d* most likely probabilities in each letter position are checked. The computational overheads of the VA are thus reduced, as only a $d \times l$ trellis needs to be traced, but the performance degrades as *d* is reduced.

3.1.5 Predictor-Corrector algorithm

Shinghal and Toussaint (1979b) further improve on the MVA by combining it with an efficient dictionary lookup algorithm (DA). This is called the Predictor-Corrector Algorithm (PCA).

The dictionary is partitioned into sub-dictionaries of same-length words. Each sub-dictionary is then sorted by *value*, where the *value* is calculated by combining the transition probabilities of the letters of the word. This *value* was found to be nearly always unique for any sub-dictionary. A binary search is used to see if the output word from the MVA exists in the dictionary. If so, then that is taken as the required output. If not, then the DA is employed. A *score* is calculated for the nearest *f* words to where the output word was expected. *f* is a heuristic set by the user. The *score* is calculated by combining the transition probabilities between the letters with the confusion probabilities. The word with the largest *score* is the required output.

Shinghal (1983) describes a further enhancement to the PCA, where the *n* most likely words from the MVA are checked in decreasing order to see if they exist in the dictionary. If none of them exist, the DA is employed. The value of *n* is determined by experiment.

3.2 Ambiguous Letter-Segmentation

The problem with all of the systems based on the VA is that they do not allow for incorrect letter segmentation. For example, if the word *dog* is misrecognized as *clog* then the algorithm cannot produce the correct word and the system will fail. An ambiguous letter-segmentation retains all of the possible segmentation points within the letter graph representation.

For an ambiguous script recognition system, the contextual disambiguation process takes a letter graph as its input, and produces as its output a list of words which have been checked, in some way, against a lexicon of dictionary information. One simple way to do this is to trace every path through the graph and check each resulting word against a list of valid words. This is not practical since the letter graph is usually very large with many paths. Even the most efficient searching algorithms will take a significant amount of processing time to verify a very large list of words. (Commercial spell-checking software cannot be used since, for efficiency, the algorithms they employ usually rely on the fact that the input words will either be correct, or very close approximations to real words. Completely random sequences of letters may often be accepted as valid words, e.g. UNIX† spell.) A much more efficient technique is needed.

Two main practical techniques are described and compared below.

3.2.1 Binary n-gram graph reduction

A list of valid *n*-letter sequences is produced by analysing a dictionary of allowable words. This list can be used to reduce the size of the letter graph by removing invalid portions. Section 7.1 gives details of one such method.

3.2.2 Dictionary tree

A dictionary or word-list is restructured in the form of a tree, based on the *trie* structure suggested by Knuth (1973). This is shown pictorially in Figure 3; where the tree represents the word list {a, an, and, at, be, bet, but, by}. Each of these words can be found by tracing a path from left to right. The '@' symbol represents the start of a word and the '#' symbol represents the end of a word.

The *trie* can be used as an efficient structure to store a dictionary for a variety of applications, but is especially applicable for contextual post-processing of script recognition systems. The DVA proposed by Srihari *et al.* (1983) used a dictionary *trie* at the same time as the Viterbi lattice is traced to guarantee that a valid word is found. Bozinovic and Srihari (1982) combined a stack-decoding search algorithm with a *trie*-structured dictionary with a small dictionary of 1027 words. Bozinovic and Srihari (1989) used a similar approach but added a depth of search heuristic to limit the computation needed.

† UNIX is a trademark of Bell Laboratories

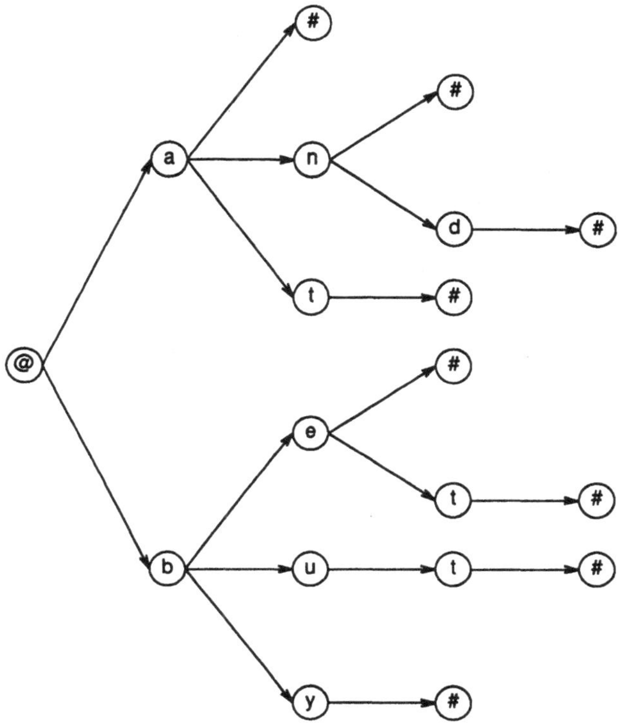

Figure 3. Dictionary Tree - Pictorial Representation.

Our method shows that a letter graph, such as that shown in Figure 1, can be efficiently traced using a recursive procedure. Such a procedure might take the head-node of the graph on which it is to act as its parameter and call itself recursively, passing each sub-graph that the head-node points to as a parameter. This will carry out a depth-first trace of the graph. Using a tree structure, the dictionary can be traced simultaneously. As each arc of the letter graph is traced, if the corresponding arc exists in the dictionary tree, then the word is valid *up to that point*. When the end-of-word marker is reached in the graph and tree, the word traced out exists in the dictionary. An invalid word in the letter graph will be rejected *as soon as* an arc cannot be found in the dictionary tree. It is not necessary to continue tracing the graph past this point. This is advantageous since it limits the time taken to search the dictionary, and allows the dictionary to be very large without seriously reducing the performance.

All valid words that exist in the letter graph can be produced in this way with a weighting of their likelihood. No information is thrown away at this stage which may prove useful later on, for example, if more than one recognition routine is used in parallel and the outputs combined.

4. Computer Representation of the Dictionary Tree

A dictionary structured in this way can be accessed very quickly. It has the disadvantage of requiring a large amount of memory, since the data structure remains resident while the program is running. This is probably the reason for its rejection in the past, but as computer memory is becoming larger and more readily available this is no longer a major problem. We consider a number of different ways to construct this data structure within a computer program.

4.1 Discussion of Possible Data Structures

Knuth (1973) suggests a static data structure for the dictionary tree. This consists of a flag to indicate whether a word can end here and an array of 26 element integer arrays together with 26 boolean flags. Each column position points to the column containing the next letter in a valid word, and a corresponding flag to indicate if a word can end at that point. This array is very sparse and would take up a large amount of memory. Nowadays, dynamic memory allocation allows more efficient data structures to be used.

A simple dynamic data structure might consist of a tree of linked nodes, where each node contains 26 pointers to possible successor nodes (see Figure 4). In this particular layout, the letter is implied by the position of the pointer in the array. Such a structure will be extremely sparse and have huge memory requirements.

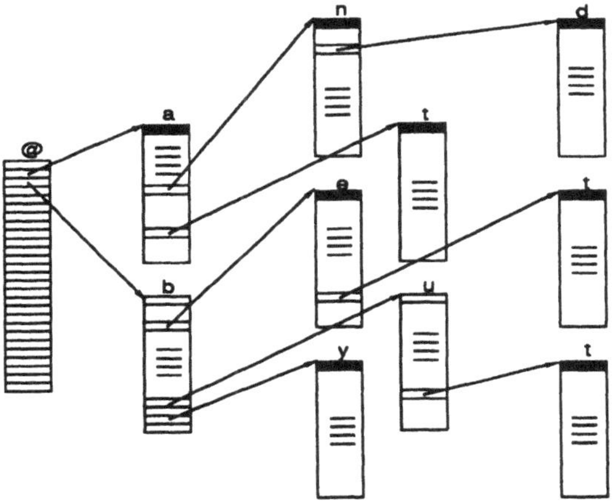

Figure 4. Dictionary Tree - Logical Representation.

A more compact dynamic structure is shown in Figure 5. In this diagram each large square represents a node of the tree consisting of a key letter, at a certain level, and two pointers. The upper pointer points to a list of letters, at the next level, that are permitted to follow this letter. The lower pointer points to the remaining list of letters, at the current level, which can occupy the same letter position. This diagram is a representation of the tree shown in Figure 3.

Figure 6 shows an internal representation for the letter graph in Figure 1. A node consists of a letter and a pointer to a list of arcs which, in turn, point to possible successor nodes.

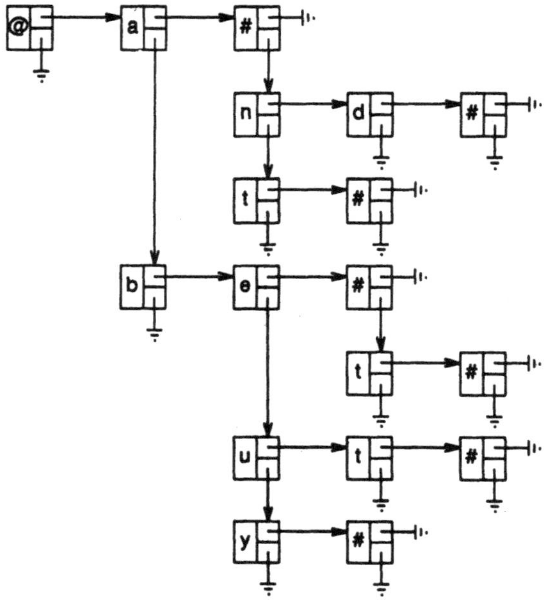

Figure 5. Dictionary Tree - Physical Implementation.

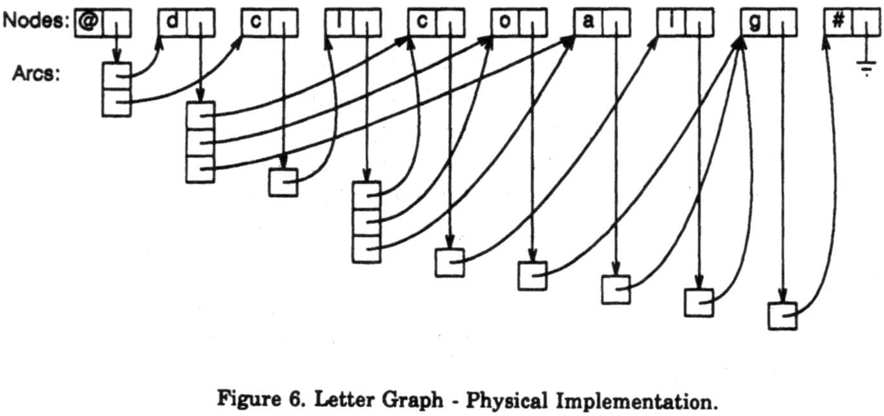

Figure 6. Letter Graph - Physical Implementation.

4.2 Possible Enhancements

The final choice of internal representation involves consideration of the trade-off needed between access speed and memory requirements. There are several different structures which can be used depending on the most important criterion.

4.2.1 Increased speed of access

We define the list of letters, at the next level, which can follow a particular letter in the dictionary tree as its *sub-list*. The search method for a structure similar to Figure 5 can be speeded up by reducing the time taken to search for a letter in a sub-list. The sub-list of letters can be ordered in several ways. If the letters are in a random order, it is necessary to search to the end of the list if the key cannot be found. However, if the ordering is known it is only necessary to search as far as the expected position of the letter.

Several different orderings are possible. Alphabetical ordering has the advantage of allowing the ASCII code to imply position in a list. Sorting on frequency of occurrence in English, or frequency of generation from the recognition system may produce a faster search, but may require extra memory to store a representation of the order. Actual speed will be dependent on the letter graph itself; a graph which contains many valid words will take a different amount of time to trace than a similarly sized graph with many invalid words. For example, a list optimized for finding a commonly occurring letter will not be optimized to show that an infrequent letter is not present in fact, it will be the worst possible ordering.

4.2.2 Reduced size of data structure

The size of the data structure can be reduced by using *tail-end compression*. If the end of a word is unique and is not common with any other word, then the dictionary structure can be reduced by constructing a special *end-of-word* node which contains the rest of the word as a string, without the need for extra nodes and pointers (see Figure 7). The disadvantage of this structure is that the dictionary cannot easily be maintained. For example, to add an extra word may involve unpacking the tail of an existing word, adding the new word, then re-packing the two new tails. Also, since the code needed to search a tail will be different to that needed to search the tree, switching lookup code would degrade the speed of lookup.

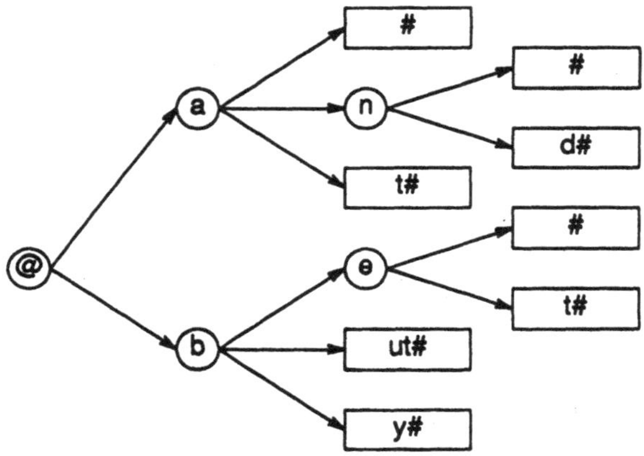

Figure 7. Dictionary Tree - Tail End Compression.

4.2.3 Improved performance

Our technique was originally developed for use as a post-processor for a cursive script recognition system. Often handwritten script samples have very poor quality sections where either the word has been written badly, or has been poorly digitized. A recognition system may not be able to identify, in any way, what has been written in these sections and may insert a *wild card*, representing every possible character, into the letter graph in that place. Alternatively, it may be clear that a segment has, for example, a descender, but no other information is obvious, so a descender subset wild card, representing every letter with a descender, could be inserted. If the poor section occurs at the end of the word, the performance of the tree structure described above will not be adversely affected, since most of the paths through the letter graph will be rejected before the highly ambiguous section is reached. However, if it is at the beginning of the word, then there will be a large amount of ambiguity at the beginning of the graph, resulting in a large amount of wasted computation.

If the dictionary tree was also structured in reverse, starting at the end of words and working towards the beginning, it would be possible, in these cases, to verify the letter graph backwards, and so speed up the process. Similarly, if both ends of the original data sample are poor, it may be possible to work from the middle outwards. An ideal data structure might have multiple linkages, starting at each letter position, so that the dictionary search would always begin at the least ambiguous part of the graph. It is intended that the properties of such a structure will be investigated in future implementations.

It should be noted that a dictionary lookup system based on a letter graph will fail if the correct letter is not in the letter graph hence the need for wildcards when no letters are obvious candidates. A fixed letter-segmentation recognition system does not suffer from this problem since it is possible to change one letter to another.

5. Implementation Selected

For our implementation, we initially chose a simple approach, for ease of coding, to ascertain whether this data structure was suitable in practice. The data structure is an exact representation of Figure 5. Lists are ordered alphabetically, and the tree is linked only from start to finish. It was discovered that its performance and size were perfectly adequate for our requirements, even with very large ambiguous letter graphs.

To test the performance of this implementation, a 210,000 word dictionary was used. This is probably much larger than would be needed in a practical system. A substitution set was used to generate simulated letter graphs of possible output from typed input words. A test set of 2,100 letter graphs was verified using the system running on a SUN 3/160 with 8 Mbytes of main memory, timeshared with other users but with light loading. The average real time needed per word was 0.45 seconds, and the actual CPU time used by the system averaged 0.19 seconds per word. In a practical system the lookup process would be quicker, since the times quoted above include routines to set up the data structure for the dictionary, to time the program and to calculate and print various analytical results.

6. Size of Data Structure

The size of the dictionary data structure can be measured by counting the number of nodes in the tree. Experiments have been carried out using different sized word lists to investigate the effect of dictionary length against data structure size. As can be seen from the graph in Figure 8, this is roughly a linear function of dictionary size.

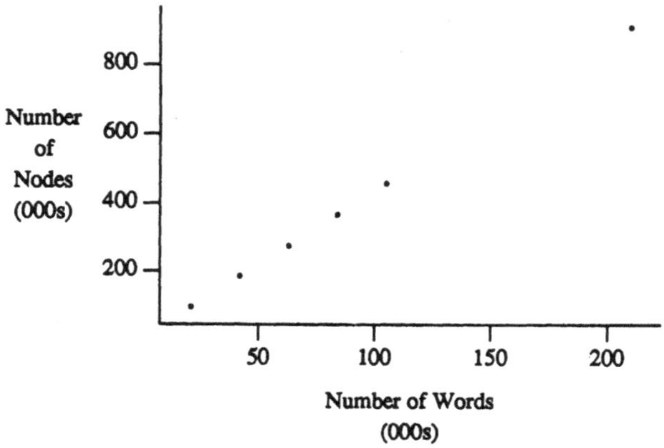

Figure 8. Size of Data Structure *vs* Dictionary Size.

For example, it seems that a dictionary of 60,000 words will probably be sufficient for a usable script recognition system. (This would allow for a vocabulary of about 20,000 root words with plurals and verb endings.) Assuming that a node consists of a character and two pointers, it can be seen from the graph that such a dictionary would require approximately 2.4 Mbytes of memory in our implementation. This is not an unreasonable amount of memory to expect in modern computer workstations.

The size of the data structure for a particular dictionary is dependent on the *compactness* of the dictionary, i.e. whether the words have common roots. Adding a word which has the same root as a word already in the dictionary increases the size by less than adding a unique word. This means that the dictionary can contain all participles of verbs, plurals etc without drastically increasing its size, avoiding the need to construct these words by a rule-driven system.

To investigate the effect of compactness on dictionary size, two sub-dictionaries of 21,000 words were selected from a word list of 210,000 words; one by selecting the first word from each group of 10, the other by selecting the first 10 words from each group of 100. As expected, the first of these was less compact and used 141,138 nodes, the other more compact dictionary used only 98,981 nodes.

7. Comparison with Binary 4-Gram Graph Reduction

7.1 The Graph Reduction Process

This technique makes use of the existence or nonexistence of four-letter sequences in English. Higgins (1985) reported that four is the ideal length of gram to use, since only approximately 5% of 4-grams are valid in English, and the number of possible grams, $26^4 = 456{,}976$, can be reasonably stored as a binary array occupying just under 56 Kbytes of memory. A much larger percentage of 3-grams are valid, and 5-grams would require about 1.5 Mbytes of memory for the binary array without much gain in context.

The letter graph is supplemented by adding an extra start and stop node at the beginning and end of the graph. This is so that the opening two and three letter sequences can be checked using the same 4-gram approach.

The graph reduction process can be implemented in many different ways, see Whitrow and Higgins (1985). We have selected an efficient technique, which uses a similar letter graph structure and tracing algorithm to the dictionary tree search algorithm, for the purposes of comparison. This technique involves recursively tracing each path through the letter graph, maintaining pointers to the last four letters accessed. At each step the current four letter sequence is compared against the list of valid 4-grams. If the sequence *is* valid, then the arcs connecting the letters are marked. The three arcs are marked differently depending on which letters of the sequence they connect. An arc which has been in each of the three connecting positions of a four letter sequence is marked as *used*. After the traversal is complete, the graph is stripped of all *unused* links, the arc-markings are cleared, and the process repeated until no more links are removed. In this way a much reduced letter graph is produced, which can be more readily checked against a dictionary using a straightforward approach.

7.1.1 Speed of reduction

The speed of graph reduction for the technique described above is dependent on the number of possible paths through the graph, which is in turn dependent on the length of the word and on the *confusion level* of the letter graph. This is defined to be the number of letter options at each letter position. The graph in Figure 9 shows how the time taken for the reduction greatly increases as the confusion level increases. The graph shows the time taken when reducing various possible letter graphs generated from the input word "test", with increasing ambiguity at each letter position.

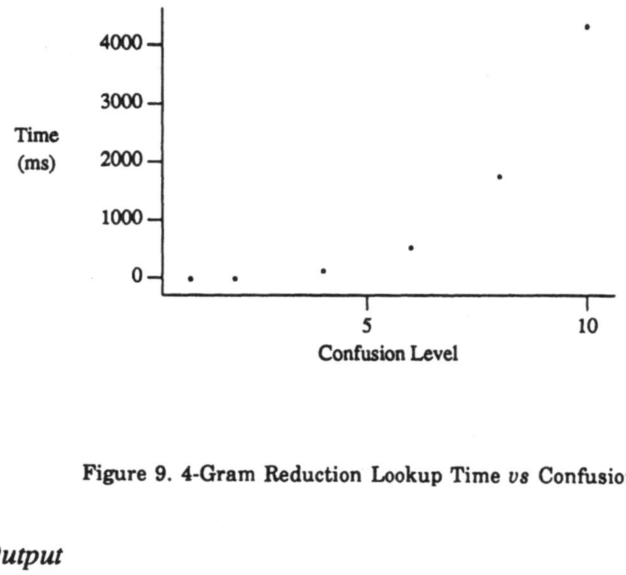

Figure 9. 4-Gram Reduction Lookup Time *vs* Confusion.

7.1.2 Output

4-gram graph reduction produces a much reduced letter graph, but this will usually still contain invalid words when all of the paths are traced. In a sequence of letters each individual gram may be permissible but this will not guarantee that the whole word is valid. Also, an arc in the letter graph can only be deleted if *no* valid gram uses it. A remaining arc will then still allow invalid grams to pass through it. (Figure 10 shows a sub-graph of a letter graph, where the gram *ebcf* may be invalid but *abcd* valid. The arc marked '*' must therefore remain in the graph, so the invalid gram will still be present when the graph is traced.) It is therefore still necessary to check each word against a dictionary to guarantee validity. This will incur the overhead of additional processing time.

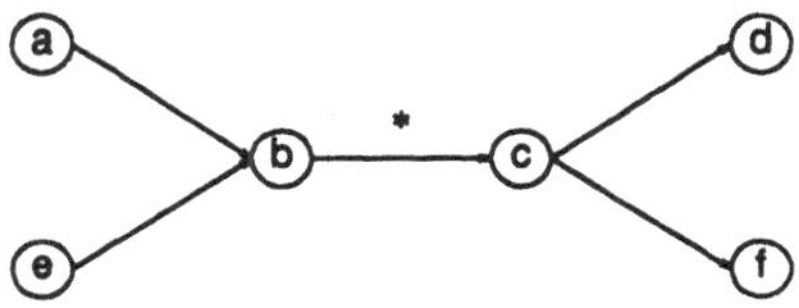

Figure 10. Arc Deletion.

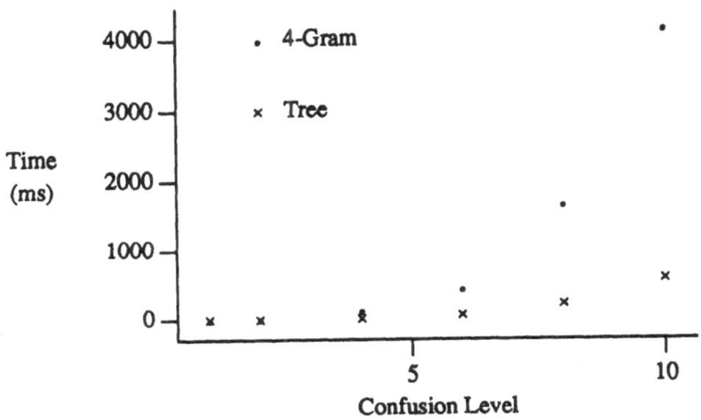

Figure 11. Comparison - Lookup Time *vs* Confusion Level.

7.2 Speed Comparison with Dictionary Tree

The graph in Figure 11 shows the time taken for the graph reduction and the dictionary lookup, plotted against confusion level of the input graph, again for the word "test". It can be seen that with very small graphs the two techniques are comparable, but, as the ambiguity increases, the dictionary lookup is considerably faster. It should also be remembered that the 4-gram output still needs to be verified against a dictionary.

8. Conclusions

Whereas a large amount of work has been done in the field of contextual post-processing of text recognition systems, not many of the techniques suggested successfully address the problems caused by ambiguous letter-segmentation, which are common with cursive script. The tree-based dictionary lookup technique described here is an ideal application of letter and word context for the output from text recognition systems and allows for ambiguous letter-segmentation of the script. It is extremely fast and efficient, can be used with a large dictionary or word list and produces all possible output words, with no loss of information that can occur when arbitrary cut-offs are applied. The dictionary contains every word which can be recognized by the system, without the need for prefix and suffix generation. It is simple to add extra words to the dictionary without greatly increasing its size. The technique is especially appropriate for very ambiguous letter graphs, typically produced by cursive script recognition systems. Wild card substitution can be used in areas where the script is very poorly written or digitized.

The performance is superior to n-gram graph reduction techniques, and is in fact a super-set of the n-gram approach, since all possible values of n are effectively applied simultaneously.

References

Bozinovic, R. and Srihari, S.N., "A String Correction Algorithm for Cursive Script Recognition," *IEEE Transactions on Pattern Analysis and Machine Intelligence* 4 no. 6, November 1982, pp. 655-663.

Bozinovic, R. and Srihari, S.N., "Off-Line Cursive Script Word Recognition," *IEEE Transactions on Pattern Analysis and Machine Intelligence* **11** no. 1, January 1989, pp. 68-83.

Duda, R.O. and Hart, P.E., "Experiments in the Recognition of Hand-Printed Text: Part II - Context Analysis," *AFIPS Conference Proceedings*, **33**, 1968, pp. 1139-1149.

Forney, G.D. Jr., "The Viterbi Algorithm," *Proceedings of the IEEE* **61** no. 3 March 1973 pp. 268- 278.

Hayes, K.C., "Reading Handwritten Words Using Hierarchical Relaxation," *Computer Graphics and Image Processing* **14** 1980 pp. 344-364.

Higgins, C.A. and Whitrow, R., "On-line Cursive Script Recognition," *First IFIP Conference on Human-Computer Interaction - INTERACT 84*, London, 1985, pp. 139-143.

Higgins, C.A., "*Automatic Recognition of Handwritten Script*," PhD Thesis, CNAA, 1985.

Hull, J.J. and Srihari, S.N., "Experiments in Text Recognition with Binary n-Gram and Viterbi Algorithms," *IEEE Transactions on Pattern Analysis and Machine Intelligence* **4** no. 5 September 1982 pp. 520-530.

Knuth, D.E., in *Sorting & Searching, The Art of Computer Programming*, **3** Addison-Wesley, 1973, pp. 481-487,.

Lowerre, B. and Reddy, R., "The HARPY Speech Understanding System," in *Trends in Speech Recognition*, ed. W. Lea, 1980, pp. 340-360.

Neuhoff, D.L., "The Viterbi Algorithm as an Aid in Text Recognition," *IEEE Transactions on Information Theory* March 1975 pp. 222-226.

Peleg, A., "Ambiguity Reduction in Handwriting with Ambiguous Segmentation and Uncertain Interpretation," *Computer Graphics and Image Processing* **10** 1979 pp. 235-245.

Riseman, E.M. and Ehrich, R.W., "Contextual Word Recognition Using Binary Digrams," *IEEE Transactions on Computers* **20** no. 4 April 1971 pp. 397-403.

Riseman, E.M. and Hanson, A.R. "A Contextual Post-processing System for Error Correction Using Binary n-grams," *IEEE Transactions on Computers*, **23** no. 5 1974 pp. 480-493.

Shinghal, R. and Toussaint, G.T., "Experiments in Text Recognition with the Modified Viterbi Algorithm," *IEEE Transactions on Pattern Analysis and Machine Intelligence* **1** no. 2 1979a pp. 184-192.

Shinghal R. and Toussaint, G.T.,"A Bottom-up and Top-down Approach to Using Context in Text Recognition," *International Journal of Man-Machine Studies* **11** 1979b pp. 201-212.

Shinghal, R., "A Hybrid Algorithm for Contextual Text Recognition," *Pattern Recognition* **16** no. 2 1983 pp. 261-267.

Srihari, S.N., Hull, J.J. and Choudari, R., "Integrating Diverse Knowledge Sources in Text Recognition," *ACM Transactions on Office Information Systems* **1** no. 1 January 1983 pp. 68-87.

Viterbi, A.J., "Error Bounds for Convolutional Codes and an Asymptotically Optimum Decoding Algorithm," *IEEE Transactions on Information Theory* **IT-13** April 1967 pp. 260-269.

Whitrow, R. and Higgins, C.A., "The Application of n-Grams for Script Recognition," *Proceedings of the Third International Symposium on Handwriting and Computer Applications*, Montreal, Canada, July 1987, pp. 92-94.

Computer Processing of Handwriting
Eds. R. Plamondon & C. G. Leedham

A TASK ORIENTED APPROACH TO READING AND TO HANDWRITTEN TEXT RECOGNITION ¶

CHRISTOPHE PARISSE, VICTOR ROSENTHAL,
ABDELMALEK IMADACHE, EVELYNE ANDREEWSKY
and FRANCOISE COCHU
INSERM-TLNP, Pavillon Claude Bernard, Hôpital de la Salpêtrière, 47, Bd. de l'Hôpital, 75013 Paris, France

This chapter examines the relationship between reading and the machine recognition of handwriting and develops arguments in favour of a task oriented approach to these two research problems. This approach leads us to suggest a model of the initial stages of reading which we applied to the problem of machine recognition of unrestricted handwriting. The model specifies a sequence of operations defined by the means of analyses which gradually become available in the course of reading. They consist, *inter alia*, of primary coarse-grained global analyses followed by local fine-grained verification. We believe that this principle of gradual approximation, complemented by syntactic and lexical-semantic analyses, is appropriate to deal with the problem of the inherent instability of handwriting. A computational experiment conducted on three very different samples of script is described. The experiment was designed to evaluate the suitability of primary coarse-grained analyses for the recognition of unrestricted handwriting and yielded results which were interpreted as encouraging for the present approach.

1. Introduction

For a normal literate individual familiar handwriting of good quality is virtually as legible as print. Even unfamiliar handwriting does not necessarily pose a difficulty: we often read a letter from an unknown correspondent with no sense of any particular effort. It should be borne in mind, however, that

¶ This work was supported by INSERM, CRE No. 88/1802 and by a grant from the University Pierre et Marie Curie.

even very good quality handwriting shows marked irregularities (in letter shape, segment size, spacing, direction of stroke, etc...) as compared to the typset and impression stability of print. Although very little comparative work has been undertaken to date, it is generally assumed that most phenomena reported to occur for printed material also take place in the reading of handwriting. In the past, many psychological experiments on reading or word recognition were devised using a handwritten format, with print being used only once the experiment had been fully set up.

The widespread assumption that it takes no fundamentally different skills to read print or handwriting provides excellent reasons to be puzzled by the *small effect* of script variability on reading performance. Psychologists are used to taking this observation for granted and thus have never seriously attempted to exploit its implications for theories of reading. Computer scientists, on the other hand, who have seldom been interested in psychological evidence - despite the obvious relationship between script recognition and reading - have tackled the problem of the machine recognition of handwriting with the means at hand. Still, the puzzle of the small effect of script variability remains unresolved. How is it that even very important variations may have a negligible effect on the speed and accuracy of reading while at the same time irregularities which to the human eye appear unremarkable often have dramatic effects on the machine recognition of handwriting?

Unless one takes this observation as an example of the irrationality of nature, this striking contrast between the performance of natural and artificial systems invites one to rethink the usual way of conceiving reading and its relationship to script recognition. In particular, the ability to perceive regularity in ever variable patterns, to recognize a pattern we have never seen before as an instance of a familiar gestalt (despite an infinity of locally different forms it may potentially assume), appears to be inherent in reading skills. An explanation of this ability may thus be necessary for the development of a plausible theory of reading.

2. A Task Oriented Description of Reading

What this explanation may require and how it might affect theories of reading and work on the machine recognition of handwriting are questions we shall seek to clarify as we proceed. Needless to say, we do not intend to suggest that the mechanisms of reading are designed in such a way as to be minimally affected by the variability of script. Rather, we shall examine what the earlier-mentioned *small effect* of script variability on reading performance implies for the structure of *reading tasks*. By *task structure* we are referring to a logical sequence of steps, each of which is defined by a subgoal, that always have to be achieved - whether by a human mind or by a machine - for a given task to attain its goal and accomplish what it is supposed to (see Marr, 1982; Rosenthal, 1988). In the present context, this structure would consist of a sequence of steps leading to the accomplishment of what reading is supposed to do. It would have to be defined in such a way as to allow the reading system to be relatively insensitive to a broad range of variations: as long as relevant patterns satisfy certain, as yet to be identified, conditions, what may be an important irregularity to an illiterate eye is but negligible variation to a literate eye.

In other words, we need to understand what information processing structure grants insensitivity to a range of "irrelevant" variations. Such a structure may, of course, not be specific to written input; it may apply to many tasks of visual perception. Yet even if there are grounds to place a certain degree of commonality of mechanisms underlying various forms of visual perception, there are also important reasons to view reading mechanisms as domain specific. Note, for instance, that written words are very poor visual patterns: they have no volume, no texture, they are written against an extremely poor background (plain white paper provides no cues to the identity of written words). Topographical variability further increases the difficulty of dealing with such poor perceptual patterns on purely visual grounds. On the other hand, written words have a very characteristic "look" for a literate eye; an understanding of what this "look" consists in may be very useful for the present purpose. Accordingly, it seems most advisable to tackle the recognition of handwriting as a problem of its own, rather than as an instance of a general problem of pattern recognition. A consideration of all sorts of information which are specific to written language would play an important role in this approach.

Consider, for instance, that reading skills involve the knowledge of the language in which a text is written. If written words are recognized as meaningful patterns it is because they are conventional codes built upon the knowledge of what they are designed to encode. Conceivably, this knowledge ought to play an important role in reading and be extremely useful to, if not necessary for, machine recognition of written texts. But if some linguistic knowledge is to contribute to the overall process in a significant way it has to apply quite early in the course of processing, and it must determine some of the initial steps of the whole sequence, steps which should be executed in order to recognize words, not once recognition is completed. This raises another important question into which we shall inquire in what follows. The inquiry will lead us to combine evidence from psychological studies of reading and of acquired dyslexia with evidence from work on the machine recognition of handwriting. Our goal in this chapter is twofold: first we shall examine the structure of reading tasks and evaluate its implications for machine recognition of handwriting; second we shall define the primary steps of the recognition process and evaluate the adequacy of this definition by means of a computational experiment.

3. The Psychology of Reading

3.1 In Search for Units of Visual Analysis

From its beginnings in the late 19th century, the psychological study of reading evolved around the debate concerning the definition of the units of analysis in visual word recognition: is word perception mediated by letter, syllable, or full-word units (Cattell, 1886; Huey, 1908)? As the research went on, the debate became increasingly complicated due to a myriad of discoveries concerning the effects of semantic context, orthography (e.g. spelling-to-sound regularity), lexical frequency and familiarity, and syntax and phonology on word recognition (see Coltheart, 1987). The emergence of relatively complex and sophisticated models such as dual route (e.g. Morton, 1969; Patterson and Morton, 1985) or multilevel parallel processing models (e.g. Greenberg and Vellutino, 1988; McClelland and Rumelhart, 1981) which apparently reconcile the proponents of global (full-word oriented) and local (e.g. letter oriented) units of processing did not result in an end to the controversy; it simply provided additional issues.

By units of visual analysis we are referring to elaborate clear-cut patterns (e.g. letters, words) having all the qualities of a distinct image. What characterizes the unit-based approach, apart from the unit size issue, is that visual analysis is conceived as a process of checking a target (visually relevant unit) against an inventory of known, mentally encoded patterns. The target is recognized if it matches an item in the inventory, otherwise it is unknown. (Note that in this account recognition is an all-or-nothing process.) Notwithstanding that a comparison must be carried out between the stimulus and known patterns, this approach either skips the process of visual analysis or implicitly reduces it to a sort of superposition of retinal image and mental image. (For computer scientists this view is reminiscent of the classical method of pattern matching.) The primary question which arises in this context is how the visual system copes with typographical variations (font type and size, style and spacing)? How does it recognize unknown fonts? As far as print is concerned, it could be argued that minor (but systematic) transformations of target image would suffice. But what about handwriting which would require major unsystematic transformations? Would it not make the reading of handwritten texts much more difficult than it is?

3.2 Methodological Restriction

With these questions in mind we may turn now to a discussion of psychological studies on reading. Two preliminary comments may however be necessary. First, it should be noted that a considerable amount of evidence in this area arises from tachistoscopic studies which almost exclusively bear on printed words, mostly upper-case printed words. The choice of upper-case letters is understandable if one wants to preclude any facilitation due to the shape of the word (one can see that **BLACKBOARD** is harder to read than **blackboard**) or to avoid having some letters more easily recognizable than others. Still, taking this methodological step makes sense only if word shape can be considered irrelevant or undesirable. It may indeed be so in certain tasks. But when reading and word recognition are involved, one needs good reasons to assume that word shape is irrelevant. In a similar vein, the choice of print may also be justified on methodological grounds: one may consider that handwritten words are not all equally recognizable due to the inherent instability of handwriting. But, again, this precaution may only be valid for some tasks. When it becomes a rule it is tantamount to assuming that the recognition of printed words best exemplifies reading skills. If stability of print, however, is unessential for reading skills then the psychological study of reading has eliminated from its field of inquiry a potentially important source of information.*

Second, it should be recalled that roughly twenty years ago the psychological investigation of reading split up into two areas of research with hardly any inter-communication between them: on the one hand, the work on word recognition; on the other, the study of text comprehension with relatively lit-

---

* This point is not in contradiction with the earlier proposal that it takes no fundamentally different skills to read print and handwriting. It can in fact be argued that the ability to read print represents a special case of reading, the ability to read handwriting being a more general or more characteristic example of these skills.

tle interest in perceptual processes. In this context, the studies actually related to the mechanisms of reading were confined to word recognition and perception, mostly of single words. As a consequence, the scope of semantic context had to be restricted to the facilitatory effects of semantic or associative priming on word recognition (the presentation of, say, the word *'table'* shortly before the word *'chair'* speeds up the recognition of the latter, relatively to the word *'smoke'*, say). Similarly, the scope of syntax has been reduced to the distinction between open- and closed-class words. Thus, even the studies based on proofreading (e.g. Healy, 1980; Haber and Schindler, 1981) and on the letter cancellation technique (Corcoran, 1966; Healy, 1976), which of necessity used real texts, were limited, as far as syntax is concerned, to the demonstration of differences in the status of the two vocabulary classes (i.e. closed-class *vs* open-class words).

3.3 Eye-Movement Studies

Notwithstanding these reservations, psychological inquiries offer a wide range of findings that any reasonable theory of reading must take into account. Some of these findings stem from studies based on the technique of tracking eye fixations during reading. The technique is based on the discovery reported in 1879 by Javal that eye movement behavior during reading does not consist of steady, letter-by-letter-like movements but of saccades and return sweep movements. This observation raises immediately two questions: what is fixated and what determines the position of the next fixation? With the advent of the modern eyetracking apparatus it became possible to register the exact position of each fixation and the exact length of saccades. It is important to keep in mind that, although the span of foveal vision is very narrow (around 3 characters), many words are fixated only once in a more or less central spot, some words, even relatively short ones, are fixated in two different locations (regardless of backward sweeps) and some words are skipped altogether (see Rayner and Pollatsek, 1987). Figure 1 shows the sequence of gaze movements of a subject reading a sentence.

An analysis of Figure 1 lets one understand the finding that seems the most important in the present context: since saccade length generally exceeds the span of foveal vision it is parafoveal vision that determines the location of future eye fixations (Loftus, 1983; Balota, Pollatsek, and Rayner, 1985) and whether to skip a word (Rayner and Pollatsek, 1987). Yet, parafoveal vision normally does not enable one to overtly recognize a word. A problem thus seems to arise here. In order to know where to look one has to have some knowledge of what is going to be looked at. But how can one know what is going to be looked at before recognizing it?

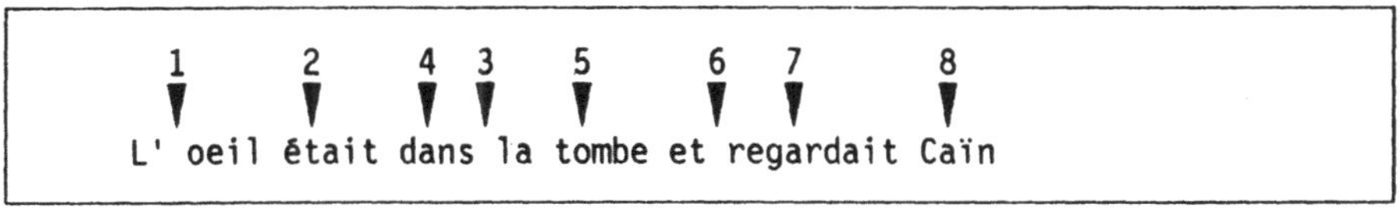

Figure 1. Fixation points of a subject reading a sentence. The data are based on a single record of eye movements of a subject. The gaze moves by saccades from one location indicated by ▼ to another. Each ▼ shows the centre of fixation, the digits indicate the sequence of movements. The inversion of 3 and 4 shows a backward sweep of the gaze. (Adapted from Lévy-Schoen and O'Regan (1989)).

3.4 A Task Oriented Approach To Reading

We submit that this observation poses a problem only within the confines of the clear-cut unit recognition framework defined on an all-or-none basis. Parafoveal vision appears to be a mystery if one assumes that what has not been overtly recognized has not been seen and what has not been seen cannot provide information regarding just what it is. The present suggestion is, however, different. Specifically, we propose that, in parafoveal vision, the system carries out a preliminary processing of a stimulus word which somehow "gives it an idea" about the morphology of this word. This "idea" will be precise enough to determine the interesting point of verification (the location of the next fixation) but in most cases it will be too rough for identification. Or, more generally, this finding of the parafoveal determination of saccades suggests the existence of a *primary coarse-grained global analysis* of a word which precedes a *local fine-grained verification*. In a way, we are no longer concerned with units of perceptual analysis. For if the primary analysis amounted to a fine clear-cut identification then why is parafoveal presentation insufficient for word recognition? Why bother checking some supposedly critical spots if identification has already taken place? On the other hand, if identification has not taken place why might locally limited checking most often suffice, why certain words are merely skipped?

Note that the present point is not merely to suggest a division of labor between parafoveal and foveal vision. This division appears only as a manifestation of an important general principle whereby primary coarse-grained analyses prepare and enable subsequent fine-grained analyses. In this sense, this principle should hold for and thus be detectable in all instances of reading behavior. The process of reading is considered here as involving a series of steps defined by operations. Each operation has its own purpose and yields a result necessary for the next operation to take over and carry on the job (see Marr, 1982; Rosenthal, 1988). This sequence of steps or operations constitutes the task structure of reading. Let us characterize this view as a *task oriented approach to reading*. This approach takes it that information processing unfolds in time and in the case of complex tasks involves a certain succession of operations. Such a succession is necessary in part because not all sources of knowledge or means of analysis are available from the outset. The present task oriented approach aims to specify the means of analysis of written or printed words which gradually become available in the time-course of reading. It goes without saying that these means determine the operations which will have any chance of being productive.

Reading, as noted earlier, cannot be merely reduced to a problem of visual processing; it is a complex process which involves linguistic knowledge. Normally, when one reads a text one is actively involved in the on-line understanding of what is being read. Parafoveal processing is thus most likely to interact with or otherwise take advantage of ongoing comprehension. In some cases, the prior semantic context may somehow pretune and/or complete the parafoveal analysis of a word, thereby permitting to skip it on the next saccade. This is particularly true for speed reading wherein the amount of fixated words is considerably smaller than in normal reading (see Jacobs, 1987). Furthermore, it is clear that syntactic analyses also are implicated in the job of parafoveal vision. For instance, Carpenter and Just (1983) reported that while reading a text their subjects fixated on 83% of content words and skipped 62% of function words, especially words like: *the, of, a*.

3.5 Evidence from Letter Cancellation Studies

It is interesting to note that these findings converge with evidence arising from studies using different techniques. For instance, in experiments based on letter cancellation while reading a text, it is generally found that letters in function or closed-class words are more often missed than letters in content words (Smith and Groat, 1979). Furthermore, silent letters tend to be omitted more often than pronounced letters (Corcoran, 1966; Cochu, 1988). Since it cannot be decided whether a letter is to be pronounced prior to having a certain idea about the identity of the target word, this finding is in keeping with the suggestion that local (e.g. letter level) verification is pretuned by hypotheses provided by primary analyses. In Cochu's study there were several groups of subjects ranging from beginning readers (children after 6 month's training) to adult skilled readers. Similar results were found in all groups, letters in texts were as invisible for beginning as for proficient readers, thereby suggesting that the mechanism envisioned may emerge very early in the course of acquisition of reading skills, or otherwise be critical for reading.

3.6 Further Neuropsychological Evidence

The present position is further corroborated by evidence arising from neuropsychological investigations of certain forms of pathology of reading due to brain damage. It is often observed that brain damage selectively impairs certain cognitive functions while leaving others apparently intact. For instance, a patient may present a severe impairment of spoken language but no observable deficit in the comprehension of written language. Furthermore, even the pathology of a single function may present various combinations of deficit and preservation. The pathology of reading offers indeed a panorama of selective disruptions. Consider: in so-called *literal alexia*, a patient may be able to read a word but be unable to name any of its letters (Hécaen, 1967). Letter recognition, in the sense of enabling explicit identification, thus does not appear as a necessary condition for word recognition. In a thorough investigation of a patient who was manifestly unable to read aloud or understand any written material (words, sentences), Albert, Yamadori, Gardner, and Howes (1973) found intriguing evidence for preserved aspects of reading behavior. When presented with a list of written words (e.g. cat, dog, hat, pig, wolf) and instructed to point to the odd word without attempting to read it, the patient selected the odd word without error, although he was unable to read those words aloud when presented individually nor to describe their meanings. Moreover, although unable to read, the patient was able to match correctly written words to pictures. This apparently startling behavior makes, however, perfect sense in the light of the suggestion that perceptual processing of written or printed words starts with primary coarse-grained global analyses and is followed by fine-grained local verification. The primary analyses provide certain information about the target word. This information is too rough to achieve lexical identification but it is precise enough to trigger preliminary semantic processing of the target and to enable local verification. If pathology of reading in Albert et al.'s patient can be characterized as an interruption of this sequence of steps (though not of processing), such that local verification cannot take place or otherwise be effective, then we have here a rational explanation of the observed startling behavior. Even very rough semantic processing may suffice to point to a semantically odd word in a list. In a way, what we find in the present case are partial products of processing that normally do not manifest themselves in behavior but undergo further elaboration.

The observation of such partial products in *deep dyslexia* will better illustrate this point. Deep dyslexic patients are characteristically unable to identify letters and to read aloud pronounceable non-words. They preserve nevertheless the ability to read words, though to a variable degree: nouns, verbs and adjectives are read the best whereas conjunctions and articles are almost never read out loud. In reading aloud content words, they quite often make semantic paralexias (e.g. reading *priest* for *church*) and sometimes visual (e.g. *deep* for *deer*) or derivational (e.g. *registered* for *register*) errors (Coltheart, Patterson and Marshall, 1980). One would thus hardly find a better example of interruption taking place after the primary coarse-grained analyses wherein perceptual-morphological, syntactic and semantic aspects are all intermingled. Consider: if the primary analyses are too coarse for letter or full-word recognition, then these analyses can do nothing for non-words. If the information they provide is too rough for word identification and cannot be checked locally, then it is perfectly understandable that various errors occur (e.g. visual and derivational). And when the information is too rough, the semantic processes cannot but remain approximative, and hence yield a certain amount of semantic paralexias. And although the rough information provided by primary analyses is sufficient for the identification of function words as such, the missing information can hardly be compensated for by approximative semantic processes. Function words have no semantic fields.

3.7 Conclusion

All these statements involve conditionals. It may now be useful to confront the logic of the present position with blind machine logic. We do not intend to implement, as it were, a psychological model and see how it fares. Rather, the purpose of this confrontation is to examine whether our tentative and partial definition of the logical structure of reading tasks, be they performed by man or machine, makes a sound research program. Let us now sum up our argument so far.

We argued that typeset and impression stability of print may be unessential for reading skills. What appears to be inherent in these skills is the ability to perceive regularity in variable and visually poor patterns, to recognize a pattern as an instance of familiar gestalt in spite of its being at variance with all known instances of this gestalt. The irregularity of even good quality handwriting makes clear-cut patterns such as letters or words implausible candidates for units of initial visual analysis. The examination of psychological evidence based on studies involving almost exclusively printed words casts doubt on the adequacy of the concept of unit as a vehicle for the initial analysis of script. What rather appears to operate in the initial stage of visual text processing is a sort of coarse-grained global analysis. It does not take a word or a letter as a clear-cut pattern nor does it eventuate in word recognition. Nevertheless, it provides rough information about the shape of the target word, sufficient information to direct local fine-grained verification to a certain critical spot (e.g. a particular letter).

We also insisted that reading cannot be abstracted from the reader's knowledge of the language and noted that this knowledge actively contributes even to the very process of visual word recognition. On the one hand, syntactic and semantic knowledge allow the system to directly focus on plausible candidate-words; on the other, as soon as certain rough information about the target is provided by the preliminary process, semantic and syntactic analyses can come into play. Note that syntactic analysis does not necessarily have to be exhaustive. What may suffice at this stage is morpho-syntactic disambiguation necessary for the determination of the lexical identity of a given word and thereby of its syntactic class (see Andreewsky, Debili and Fluhr, 1980; Parisse, Andreewsky and Imadache, 1989).

4. Machine Recognition of Handwriting

Do these arguments concerning the structure of reading tasks apply to the work on machine recognition? Is the concept of basic units of analysis inadequate for this work?

4.1 Units of Analysis and Recognition

In fact, from its very beginnings in the 1960's, the work on the machine recognition of handwriting has similarly been dominated by the question of units, primarily: individual characters. But since letters are interconnected in handwriting, in order to recognize individual characters it may first be necessary to segment words into letters. The segmentation of handwritten words into letters proved to be extremely difficult: how to segment words prior to the identification of letters and how to identify individual letters embedded in a complex unsegmented pattern? This puzzle turned out to be one of the most discouraging problems in the field of pattern recognition in the late 1960's and in 1970's. Yet it never prompted a serious examination of the relevance of letters as basic units of recognition. It was direct segmentation which had been thought to be the source of misfortune, not the status of characters as individual units. Various indirect solutions which have been attempted proved relatively complex and not necessarily reliable, e.g. analysis by synthesis, structural and syntactic methods, etc... (see Eden and Halle, 1961; Lingden, 1965; Harmon, 1972; Sayre, 1973). Subsequent methods have avoided segmentation using statistic, structural or syntactic algorithms but they imply some form of presegmentation. They achieve letter recognition at the end of a long chain of preprocessing, feature extraction, reference lines' finding, etc... (see Suen, Berthod and Mori, 1980; Mantas, 1986). A recent example of this approach is the work reported in Srihari and Bozinovic (1987) who use contextual probabilities of character strings defined on the basis of topological features. Yet despite these improvements, the question of recognizing unrestricted handwriting on the basis of character identification is far from being resolved. And so far there is no off-line system whose recognition rates might be viewed as satisfactory.

On the other hand, certain studies undertaken in the early 1960s were directly oriented to the recognition of entire handwritten words, i.e. the word was treated as a single complex symbol (see Frishkopf and Harmon, 1961). However the patterns to analyse turned out to be of immense complexity and this sort of global approach impracticable, except for extremely restricted use (see Farag, 1979).

One can thus hardly conclude in this context that the concept of *a priori basic units of analysis* proved particularly productive in machine recognition.

4.2 A Task Oriented Approach to Machine Recognition

In the preceding sections we made a distinction between primary coarse-grained and subsequent fine-grained analysis. This distinction emerged from a task oriented approach to reading and we marshalled arguments as to the psychological validity of this model of the initial stages of reading. We need now to evaluate the putative validity of this suggestion in the field of machine recognition of handwriting. But in order to do so we have first to specify this suggestion so as to make it applicable to machine recognition. The primary analyses ought to be coarse enough to be unaffected by minor variations in word shape, much as human readers are. They have, on the other hand, to be precise enough to yield results which enable the use of other sources of knowledge (syntax, lexical semantics, and fine-grained morphological analysis). This primary step should also yield results compatible with the earlier observation of the little difference between reading print and good quality handwriting. Furthermore, these analyses should apply in the same way to any handwriting, whether of good or poor quality. The results may be mediocre for poor quality handwriting but they are likely to be mediocre for human readers as well.

These requirements are not easy to meet, however. Coarse-grained analyses are necessarily global: they run on the full word. We are thus confronted with the problem of complexity similar to that encountered by the studies directed at full-word recognition. One of the most difficult problems in machine recognition of non geometrical patterns is how to bridge the gap between the status of topological description and that of meaningful pattern. The idea of a primary coarse-grained analysis seems very compelling in this context. If we can get directly an approximation of the category to which the input belongs we have at our disposal hypotheses about its identity. And once we have these hypotheses we are no longer dealing with a pure topological description, we know what we have to check. But to this end we have to be able to extract certain stable global characteristics of handwritten words, without any prior knowledge about their identity. Most global methods (Karhunen-Loeve, Fourier or Hadamard transforms, curvature analysis, skeleton extraction, etc...) are blind with respect to the distinction between essential and unessential variations. They are also very sensitive to local variations of individual handwriting. And even if we succeed in finding an appropriate global approximation we are confronted with a ticklish technical problem. Once we obtain the right approximation we have to compare it with known patterns. The trouble is that the signal we are dealing with - handwriting - is bi-dimensional and there are no really effective methods of comparing bi-dimensional signals.

Two general principles emerge from these considerations. First, we have to avoid any global transformations which do not take into account the graphic specificity of the signal and which do not make a sense as a simplified description of this signal. Second, we have to obtain a unidimensional datum to enable efficient comparisons. The method we actually used will be described below. In a nutshell, it consists in following separately the major variations of the external upper and lower contour of a word. We thus obtain two separate unidimensional data characterizing the global (as contrasted to local) variations of a signal's outline.

5. Evaluation of the Hypothesis of Primary Coarse-Grained Analyses: Description of a Computational Experiment

The computational experiment we are going to report is a part of a larger project concerning the development of a system designed for the recognition of unrestricted handwriting (Parisse, Rosenthal, and Andreewsky, 1987; Parisse, Imadache, and Andreewsky, 1989). The projected system comprises preliminary coarse-grained analyses followed by local fine-grained verification and syntactic and lexical-semantic processing. The following report bears only on the work conducted for the first part of the project. It resulted in the definition of two complementary algorithms for coarse-grained analysis which broadly satisfy the above-mentioned conditions. These algorithms apply to the external contour of words. They involve two distinct stages which can be summarized as follows.

5.1 Methods

The first stage consists in extracting from a word image significant points which follow its contour from left to right and make up its outline. This provides us with an ordered unidimensional sequence of point coordinates. It is obtained in the following way. Once the image of a scanned word is isolated (as an object between two clearly distinguishable blank spaces) it is possible to seek its left and right extremities (i.e. the two pixels belonging to this object and having respectively the smallest and the largest value on the abscissa). The upper and lower contours of the word are extracted by moving from left to right, starting at the left extremity (the beginning of the word), along its external upper and lower limits on the y axis. This operation results in two separate series of points, one for the upper and one for the lower contour. If a word is not written in an entirely connected fashion (i.e. if there are blank spaces between some letters or clusters of letters) this method is applied to all segments and the resulting upper and lower contours are concatenated to represent a full word. This reduces the impact of variations in the internal connectivity of unrestricted handwriting.

In the present experiment we used two variants of this method which differed as to the precision in following the variations of contour. One variant consisted in following the trajectory from left to right (ignoring backward turns) and taking exclusively the points having the largest value on the ordinate (the smallest value for the lower contour). We shall call it the rough method. The other variant, let us call it the fine method, consisted in following all variations of the trajectory including concavities and backward turns.

The upper and lower contours provide us with means of comparing significantly related data which are meaningful for the present analysis. These means are independent of any individual handwriting and of any *a priori* idea about a word. However, before attempting a comparison a second stage of processing is necessary whereby the series of points (extracted by the two methods) standing for the upper and the lower contour are vectorized. The use of vectors instead of points allows us to compare only the major variations of contour (smoothing). The data are thus simplified and lend themselves to easy and more effective comparison. In this way significant elements (vectors) which represent segments of a word's contour are compared, not just points (see Parisse, 1989 for further details of the whole method). Figure 2 provides an example of a scanned word, the contour points extracted by the two methods and their final vectorized format.

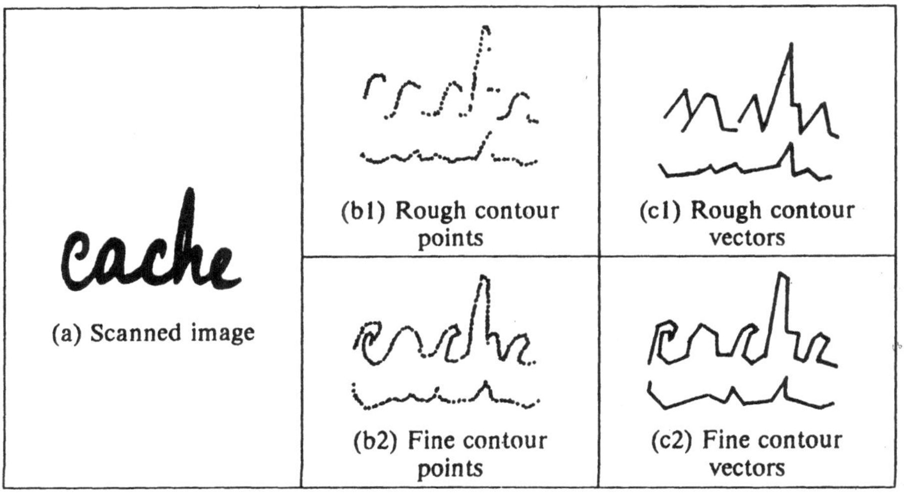

Figure 2. The scanned image of a word, its upper and lower contour points extracted by the rough and fine method, and their final vectorized form.

The comparison involves two samples of handwriting of a given writer (who writes the same text twice). One sample is the reference base. Once it is scanned, all its words are labelled manually, processed by the two (rough and fine) methods, and the series of points thereby extracted are vectorized. The other sample is the test sample. It is processed in the same way except that it is not labelled. The comparison is conducted using a dynamic time-warp algorithm. It runs between each word of the test sample (i.e. its vectorized upper and lower contour) and all words of the reference base (their vectorized upper and lower contours). This yields for each target word a list of candidate words ordered on the basis of proximity (the closest on the top). Since for each word the contours are extracted by two methods there are in fact two independent comparisons, one for the rough vectorized contour and one for the fine contour. Figure 3 provides an example of two scanned words and their vectorized contours which undergo comparison.

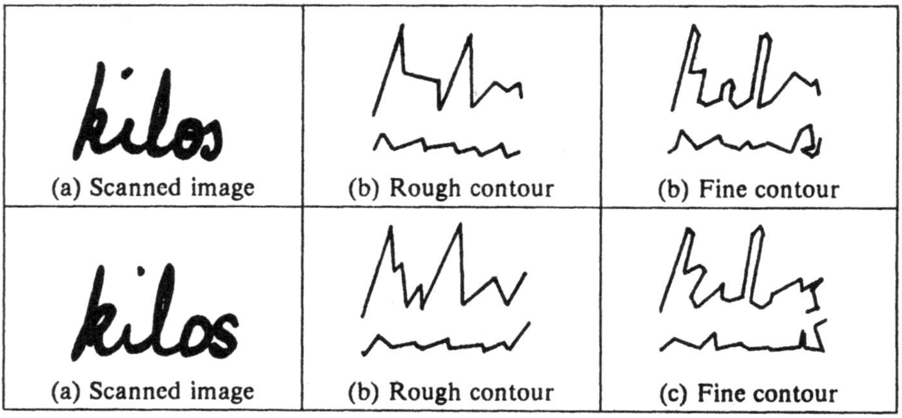

Figure 3. Two scanned words and their vectorized upper and lower contours. The comparison is conducted between the rough contours and between the fine contours.

The experiment was conducted under the extremely stringent conditions of comparing a sample of five-letter words which sometimes differ only by a single but similar letter (e.g. consider the handwritten *a e*). The use of words of the same length creates an additional difficulty since length constitutes a very distinctive source of information. Thus, in a way, the five-letter sample becomes equivalent in difficulty to a much larger sample. The experiment bore on a sample composed of 976 five-letter words selected from a 12,000 word dictionary. Two individual handwritings, one of good the other of mediocre quality, and a sample of print of a dot matrix printer were submitted to a recognition test. Specifically, each subject wrote 976 five-letter words for the reference base. The test sample consisted of 150 five-letter words written again by the subject (or re-printed by the dot matrix printer). Figures 4 to 6 present specimens of the three samples.

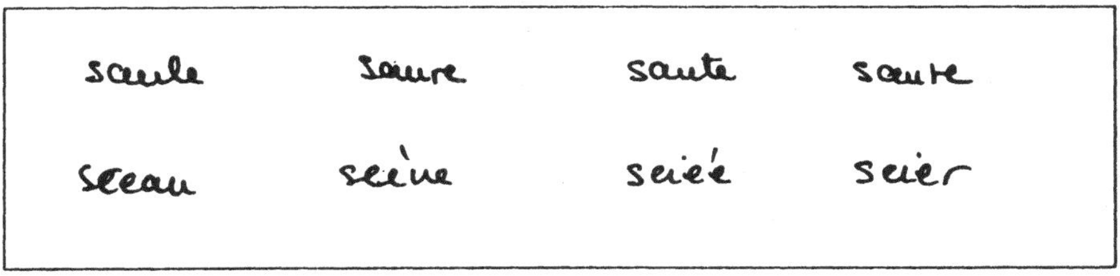

Figure 4. Sample of mediocre quality handwriting.

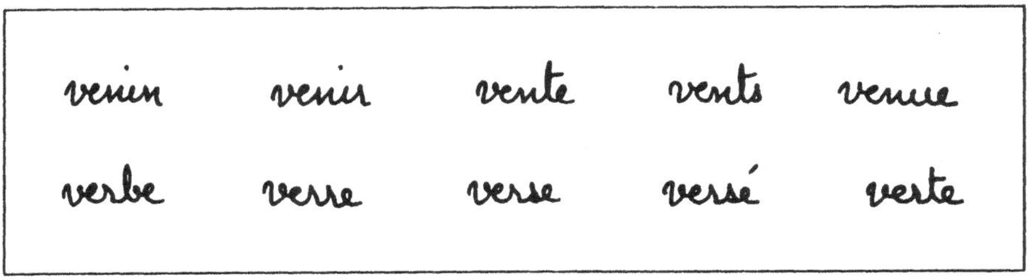

Figure 5. Sample of good quality handwriting.

véler	velue	vende	vendu	venin	venir
vente	vents	venue	venus	verbe	verre

Figure 6. Sample of dot matrix print.

5.2 Results and Discussion

Figure 7 summarizes the results in percentages of correct words occurring on the top of the list of candidates and among the first five candidates.

These results of the primary coarse-grained global analyses appear encouraging as they are sufficient for local fine-grained morphological verification and linguistic (syntactic and lexical-semantic) analyses to take over the job. Note that Figure 7 does not go beyond the five top choices. If we take the list of 10 or 15 top candidates there are still better chances to have the correct word in the list. And of course the subsequent local and linguistic analyses can run on any number of candidates, all the more so since in case of a normal text the list of candidates is easily reduced by the elimination of syntactically implausible candidates.

	Rough contour		Fine contour	
	First choice	Five top choices	First choice	Five top choices
Mediocre quality handwriting	41.1	70.8	43.2	66.6
Good quality handwriting	62.1	82.8	49.5	70.2
Dot matrix printer	76.7	99.4	71.2	94.9

Figure 7. Percentage of correct recognition of 150 words.

What is particularly interesting in light of the earlier discussion is the finding that the difference of recognition rates for the three samples is not overwhelming (although it is considerable). The system does not collapse when presented with mediocre quality handwriting, there is an obvious gradation of results, and, although the dot matrix print is best recognized, the primary coarse-grained analyses are insufficient for 100% recognition of print. In other words, the present results are consistent with the previously spelled out definition of the purpose of primary coarse-grained analyses.

Another result which deserves attention is the finding that the two methods yield quantitatively similar results (except for good quality handwriting where the rough method is clearly more effective). However, a careful qualitative analysis of the results reveals that the two methods are in fact complementary: items which are poorly recognized by one method are often better recognized by the other and vice versa.

It is finally worth mentioning that the present work tackles the basic problem of off-line text recognition, whatever its graphical quality. We tested the method on very different samples (including poor quality handwriting) and gathered results comparable to those described by other authors using specific methods for each type of script. In this sense our approach differs significantly from the general trend of dealing separately with each type of script.

6. General Conclusion

We marshalled throughout this chapter arguments for a task oriented approach to reading and machine recognition of handwriting. This approach led us to suggest, mainly on the basis of psychological evidence, a model of the initial stages of reading which we applied to the problem of machine recognition of unrestricted handwriting. The model specifies a sequence of operations defined by the means of analyses which gradually become available in the time-course of reading. They consist, *inter alia*, of primary coarse-grained global analyses followed by local fine-grained verification. We believe that this principle of gradual approximation is more appropriate to deal with the problem of the inherent instability of handwriting than the classical unit-based approach.

The goal of coarse-grained global analyses is not to identify the input as a specific entry of a dictionary, but rather to provide enough information to generate hypotheses about its identity. As a result, the input becomes a meaningful category that lends itself to finer analyses on the basis of lexical, syntactic and semantic knowledge which is specific to this category. At the morphological level, these analyses involve local fine-grained verification. The nature of the verification depends on the types of hypotheses generated by the primary analyses. In this sense local verification is neither defined a priori nor need it be exhaustive. As a result of the primary approximation only certain elements are actually critical: identification can be achieved by selectively checking these elements. In this manner the complexity of the whole process is considerably reduced, and the diversity of potential verifications makes the result more reliable.

We implemented a method based on the hypothesis of primary coarse-grained analyses and conducted a computational experiment to evaluate its capacity to deal with unrestricted handwriting. The results appear encouraging as they eventuated in data suitable for local verification. In that we hope to have demonstrated not only the viability of this hypothesis but also the interest of considering both the reader's and machine's side of the story.

Acknowledgements

We thank the editors and two anonymous reviewers for helpful suggestions and comments on an earlier version of the manuscript. We are indebted to Ralph Eisenberg for help in preparation of the English text.

References

Albert, M. L., Yamadori, A., Gardner H., and Howes, D., "Comprehension in alexia", *Brain* **96**, 1973 pp. 317-328.

Andreewsky, A., Debili, F., and Fluhr, C., "Apprentissage, syntaxe, sémantique lexicale", *Revue du Palais de la Découverte*, **83**, 1980 pp 17-40.

Balota, D. A., Pollatsek, A., and Rayner, K., "The interaction of contextual constraints and parafoveal visual information in reading", *Cognitive Psychology* **17**, 1985 pp. 364-390.

Carpenter, P. A., and Just, M., "What your eyes do while your mind is reading", in *Eye movements in reading: Perceptual and language processes*, K. Rayner (Ed.), New York: Academic Press, 1983, pp. 275-307.

Cattell, J. M., "The time it takes to see and name objects", *Mind* **11**, 1886 pp. 63-65.

Cochu, F., "*Différence de status des mots pleins et fonctionnels dans la lecture et son apprentissage. Aspects normaux et pathologiques.*" Unpublished DEA dissertation. Université Paris V, 1988.

Coltheart, M. (Ed.), *Attention and Performance XII*: The Psychology of Reading, Hillsdale, N. J., Lawrence Erlbaum, 1987, 707 pp.

Coltheart, M., Patterson, K., and Marshall, J. C. (Eds.), *Deep dyslexia*, London: Routledge & Kegan Paul, 1980, 490 pp.

Corcoran, D. W. J., "An acoustic factor in letter cancellation", *Nature* **210**, 1966 p. 658.

Eden, M., and Halle, M., "Characterization of cursive writing", *Proc. 4th London Symposium on Information Theory*, C. Cherry, (Ed.), London: Butterworths, 1961, pp. 287-289.

Frishkopf, L. S., and Harmon, L. D., "Machine reading of cursive script", *Proc. 4th London Symposium on Information Theory*, 1961, pp. 300-316.

Greenberg, S. H. and Vellutino, F. R., "Evidence for processing of constituent single- and multiletter codes: support for multilevel coding in word perception", *Memory and Cognition* **16-1**, 1988 pp. 54-63.

Haber, R. N. and Schindler, R. M., "Error in proofreading: evidence of syntactic control of letter processing", *Journal of Experimental Psychology: Human Perception and Performance* **7**, 1981 pp. 573-579.

Harmon, L. D., "Automatic recognition of print and script", *IEEE Proc.* **60**, 1972 pp. 1165-1176.

Healy, A. F., "Detection errors in the word the: evidence for reading units larger than letters", *Journal of Experimental Psychology: Human Perception and Performance* **2**, 1976 pp. 235-242.

Healy, A. F., "Proofreading errors on the word the: new evidence on reading units", *Journal of Experimental Psychology: Human Perception and Performance* **6**, 1980 pp. 45- 57.

Hécaen, H., "Aspect des troubles de la lecture (Alexies) au cours des lésions cérébrales en foyer", *Word* **23**, 1967 pp. 265-287.

Huey, E. B., *The psychology and pedagogy of reading*, New York, Macmillan, 1908.

Jacobs, A. M., "On localization and saccade programming", *Vision Research* **27**, 1987 pp. 1953-1966.

Javal, E., "Essai sur la physiologie de la lecture", *Annales d'Oculistique* **82**, 1879 pp. 242-253.

Lévy-Schoen, A., and O'Regan, J. K., "Le regard et la lecture", *La Recherche* **211**, 1989 pp. 744-753.

Lingden, N., "Machine recognition of human language", *IEEE Spectrum* **III**, 1965 pp. 104-116.

Loftus, G. R., "Eye fixations on text and scenes", in *Eye movements in reading*, K. Rayner (Ed.), New York: Academic Press, 1983, pp. 359-376.

Mantas, J., "An overview of character recognition methodologies", *Pattern Recognition* **19-6**, 1986 pp. 425-430.

Marr, D., *Vision*, New York: W. H. Freeman, 1982, 397 pp.

McClelland, J. L., and Rumelhart, D. E., "An interactive activation model of context effects in letter perception: Part 1. An account of basic findings", *Psychological Review* **88**, 1981 pp. 375-407.

Morton, J., "Interaction of information in word recognition", *Psychological Review* **76**, 1969 pp. 165-178.

Parisse, C., *Reconnaissance de l'écriture manuscrite: analyse de la forme globale des mots et utilisation de la morpho-syntaxe*, Unpublished doctoral dissertation, Université Paris-Sud, Orsay, 1989, 170 pp.

Parisse, C., Andreewsky, A., and Imadache, A., "Un système d'apprentissage de relations lexicales-sémantiques pour la reconnaissance de l'écriture manuscrite", *Interfaces* No **86**, 1989 pp. 9-12.

Parisse, C., Imadache, M., and Andreewsky, E., "MEPHISTO: système de reconnaissance de l'écriture manuscrite", *Actes du 7ème Congrès AFCET-RFIA*, Vol. 2, Paris, 1989, pp. 955-974.

Parisse, C., Rosenthal, V., and Andreewsky, E., "An approach to machine recognition of handwritten texts", *Proc. 7th Int. Congress Cybernetics and Systems*, Vol 1, 1987, pp. 185-188.

Patterson, K. E., and Morton, J., "From orthography to phonology: an attempt at an old interpretation", in *Surface dyslexia*, K. E. Patterson, J. C. Marshall and M.C. Coltheart, (Eds.), London: Lawrence Erlbaum, 1985, pp. 335-359.

Prinzmetal, W., Treiman, R., and Rho, S. H., "How to see a reading unit", *Journal of Memory and Language* **25**, 1986 pp. 461-475.

Rayner, K., and Pollatsek, A., "Eye movements in reading. A tutorial review", in *The psychology of reading*, M. Colheart, (Ed.), London: Lawrence Erlbaum, 1987, pp. 327-362.

Rosenthal, V., "Does it rattle when you shake it? Modularity of mind and the epistemology of cognitive research", in *Perspectives on cognitive neuropsychology*, G. Denes, C. Semenza and P. Bisiacchi, (Eds.), London: Lawrence Erlbaum, 1988, pp. 31-58.

Sayre, K., "Machine Recognition of Handwritten Words: A Project Report", *Pattern Recognition* **5**, 1973 pp. 213-228.

Smith, P. T., and Groat, A., "Spelling patterns, letter cancellation and the processing of text", in *Processing of visible language*, P. A. Kolers, M. E. Wrolstad and H. Bouma, (Eds.), Vol. 1, New York: Plenum Press, 1979, pp. 309-324.

Srihari, S. N., and Bozinovic, R. M., "A multi-level perception approach to reading cursive script", *Artificial Intelligence* **33**, 1987 pp. 217-255.

Suen, Y. C., Berthod, M., and Mori, S., "Automatic Recognition of Handprinted Characters - The State of the Art", *IEEE Proc.* **68-4**, 1980 pp.469-487.

Computer Processing of Handwriting
Eds. R. Plamondon & C. G. Leedham

AUTOMATIC PERCEPTION OF THE STRUCTURE OF HANDWRITTEN MATHEMATICAL EXPRESSIONS

CLAUDIE FAURE and ZI XIONG WANG
ENST - SIG, 46 rue Barrault, 75634 Paris, cédex 13, France

A modular system for segmenting handwritten mathematical expressions is presented. An attributed relational tree represents the hierarchical structure of the expression. The system is a set of specialist modules which perform specific tasks using data-driven or knowledge-driven strategies. The modules are defined according to the properties of graphic languages: they can not be interpreted without a specific knowledge (in this case the symbol lexicon and the mathematical syntax); they are generated in a form that helps the visual perception of meaningful components or relations without prior identification of the components (in this case the collinear arrangements and the subscript and superscript positions).

1. Introduction

The specific material that constitutes mathematical expressions, leads to the splitting of the problem of their interpretation into three sub-problems: handwriting recognition, two-dimensional structure recognition and syntactic analysis. Mathematical expressions differ from classical textual data in the "dimension", the vocabulary and the syntax. Textual data are written and read by scanning the writing line, from left to right for example. This implies a level at which the writing has a uni-dimensional structure. This is no longer true for most mathematical expressions: there is a main writing line and also several secondary writing lines. When writing and reading an expression, these lines may be visited once or several times; the expression is not linearly scanned. The mathematical vocabulary constitutes a lexicon containing symbols of different alphabets (Latin, Greek...) and words (cos, tan, min ...). The syntax is well known and follows formal rules. Unlike in compilers where the parser processes a sequence of symbols, the syntax analysis of handwritten mathematical expressions is performed on two-di mensional patterns. The present system is mainly aimed at segmenting handwritten mathematical expressions in order to provide their structural descriptions. The relationships between the structural description and the knowledge related to the lexicon and the syntax will be pointed out.

The well known works on mathematical expressions are now over 10 years old: Anderson (1977); Chang (1970); Martin (1971). These authors essentially focus on the syntax analysis of printed or well-formed handwritten expressions. This kind of material facilitates the recognition of the lexical symbols and their spatial relationships; therefore pattern recognition becomes a problem of minor interest to the authors. More recently, Belaid (1984) presents a system for handwritten mathematical expression recognition. It integrates pattern recognition and syntax analysis. The expression is segmented into symbols that are further recognized. A model-driven process is activated in order to interpret the symbols or groups of symbols as mathematical items. This system deals with restricted lexicon and syntax. However, it shows that lexical knowledge and syntactic knowledge are complementary: if a stroke is recognized as a parenthesis or a "1" at the lexical level, then it may be possible to eliminate one candidate by checking the syntax.

The interpretation of mathematical expressions is the result of a complex process involving the treatment of different kinds of information. The whole problem-solving process can be divided into several sub-problem-solving steps. The relevant information is selected at each step of the process. The simulation of a natural system by an artificial one requires a knowledge of which parts of the information are processed, and how they are processed, in order to solve a given sub-problem. In our approach, the structural description of an expression is a sub-problem of its interpretation. It will be solved prior to the recognition of all the constituent symbols of the expression. This approach is based on knowledge of visual perception skills and on the verbal data obtained by reading a mathematical expression aloud. Pattern perception is illustrated with a well known example: a digit 2 is composed of concatened small 2's. In this example, the digit 2 is identified prior to the recognition of the small 2's. Such examples have been used by psychologists to conclude that a pattern is perceived prior to the recognition of its components, see Ullman (1986). When reading a mathematical expression aloud, the verbalisation order of the symbols depends on the syntax and does not require the previous scanning of all of the constituent symbols.

This leads us to suppose that the expression is segmented into mathematical items in the course of the reading process before all of the constituent symbols are recognized by attentional processes. For example, the reading of a fraction starts with the verbalisation of the numerator symbols despite the fact that some of them may not be recognized and therefore stop the reading process; the numerator is recognized as an aggregate of graphical units and not as an aggregate of lexical symbols. Therefore, the system we present will test the statement that local structures (symbol shapes) are not always needed to detect global structures (mathematical items). In addition, it will be shown that some symbols must be recognized to ensure a mathematically or lexically valid segmentation of the expression.

This paper will focus on the system architecture and the information treatment process without providing a full description of the programs. More details can be found in Wang and Faure (1988); Wang (1989). The system is written in Pascal, and runs on a VAX 3600.

2. Structure of Mathematical Expressions

A structure is usually defined as a set of components and relationships. Morphological criteria are added in order to specify which of all the possible segmentations into components are the meaningful ones. The morphological criteria are mapped onto a set of procedures that processes the input data in order to provide their structural description. These procedures may be automatic processes without cognitive control, as are those which enable us to aggregate items that satisfy properties such as proximity, similarity or collinearity. Other procedures are driven by a knowledge related to the domain of the processed data. These two types of procedures are involved in the system for segmenting the mathematical expressions.

In our application, the structure's components must be mathematical subexpressions or lexical symbols. For example, if a fraction is a component of the expression, it can be segmented into the fraction bar, the numerator and the denominator, which can in turn be segmented into symbols or groups of symbols. This example shows the importance of hierarchical structural description. At the lowest level of the hierarchy, the structure's components are the lexical symbols. At the higher levels, the mathematical sub-expressions are represented (fraction, numerator, denominator ...).

Mathematical expressions are not natural patterns: they are constituted of arbitrarily defined symbols and follow the rules of a formal language. These patterns are drawn by a writer to communicate a message to the reader. The arbitrary choices from which the formal rules proceed, are constrained in order to generate efficient material for communication: the rules should allow good readability and non-ambiguous interpretation of the expression. Within textual data, the spaces between characters are chosen to aid the perceptual grouping from which the words emerge. In the same way, the organisation of the symbols in mathematical expressions is defined by formal rules that assist the visual segmentation into a hierarchy of mathematical items.

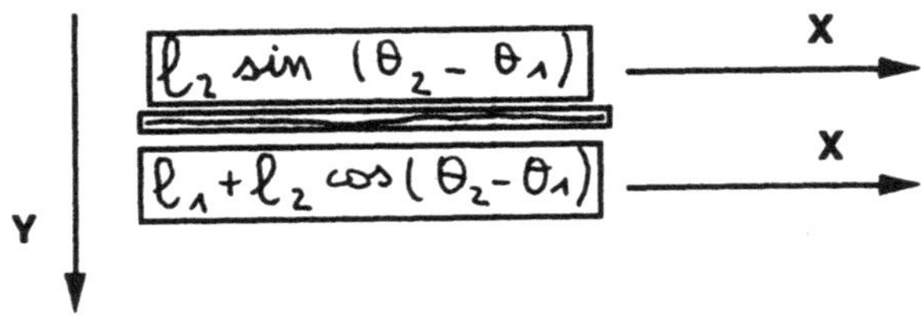

Figure 1. Collinear arrangements along the X and Y axes.

The main property that leads to the detection of symbol aggregates in mathematical expressions is the collinearity along the vertical and horizontal axes. As shown in Figure 1, a fraction is a collinear arrangement along the vertical whereas a writing line, such as the numerator or the denominator, is a collinear arrangement along the horizontal. Collinearity is a morphological criterion that will be used by the system for the segmentation of the mathematical expressions. Nevertheless, this criterion is not sufficient to perform a complete and valid segmentation of any expression - symbol recognition may be required. The situations in which a symbol recognition is performed by the system will be described later in the section presenting the segmentation process (Section 4).

The automatic segmentation is performed from the expression to the lexical symbols (from global to local). The hierarchical structural description is represented by an attributed relational tree that must be as close as possible to a goal tree. The nodes of the goal tree represent the components of the structure. The root corresponds to the expression itself and the leaves correspond to the lexical symbols. Intermediate nodes correspond to mathematical sub-expressions having varying degrees of structure. The tree has two types of branches. Relational branches are labelled with the spatial relationship between two brother nodes. Segmentation branches that link a father node to its first son, are labelled with the name of the procedure that segments the father node. An example is given in Figure 2. The vertical branches are segmentation branches, they are labelled XX for projection onto the X axis, YY for projection onto the Y axis and MY for Y-mask removal. The horizontal branches are relational branches, the label A is for "Above" and the label L is for "on the Left of". The relational branches linking the nodes that belong to the same writing line (collinear arrangement of symbols and/or groups of symbols along the horizontal) receive another label which specifies the position of the node in the writing line (subscript, superscript, main line). This step of treatment is described in Section 5. This labelling does not involved symbol recognition.

3. The System

The system processes mathematical expressions in order to solve the problem of their structural description. The data are handwritten on a digitizing tablet. The writer is asked to write in the horizontal direction; this request is not strictly respected by the writer. The system is designed to work on freely written material: the data do not fit the formal rules of mathematical expressions perfectly and the direction of the writing lines is not strictly horizontal.

A schema of the system is given in Figure 3. The system contains the acquisition module and a set of specialist modules: two segmentation modules and a line labelling module. Each of them solves a sub-problem, and each has its own goal, strategy, procedures and knowledge-base. The specialist modules operate independently, and they communicate through a shared memory. These modules have read/write access to the shared memory. This memory contains the initial data file and the representation tree which is updated throughout the process. At each step of the treatment, the configuration of the representation tree reports the progress of the problem solving process.

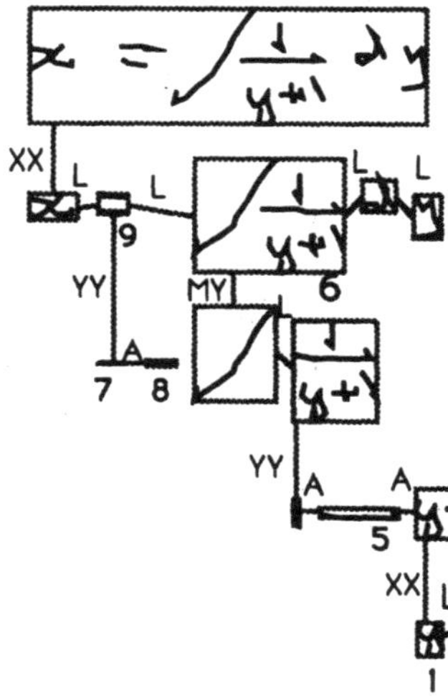

Figure 2. An example of attributed relational tree. A=Above; L=on the Left of; XX=projection onto the X axis; YY=projection onto the Y axis; MY=Y-mask removal.

The first segmentation module builds the relational tree by activating a set of procedures. This module does not take into account the writing order, and therefore it could work on off-line data such as image skeletons. The second module corrects the tree resulting from the first segmentation step. It uses different kinds of knowledge, including the writing order, available in the input data file. The third module labels the spatial relationships between the components of a same writing line. This labelling makes it possible to detect the principal writing line and the secondary writing lines that correspond to subscripts and superscripts.

The system is knowledge-based. The categories of knowledge involved in the system are:

— *mathematical knowledge*, that is the syntax of the two-dimensional expressions,

— *lexical knowledge*, related to the shapes of the lexical symbols,

— *knowledge about the writing generation process*, such as the stroke order,

— *procedural knowledge*, for example the procedures that segment the data according to morphological criteria,

— *empirical knowledge*, that the designer possesses about the system.

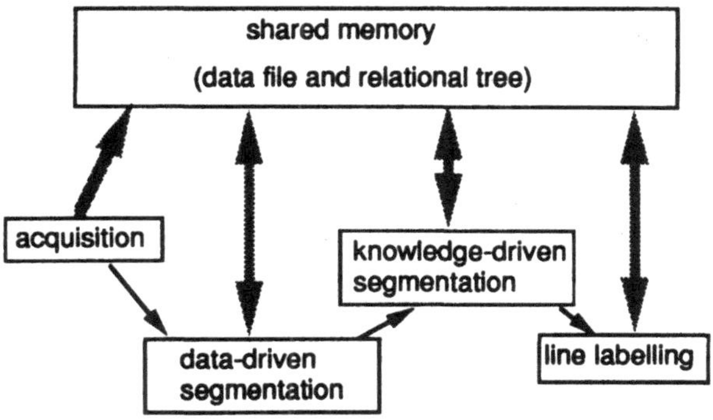

Figure 3. General architecture of the system.

4. The Two Segmentation Modules

These modules differ in the problems they solve, the knowledge they possess and the strategy they use.

4.1 First Module

The first module is aimed at building the relational tree during a segmentation process that activates a set of operators. Each operator performs a segmentation into components according to a criterion. The components and their relationships result from the application of the operator on the raw data and not the reverse: a criterion is verified on pre-segmented components. When a segmentation succeeds, the results are recorded in the relational tree. In the tree nodes, the components are represented by blocks; a block is the smallest rectangle containing the component strokes. The way in which the segmentation has been obtained is recorded by labelling the branches.

The segmentation process is data-driven. At each step of the treatment, a subset of the expression is selected as a focus of attention. This subset constitutes the data to which the operators are applied in order to perform their segmentation into several components. The components are in turn selected as foci of attention. The process is recursively activated on the data until a halting situation is reached for all of the leaves of the relational tree. The halting situation is defined by a data-dependant property: the strokes contained in a tree leaf can not be separated by the collinear arrangement detecting operators, and the number of strokes is at most n (the experimental value of n is 3). The information available to an operator is restricted to the data (a subset of the expression) on which it is applied. No contextual information (the whole expression) is available to the procedures of the first segmentation module. The operators that detect collinear arrangements have priority in the segmentation process. They project the raw data onto the vertical and the horizontal axes. When a collinear arrangement is detected, then the data is segmented into collinear components. Projections onto X and Y are activated alternatively. As an example, the case of a writing line is presented (see Figure 2). A writing line is composed of collinear symbols, or groups of symbols, along the X axis. The projection (onto X) does not find the position of the line, but merely answers the question of its existence, and finds the line components. The topological relationship of the writing line with the context is obtained directly from the process that builds the relational tree, and is available in the labelled branches. In Figure 2, the line "y+1" (nodes 1, 2 and 3), with three component symbols (leaf 1 "on the Left of" leaf 2 "on the Left of" leaf 3), results from the projection onto X of the group of symbols "y+1" (node 4) as indicated by the label XX on the segmentation branch 4-1. The relational branch 5-4 indicates that the fraction bar is above "y+1".

During the segmentation process, some situations arise in which the projection operators can no longer segment the data. Indeed, projection is not efficient when a mask hides the spaces between symbols or groups of symbols. This happens in the case of a square root symbol which hides the symbols beneath it; projections onto the X and Y directions fail to segment such a sub-expression. The separation can only be made by performing a square root recognition. The mask effect happens in other cases, most often because the symbols are written very close to each other and/or are not written in a strictly horizontal direction. For example, in Figure 2, the fraction bar and the integral symbol prevent the segmentation by projection of the block denoted 6. A set of procedures grouped under the name "mask removal operator" makes it possible to resolve these situations by splitting the mask and the remaining strokes. The mask removal is performed until a projection becomes efficient or a halting situation is reached. Mask removal implies mask detection within the block. Most of the mask situations arise from two symbols: the square root and the fraction bar. The recognition of these two lexical symbols is attempted first. When neither of these lexical symbols are recognized then the mask is detected using numerical features: the ratio between the stroke size along the X (Y) direction and the horizontal (vertical) block edge size. The mask is the stroke with the greatest relative size along one direction.

4.2 Second Module

The second module is aimed at correcting the output relational tree of the first segmentation module. The errors that occur in the tree are mainly related to specific patterns. For example, "i", "j", "=", are split into two collinear components by projection onto the Y axis when they belong to the same lexical symbol. The second segmentation module is knowledge-driven. It looks for a set of specific patterns that constitute an a priori knowledge-base, and updates the relational tree by pattern-dependant corrections. The process continues until no further corrections are possible (there are cases where no corrections are made). The set of specific patterns for which a correction is defined can be increased as the a priori knowledge is enlarged.The knowledge base of this module contains knowledge about a subset of the lexicon, the syntactic rules related to these symbols, the writing order (for example: the point is written after the body in "i" and "j"), and the symbol shapes (for example: the strokes of "E", "F" and "Σ" may not be connected). This module is structured as a hierarchy of *frames*. Each node *frame* represents a specific pattern. The related knowledge is contained in the slots of information associated with it. Figure 4 shows the *frame* hierarchy. The root *frame* contains information about the expression, such as the average symbol size and the average length of the inter-symbol intervals. In this tree, the nodes inherit the properties of their ancestors. For example, the [≡] *frame* inherits the description (size, location) of the two-strokes pattern that instantiates the [=] *frame*. The hierarchy of *frames* is scanned using a depth-first strategy.

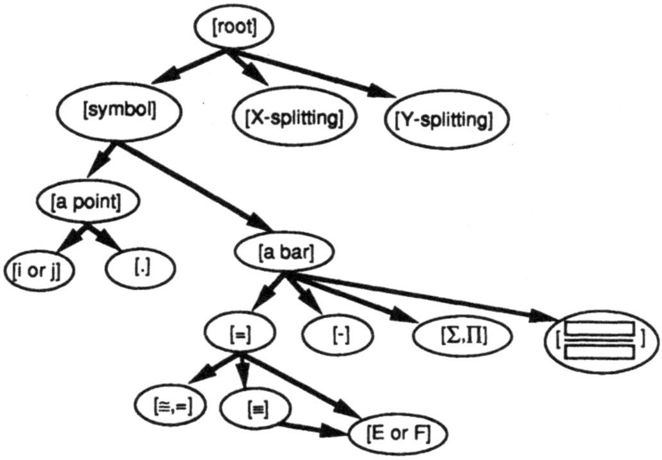

Figure 4. The hierarchy of *frames*.

Attempts are made to match the a priori given patterns with the data using a top-down strategy. When a *frame* is visited, then the represented pattern is hypothesized. The hypothesis is confirmed after the *frame* instantiation that occurs when the DESCRIPTION slot is filled in. This slot contains a prototypical description given as the conditions which have to be fulfilled in order for the hypothesized pattern to be recognized. The prototype does not allow the recognition of all the members of the pattern class, but it avoids false recognitions that could easily occur in a decision process where there is no competition between several classes (only the hypothesized pattern class is considered). The prototype is described by information about the writing order, the shape and size of the blocks and the relative position of blocks. For example, the DESCRIPTION slot of the [Σ or Π] *frame* contains the following conditions:

1. block i contains a horizontal stroke: stroke n (information inherited from the father *frame*: [a bar] *frame*)

2. there exists block j such that: (block i) above (block j)

3. block j contains one or several strokes following the stroke n in the writing order

4. block j has an almost square shape

....

The instantiation may require new information that is not directly inherited from the *frame* ancestors nor directly available within the data base (the relational tree and the raw input data file). The new information is then obtained by activating procedures which process the data. After the hypothesis validation step, the ACTION slot may call another *frame*, or procedures, depending on the result of the validation (the hypothesis may be confirmed or not). The link between *frames* and procedures is realized through IF-THEN structured production rules embedded in the slots. The right hand side of a production rule may be another production rule, a procedure or a *frame*. The procedures are clustered into TEST and ACTION sets. The TEST procedures look for the information needed to instantiate a *frame*. The ACTION procedures update the relational tree - they are the ones that perform the corrections.

The correction itself implies two treatment steps - one that updates the relational tree and one that identifies which nodes and branches must be updated. The nodes involved in the correction and their locations in the tree are not known a priori by the system. These nodes are identified during the correction process. They are organised into a configuration in the tree. The set of possible configurations is too large (and difficult to know a priori) to design a configuration-dependant correction. A complex and general procedure is programmed as a set of tests enabling adaptation of the correction to the node configuration. The second segmentation module works as an attentional process that recognizes a given pattern located in the relational tree, and then propagates the information related to this pattern in its context. Figures 5b and 8b illustrate the relational trees obtained after the segmentation of the expressions, shown in Figures 5a and 8a. In order to improve the readability, the relational trees are displayed in a printed form (Figures 5b, 7, 8b, 9).

4.3 Discussion

The automatic segmentation of mathematical expressions may require symbol recognition, such as during the mask removal operation (first segmentation step) or when the son-*frames* of the [symbol] *frame* are visited (second segmentation step). This recognition is always performed with the use of knowledge related to the recognized symbol. This remark emphasizes the fact that the symbols are always recognized in order to obtain structural information on the expression. We will illustrate with some examples how the symbol related structural information is used for segmentation.

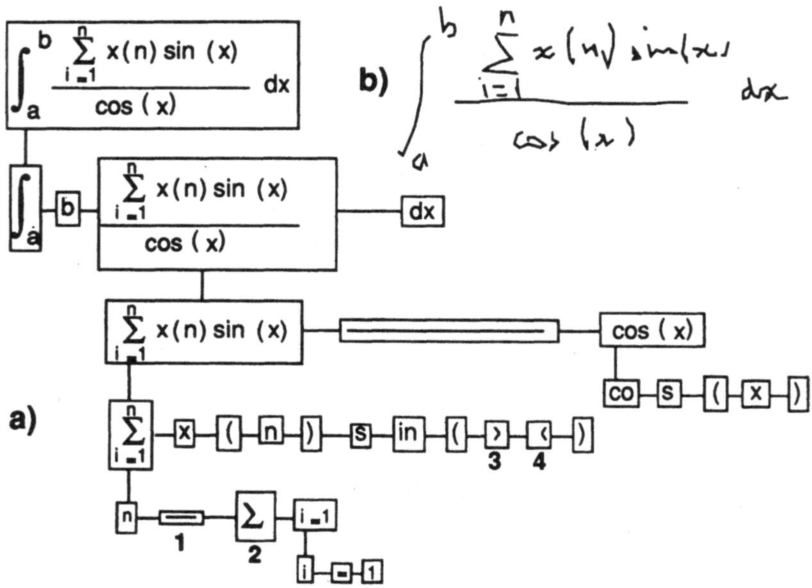

Figure 5. a) Handwritten mathematical expression. b) Relational tree after segmentation of 5a.

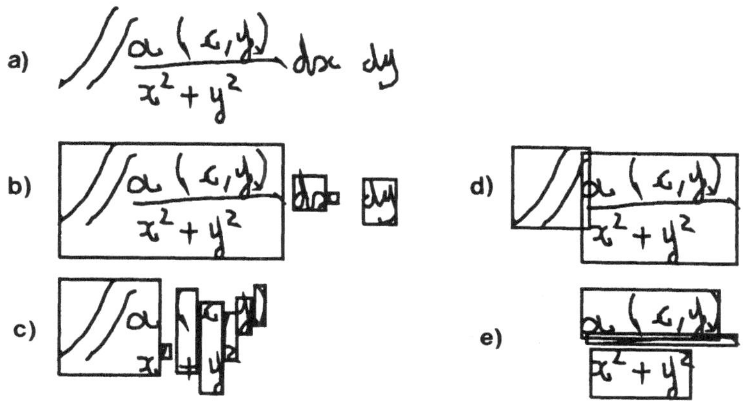

Figure 6. a) Handwritten mathematical expression. b) Segmentation by projection onto X. c) Invalid segmentation. d) Valid segmentation. e) Fraction segmentation.

We start with the mask removal case. When numerical features are used to detect the mask, then the mask removal operator splits the mask and the remaining strokes into two blocks. In the case where the mask is identified as a symbol, then the mask removal may become more complex. Invalid segmentations, such as the one illustrated in Figure 6, must be avoided. The fraction bar is identified as a mask for the block denoted 1 in Figure 6b. If the fraction bar is eliminated from this block, then the remaining strokes are segmented into components by projection onto X (see Figure 6c). The resulting components contain numerator and denominator symbols. They are linked by relational branches labelled "on the Left of". This set of components is then interpreted as a single writing line. Therefore, this segmentation is not a valid one. Such errors can be avoided by linking the mask identification to test procedures. Identification of the mask as a fraction bar activates procedures for detecting the strokes that do not belong to the fraction. In this way, these strokes are separated from the fraction block. In this example, the integral symbols are grouped into one block and the fraction into another (see Figure 6d). The segmentation of the fraction is shown in Figure 6e.

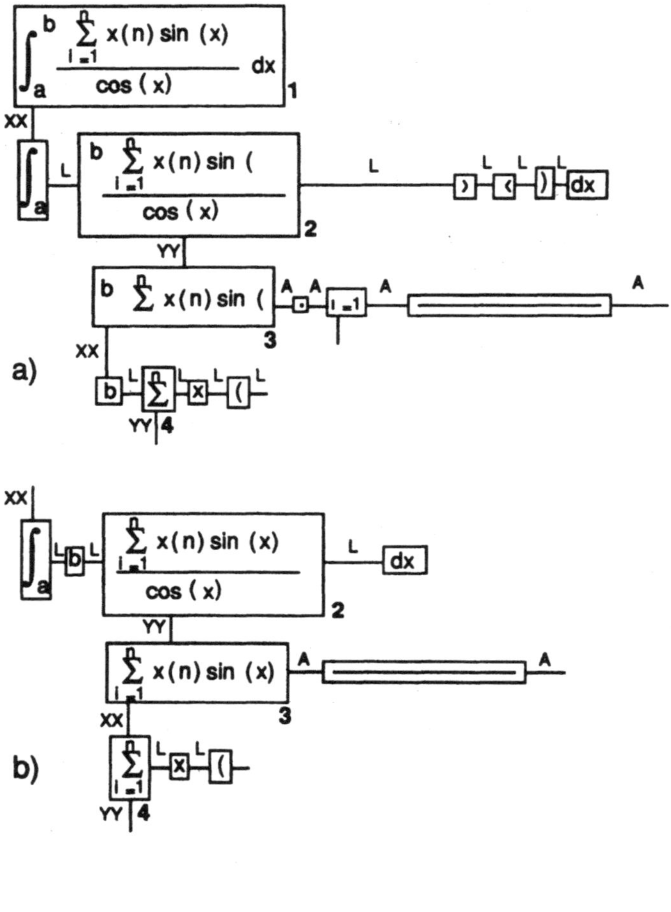

Figure 7. A part of the relational tree of 5a expression. a) After the first segmentation step. b) After the second segmentation step.

Another example is illustrated by sub-trees that are obtained after the first segmentation step (Figure 7a) and after the second segmentation step

(Figure 7b) The corresponding expression is given in Figure 5a. The fraction bar is not recognized during the first segmentation step because it does not produce a mask effect; the segmentation is continued using the projection operators. This example illustrates the fact that symbol recognition is only made on request within the first segmentation step, that is, when mask-removal becomes necessary in order for the segmentation to continue. Conversely, the second segmentation module systematically searches for fraction bars in order to correct possible errors. Examples of errors are given in Figure 7a. The fraction bar is shorter than the numerator. The projection of block 1 onto the X axis separates "x)" from the main fraction block (block 2), when "b" is included in it. The projection of block 2 onto the Y axis separates "i=1" from the numerator (block 3). During the second segmentation step, these errors are corrected. Procedures are activated by the [fraction] *frame* to detect the numerator and the denominator, by grouping the symbols, or groups of symbols, that belong to the same writing line. In this example, "i=1" and "x)" will be included in the fraction block denoted 2 in Figure 7b, and "b" will be excluded from it. Consequently, the symbols of

$$"\sum_{i=1}^{n}"$$

are grouped into a single block (denoted 4 in Figure 7b). The correction made on a given symbol or group of symbols may occur at different steps of the process, and may be achieved by different data processing operations.

Another example is illustrated in Figures 2 and 9; it corresponds to the instantiation of the [=] *frame* during the second segmentation step. Figure 2 shows the relational tree after the first segmentation step. A symbol "=" is recognized: it is composed of two strokes contained in the blocks denoted by 7 and 8. The tree is updated by deleting the leaves 7 and 8; therefore the block containing "=" becomes a tree leaf (block 9). In Figure 9a, the symbol "=" is recognized. As before, the two leaves (denoted 1 and 2) that contain the symbol components are deleted, and block 3 becomes a leaf. Moreover, the recognition of "=" activates a local parser, which detects an error: there is no symbol on the right of block 3 ("=") on the writing line ("j="). The search is then made for the right hand part by scanning the tree; it is found by climbing the tree to block 6 containing the symbol "1". The tree-updating is then performed on the sub-tree that is rooted in block 7. The root of the sub-tree is the father node of the level where a block containing the symbol on which an error is detected ("1" in this example) appears for the first time. Figure 9b shows the result of the updating. Block 3 is first deleted and then inserted on the right of "=", in the same writing line. "1" is included in blocks 5 and 4 which represent mathematical items,

$$"\sum_{j=1}^{n}" \text{ and } "j=1" \text{ respectively .}$$

We may remark that in Figure 9a, the left part of "=" is the body of "j"; the two components of "j" (the point and the body) are grouped together into a single block at another step of the correction process: when the [i or j] *frame* was instantiated. The left part of "=" becomes the symbol "j".

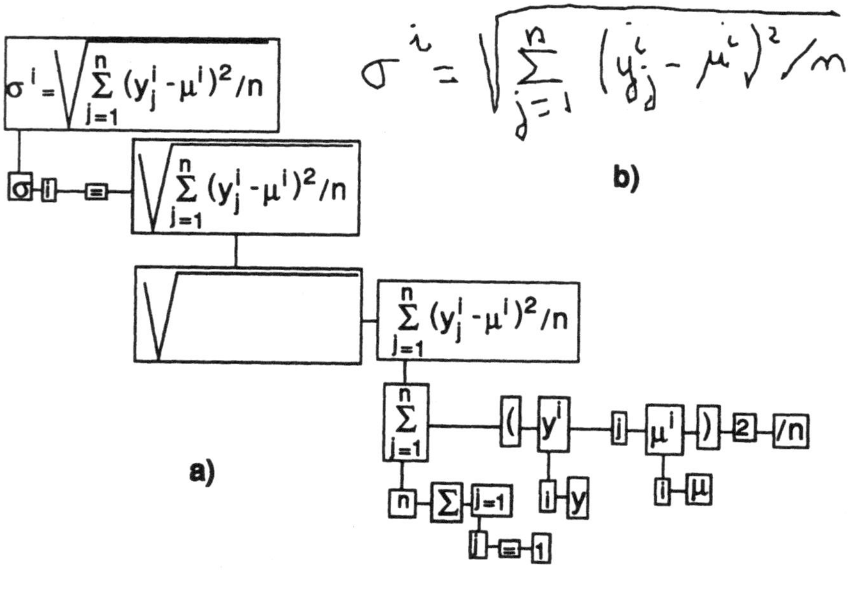

Figure 8. a) Handwritten mathematical expression. b) Relational tree after segmentation of 8a.

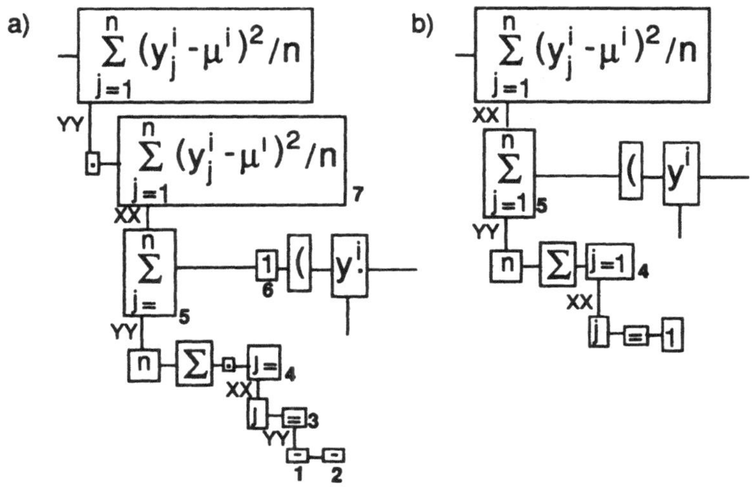

Figure 9. A part of the relational tree of 8a expression. a) After the first segmentation step. b) After the second segmentation step.

In these figures, there are some errors that are not corrected: the [Σ, Π] *frame* fails to recognize "Σ" in the Figure 5 expression, the strokes stay unconnected. The upper bar is recognized (block 1), then the [Σ, Π] *frame* is activated and the "Σ" or "Π" symbols are hypothesized. The conditions of the DESCRIPTION slot in the [Σ, Π] *frame* are partly fulfilled: the writing order of the strokes and the relative position of block 1 and block 2 match the prototype description; but conditions involving the size and the shape of block 2 are not fulfilled. This is an example where the conditions of the DESCRIPTION slot are too restrictive for a member of the hypothesized class to be recognized. In the same figure, "x" is split into two components (blocks 3 and 4). The gap between the two components of "x" produces a segmentation into two blocks by projection. There is no available knowledge in the system about the "x" pattern which could be used to correct this error. This exemplifies the incompleteness of the current system that will be resolved in future developments by the integration of a symbol recognition module and a mathematical language parser.

5. Line Labelling

The goal of this module is the labelling of the relative positions of the components of a writing line. A writing line is determined in the tree as a set of brother nodes linked by branches labelled "on the Left of". Figure 10 shows an example: this sequence of blocks may represent "$+g(x_i)^2$".

Each pair of blocks in a writing line is labelled according to the relative position of these two blocks; the labels indicate whether they are on the same line (L) or on different lines (E if the right block is written on a line above the line of the left block, S if the right block is written on a line under the line of the left block). This labelling allows us to detect subscripts and superscripts. It is achieved without using symbol recognition or knowledge of the symbol shapes. The blocks represent the symbols (or groups of symbols) in the labelling process. This work is presented by Wang and Faure (1988). Here we simply describe the main characteristics of the approach.

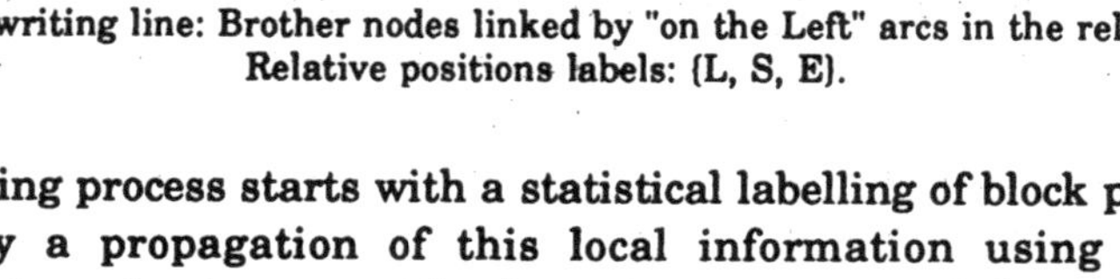

Figure 10. A writing line: Brother nodes linked by "on the Left" arcs in the relational tree. Relative positions labels: {L, S, E}.

The labelling process starts with a statistical labelling of block pairs, and is followed by a propagation of this local information using contextual constraints in order to ensure the best interpretation of the entire writing line.

The first step uses a learning set of (35 x 7) expressions: 35 expressions produced by 7 writers. These expressions are automatically segmented into blocks by the segmentation modules. Each pair of successive blocks on the X axis is manually labelled "L", "E" or "S". The representation space is two-dimensional. The features that describe a pair of blocks are x_1 and x_2 (see Figure 9). They give a full description of the relative size and position of the two blocks.

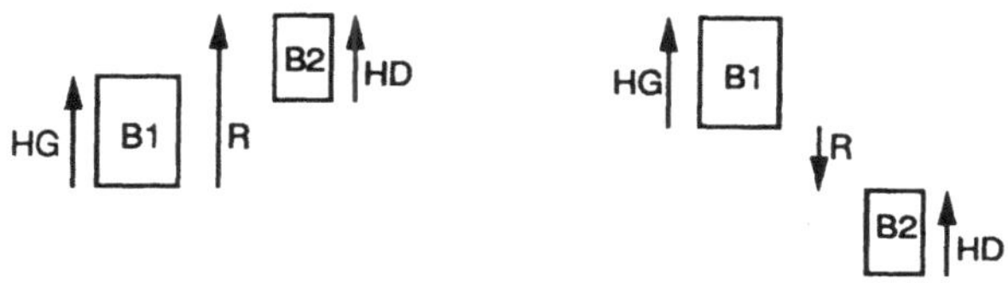

Figure 11. Relative position between two blocks. Features: x_1 = HD/HG; x_2 = R/HG.

The automatic learning is achieved using a modified version of the Ho-Kashyap algorithm, see Duda and Hart (1973). An improvement in recognition is obtained when the representation space is a priori divided into three regions: A (C) where the left (right) block is higher than the right (left) block, and B where the two blocks have almost the same height. The 3 hyperplanes (1, 3, 5 in Figure 12) which discriminate L and S classes, E and S classes, E and L classes, are obtained using the Ho-Kashyap algorithm. This method is very sensitive to the relative size of the two classes to be separated. When two classes are overlapping and differ greatly in size, the hyperplane is moved towards the smaller class, causing a prohibitive classification error for this class. In order to avoid this effect, the two classes are normalized by giving lower weights to samples of the greater class and higher weights to those of the smaller one. The resulting classification errors for the two classes are closer to each other. The hyperplanes 2 and 4 which separate the L and S classes and the L and E classes are obtained after the class size normalisation. Within each region, the labels are associated to a probability. The statistical labelling maps each pair of blocks belonging to the same writing line to three pairs (I, P(I)) where I ∈ {L,E,S}, and P(I) is the probability of the occurrence of the label I given the position (x_1, x_2) of the current pair of blocks within the representation space.

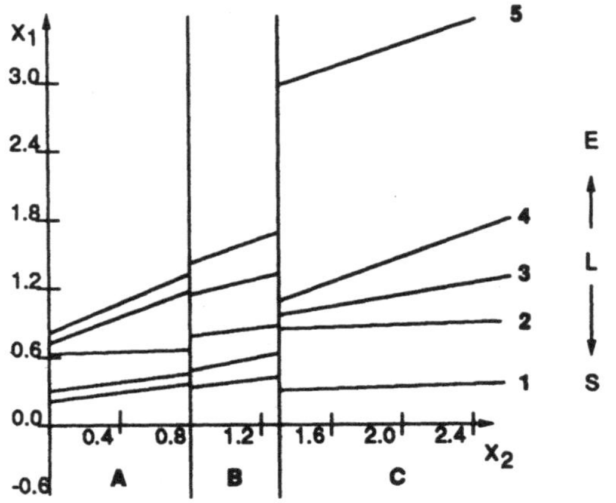

Figure 12. Representation space and the hyperplanes.

The "most likely" labelling, in which the label chosen is the one with the highest probability, leads to an error rate of 18% on a test set involving 58 expressions written by 11 writers. The low performance of this labelling is mainly a consequence of the local nature of the decision. Figure 13 shows that local information does not always allow the right interpretation when the blocks are used for symbol representation. The relative size and position of the two blocks are the same in "b+" and "y_n" when the labels are respectively L and S. Most of the errors correspond to logical contradictions such as in the case illustrated in Figure 14a, where blocks 2 and 3 are on the same line, blocks 1 and 3 are on the same line, but blocks 1 and 2 are not on the same line. Such a configuration is not acceptable; the second labelling step transforms it into an acceptable one, for example, as illustrated in Figure 14b. A set of acceptable configurations is then defined.

The goal of the second labelling step is to find the highest probability labelling that contains no logical contradictions. This is achieved by performing a contextual labelling that starts from the pairs (I,P(I)) given by the statistical labelling, and propagates constraints within the line. The context is taken into account to define the acceptable configurations and to update the probability P(I) according to the distance between the two blocks of the current pair.

Figure 13. "Same Line" (L) and "Subscript" (S) relations represented by the same block configuration.

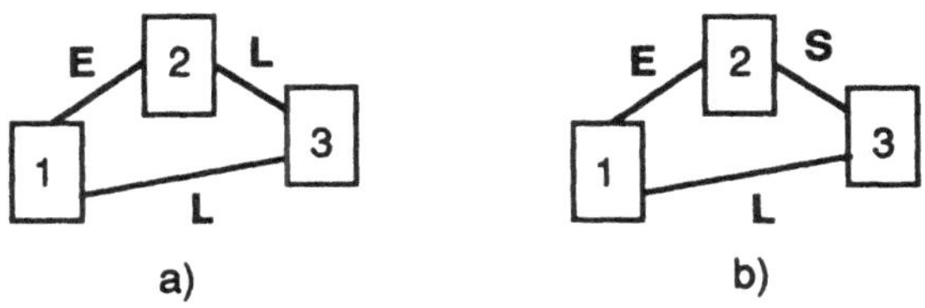

Figure 14. a) Logical contradiction between the labels. b) Acceptable labelling.

The updating of the label probabilities, P(I), $I \in \{L, E, S\}$, is based on the fact that handwritten mathematical expressions are never completely horizontal. The more distant two blocks are, the less their relative vertical positions represent their mathematical relation. The confidence in the statistical labelling decreases as the inter-block distance increases. Therefore, P(I) is replaced by

$$P_N(I) = 1/3 - (1/3 - P(I))/N,$$

where N is the distance between the pair of blocks. When N tends to infinity then $P_N(I)$ tends to 1/3. The three possible labels tend to have the same probability when the inter-block distance increases.

The search for the best final labelling is performed using a state-space represented by a tree. The root is the state where no decision has been made on the final labelling. A leaf represents a full, valid, final labelling. Intermediate nodes correspond to incomplete solutions of the labelling problem and each branch of the tree is associated with a cost that is a function of the label probabilities. The A* algorithm is used to search for the least-cost path from the root to a leaf. In order to reduce the state-space size and to speed up the algorithm, an adjustment parameter is provided: LCONT. It is defined as the shortest context length within which the contextual constraints are applied. Another adjustment parameter, DMIN, is introduced: the probability P(I) starts to vary with the inter-block distance, N, when N is greater than DMIN. The updated label probability P'(I) is then given by: P'(I)=P(I) if $N \leq$DMIN; otherwise P'(I)=P_N(I). Figure 15 shows the blocks involved in the labelling process and the values of the probabilities when the current block is block 1. Experimental results lead to the choices LCONT=6 and DMIN=3 to ensure an optimal labelling of the test set.

The error rate for the final labelling was 8.2%. Figure 16 shows the results obtained with the most likely labels without using contextual constraints (upper line) and after applying contextual labelling which corrects three errors. The figure displays the labels of successive blocks only. Figure 17 shows that the contextual labelling may fail to correct the errors; the invalid configuration is transformed to a valid one that interprets "+1" as a superscript of "y". This last line is not horizontal and the contextual information is quite poor (the line has only three symbols). Moreover, the shape of "y" stresses the upward line direction.

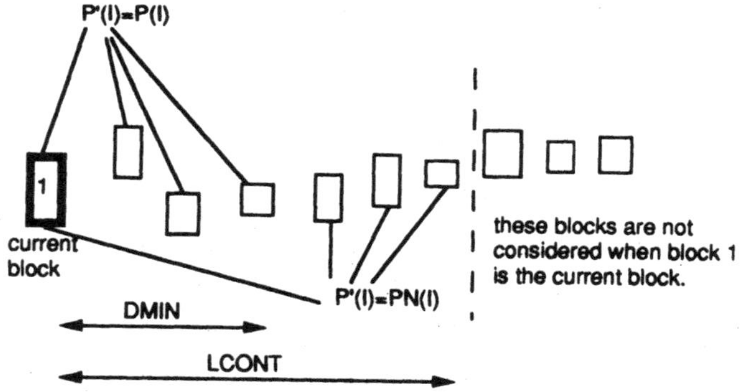

Figure 15. The parameters DMIN and LCONT and the label probabilities of the relative positions between the current block and the other blocks of the writing line.

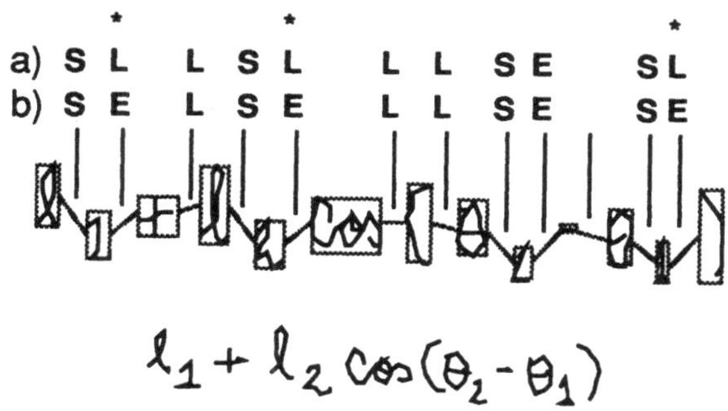

Figure 16. a) The most likely labeling. b) The contextual labelling.

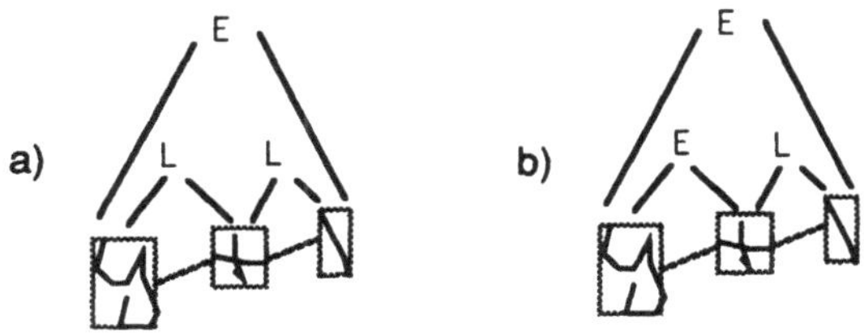

Figure 17. a) The most likely labeling. b) The contextual labelling.

6. Conclusion

This work presents the division of the problem of automatic interpretation of handwritten mathematical expressions into sub-problems. The sub-problems, and a system architecture, are described. The approach we adopt considers the processed data as a graphic language. The two-dimensional mathematical expressions follow formal rules that are defined according to visual skills in order to help the segmentation into meaningful components.

The integration of different modules seems to be the best way to deal with complex problems such as the interpretation of handwritten mathematical expressions. The system shows the necessary cooperation of data-driven and knowledge-driven processes in order to ensure a valid structural description of the expressions. This description is obtained by a set of specialist modules that perform specific tasks which do not involve recognition of the full set of symbols constituing the expression. The experimental results show the efficiency of such an approach, and point out its weaknesses. The system has not reached its final state. An improvement of the knowledge-based segmentation module is to be made; the current knowledge will be refined and enlarged as new data are processed. A more elaborate control for this module would lead to better use of the knowledge; nevertheless, the knowledge-based module will keep its structure of frame hierarchy and other major characteristics. The approach that we chose to segment mathematical expressions, leads to valuable structural descriptions. The two segmentation and the line-labelling modules are designed as components of a system which will in the future integrate a symbol recognition module and syntax analysis of the full expression.

References

Anderson, R. H., "Two-dimensional Mathematical Notation", in *Syntactic Pattern Recognition Applications*, K.S. Fu (ed.), Springer Verlag, New York, 1977, pp. 147-177.

Bélaid, A., Haton, J. P., "A Syntactic Approach for Handwritten Mathematical Formula Recognition", *IEEE Transactions on Pattern Analysis and Machine Intelligence*, **6** , no. 1, 1984, pp. 105-111.

Chang, S. K., "A Method for the Structural Analysis of Two-dimensional Mathematical Expressions", *Information Sciences* **2**, 1970, pp. 253-272.

Duda, R. O., Hart, P. E., *Pattern Classification and Scene Analysis*, Wiley Interscience Publication, 1973, 482 pp.

Martin, W. A., "Computer Input/Output of Mathematical Expressions", in *Proc. 2nd Symposium on Symbolic Algebraic Manipulation*, 1971, pp. 78-89.

Ullmann, S., "Visual Routines: Where Bottom-Up and Top-Down Processing Meet", in *Pattern Recognition by Human and Machines, Vol. 2, Visual Perception*, E. C. Schwab, H. C. Nusbaum (eds.), Academic Press, 1986.

Wang, Z. X., *Analyse de la structure des formules mathematiques manuscrites*, thèse ENST, Paris, 1989.

Wang, Z. X., Faure, C., "Structural Analysis of Handwritten Mathematical Expressions", *Proc. 9th Int. Conf. on Pattern Recognition*, 1988, pp. 32-34.

Computer Processing of Handwriting
Eds. R. Plamondon & C. G. Leedham

A COMPUTER GENERATION-MODEL OF BRUSH-USED HANDWRITTEN CHINESE CHARACTERS AND ITS APPLICATION IN EDUCATION

Jianchao ZENG[+] Xianrong ZHANG[++] Takeshi INOUE[+++] Hidehiko SANADA[++++]
and Yoshikazu TEZUKA[+]

+ *Dept. Communication Engineering, Osaka University, 2-1 Yamadaoka, Suita City, Osaka 565 Japan*

++ *Shanghai Institute of Technology, P. R. China*

+++ *Kobe University of Mercantile Marine, 5-1-1 Fukae Minamimachi, Higashinada-Ku, Kobe 658 Japan*

++++ *Faculty of Economics, Osaka University 1-1 Machikaneyama-Cho, Toyonaka City, Osaka 560 Japan*

We first propose a flexible generation-model of brush-used handwritten Chinese characters on a computer to show that the beauty of characters can be truly expressed. Based upon the generation-model, a data structure is then proposed, which can well reflect the hierarchical structures of Chinese characters and therefore is especially suitable to represent calligraphic rules for the evaluation of user-initiated brush-used handwritten Chinese characters from the viewpoint of traditional Chinese calligraphic art. Further, a support system for acquiring calligraphic knowledge is proposed. This system adopts an acquisition strategy that attempts to extract knowledge from calligraphists by monitoring and analyzing, in the form of parameters, their writing actions of characters. Finally, as an educational application, a computer-assisted instruction system of Chinese characters is described, which can evaluate characters with its calligraphic knowledge in a similar way to calligraphic teachers. The system can be used to help promote calligraphy education of brush-used handwritten Chinese characters in primary school.

1. Introduction

Calligraphy is a very important area concerned with the writing techniques of characters of both eastern and western languages. Children are taught calligraphy from primary school to high school in order to learn to write. In China and Japan in particular, where Chinese characters are used and where people have been interested in calligraphy since ancient times, calligraphy has been recognized not only as a way of learning to write, but more commonly as an art in writing, because Chinese characters are regarded as a kind of ideogram, making their writing somewhat similar to painting. From the viewpoints of efficient teaching and diffusion of calligraphy, computerization of calligraphy becomes more and more necessary and important, with the fast development of computer technology.

In this chapter, we focus on the calligraphy of Chinese characters, and deal with this problem by proposing a computer generation-model to serve as a basic tool. This model is called Structure Analysis Model (SAM). Its primary characteristic is that the beauty of Chinese characters can be truly expressed. SAM has the following four main features:

1. a newly proposed basic brush-touch pattern to allow its form and size to be flexibly modified according to brush movements during writing,
2. classification of strokes of all Chinese characters into 44 kinds and their representation by only 6 to 12 parameters each,
3. proposal of two stroke decorating-functions: a frame function for stroke orientation decision and a rounding-out function for brush-touch size decision, both of which are in the simple forms of summation of two sine functions, and
4. combination of strokes to enable brush-used Chinese characters to be generated in their truly graceful and spirited styles.

Based upon the model, a Data Structure (DS) is then proposed, which organizes the parameters of SAM so that they can reflect the hierarchical structures of Chinese characters. DS is used to represent calligraphic knowledge which will guide learners of calligraphy to write well formed characters.

Further, a Support System for Acquiring Calligraphic Knowledge (SSACK) is developed. Calligraphic knowledge exists within the minds of calligraphists. It is experience-based and therefore is generally expressed in an abstract way. Calligraphists use it almost unconsciously in their actual writing with brushes. SSACK adopts a strategy that attempts to extract knowledge from calligraphists by monitoring, recording and analyzing their writing processes in the form of parameters.

By using SAM and calligraphic knowledge represented by DS, a prototype system for Computer Assisted Calligraphic Instruction (CACI) is being developed as an attempt to computerize calligraphy. CACI can be used to help promote calligraphy education of brush-used handwritten Chinese characters in primary school.

2. A Generation-Model

Chinese characters have various kinds of complex structures. Their generation by a computer is therefore first of all faced with a memory limitation problem. It is desired that we be able to generate a character only from a small number of feature data, instead of memorizing its pattern image, and represent different kinds of possible variations of character posture, beauty and spirit through modifying only these data (Zhang et al., 1985).

2.1 Procedure of Forming Brush-Used Characters

Each Chinese character is composed of several strokes that are in turn composed of partially overlapping brush-touches. Here we define a stroke as the continuous track of a moving brush, which could consist of a single segment or multiple segments with thickness. It is somewhat similar to the definition of a component in Plamondon and Maarse (1989). By a brush-touch, we mean the shape formed by the instantaneous contact between a brush and the paper. Our model SAM is based upon this concept of brush-touch. By analyzing their structures, Chinese characters are generated in the model in the following way:

1. *Character Selection*: Input the code of a designated character and take out from the character data base a set of stroke data necessary to generate the character.
2. *Data Reading*: Read out the stroke data in written order and pass them to the corresponding stroke generation subroutine.
3. *Stroke Movement Calculation*: Calculate the values of sampling intervals, coordinate pairs of sampling points, stroking orientation and sizes of brush-touches, which are necessary for stroke generation.
4. *Generation of Brush-Touch*: Generate a brush-touch using the above values at each sampling point.

The above program flow is shown in Fig. 1.

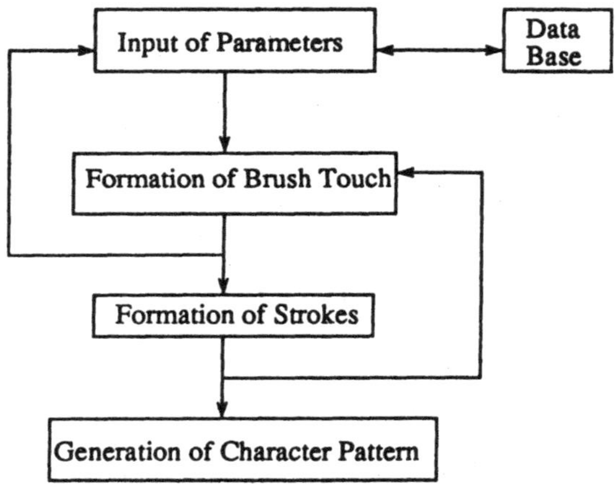

Figure 1. A Character generation model.

2.1.1 Basic Shape of Brush-Touch

Figure 2 shows a basic shape of brush-touch. Figure 2(i) shows an actual shape of brush-touch, which forms itself into a slightly long symmetrical convex triangle. However, if we directly use this triangular brush-touch in SAM, zig-zag will occur on both sides of a generated stroke unless sampling intervals are extremely small. To avoid this, here we propose a new basic shape of brush-touch in an egg form as shown in Fig. 2(ii). This brush-touch consists of two portions: a brush tip that directs against what the brush moves in, and a leading brush body front.

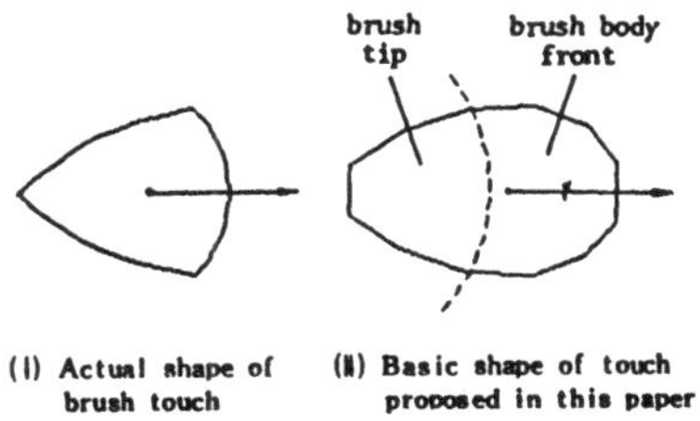

Figure 2. Basic shape of brush touch.

2.1.2 Brush-Touch Alterations

There is a proverb in the calligraphy of Chinese characters: "Stroking the center of a stroke is valuable", which means that the brush tip should always be in the center of a stroke, without leaning to either side. Actually, this holds only when the brush moves straight forward without turning. In this case, the present brush tip is always overlapped by the preceding brush-touch and therefore the new part is considered to be shaped only by the present body front as shown in Fig. 3(i). However, in the case of turning motions such as at the beginning and ending parts of a stroke, or at the stroke corners, the brush tip will play an important role and influence the stroke shapes delicately. In the actual writing of a stroke, when a brush is rotated or moved sideways, the brush tip will be slower in bending than the brush body front due to friction between the brush and the paper. Hence we propose two kinds of brush-touch alterations in SAM as shown in Fig. 4.

Figure 4(i) shows a brush-touch alteration for brush rotation. In this case, we assume that the brush body front will not change its shape, and that the brush tip will bend stronger and will be nearer to its preceding direction when its distance from the central point O becomes larger. Suppose that the preceding and present directions of a brush-touch are $C_{1,i}$ and $C_{0,i}$ respectively, then the rotation angle will be

$$\Delta C_{1,i} = C_{1,i} - C_{0,i} \tag{1}$$

We set the maximal bending of the brush tip as follows,

$$\Delta C_{0,i} = k(C_{0,i} - C_{1,i}) = -k\Delta C_{1,i} \qquad (0<k<1) \qquad (2)$$

Further, we assume that the preceding direction of the brush-touch, which is necessary to generate the next brush-touch, is

$$C_{0,(i+1)} = C_{0,i} + \Delta C_{0,i} \qquad (3)$$

which is not equal to $C_{1,i}$. In this way, once the brush rotates and bends, it can revert to its original form only after several times of gradual stretch.

Here, in Eq. 2, k is a coefficient concerning the hardness of a brush. Larger values correspond to softer brushes. Our experience suggests that a slightly softer brush would be appropriate for square-styled brush-used Chinese characters, and therefore k = 0.7 is selected in our model.

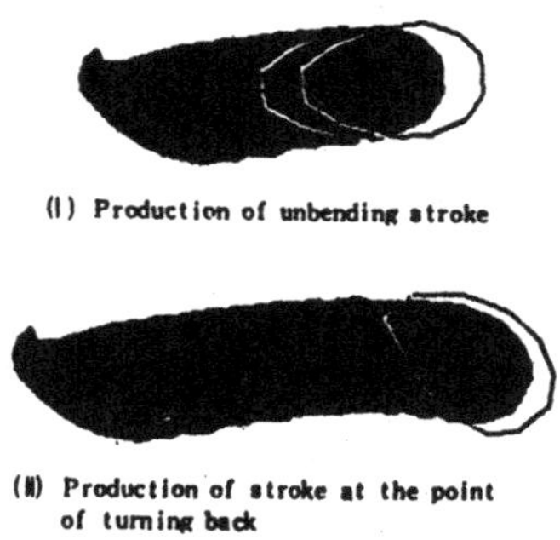

Figure 3. Brush touch and stroke generation.

The brush tip will also bend because of friction when a brush moves sideways. Suppose that the present and the preceding central points of a brush are O_i and O_{i-1} respectively; then the brush body front will not change its shape and will shift by the same distance, but the brush tip will delay to some extent and its maximal bending will be near to the diagonal line of the parallelogram, as shown in Fig. 4(ii).

All the actual brush motions can be decomposed into the above basic motions and therefore a brush-touch could have various alterations. We ensure that the Chinese characters generated using the proposed brush-touches in our model, which perform as well as actual brushes, will be much similar to those characters written with actual brushes.

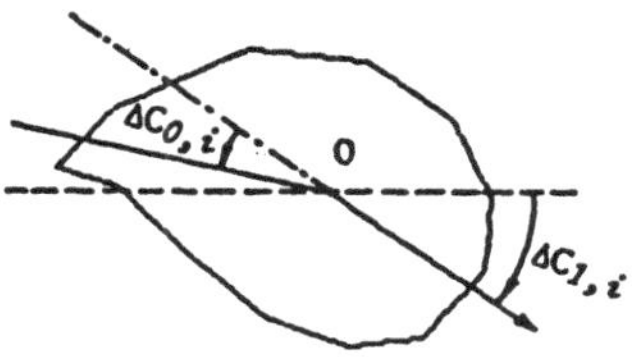

(I) Shape alteration at turning point

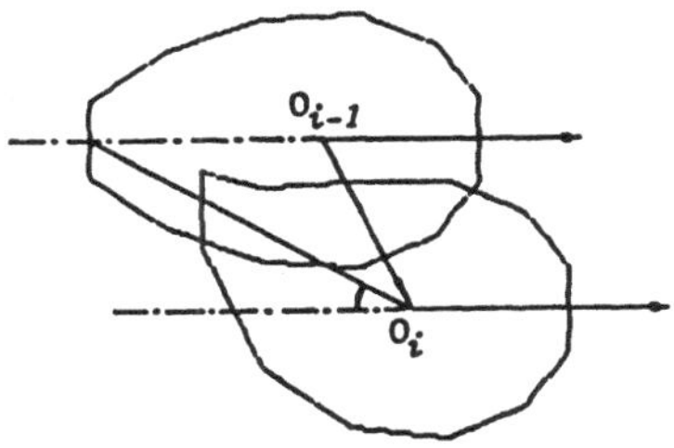

(II) Shape alteration at sideways morement

Figure 4. Brush touch alterations.

2.2 Proposal of Stroke Generation

Strokes are elements to construct Chinese characters and therefore play a very important role in representing their spirits. Ideal Chinese calligraphy is said to "show surface softness while retaining inner strength," which means that the brush tip should be hidden at the starting point of a stroke and back-turned at the end point, with the middle portion of the stroke gently changing its direction and pressure. We use the following two methods to realize these requirements.

2.2.1 Stroke Analysis and Stroke Synthesis

According to the brush motions, each stroke consists of several parts as shown in Fig. 5. These are the starting, stroking, ending, backturning and twisting parts. For example, to write a rightward stroke, one should first drop a brush leftwards and then turn back and move towards the right. To write a downward stroke, one should first drop upwards and then turn and move downwards. This inversely dropping and backturning part is called the starting part.

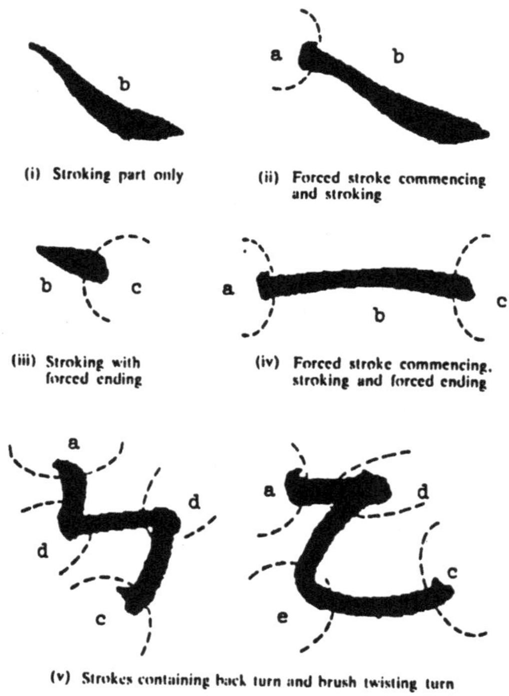

Figure 5. Stroke decomposition.

There are three ways to end a stroke:

(1) ending in the stroking direction,

(2) rejecting and

(3) turning and raising.

In SAM, only (2) and (3) are treated as ending parts. When a stroke consists of more than one segment, there will be backturning and twisting parts in the joints. A segment is defined as a simple line with thickness, which could curve but never turn. The middle portion of a stroke is called the stroking part. All kinds of strokes necessary for generating different kinds of Chinese characters could be combined from these five parts.

2.2.2 Stroke Decoration

The orientation and pressure of the stroking part of a stroke generally do not change much during the writing process, but they are extremely important in representing the spirits of Chinese characters. If we want to generate beautiful Chinese characters with spirited postures, it is necessary to be able to decorate these two elements. In SAM, two functions are therefore proposed. One is called a frame function, which is used to decorate stroking orientations of strokes, while the other is called a rounding-out function, which is used to decorate stroking pressures or the brush-touch size at each stroking point.

The frame function is mainly utilized to control the moving direction of a brush-touch at each stroking point. It corresponds to the segment differential and is therefore a function of relative position of each stroking point. Supposing that the length of a segment is L, and the curving distance between the starting point and some arbitrary point on the segment is e, then the frame function is defined as follows.

$$a_1 = f(e/L) \qquad (4)$$

If we extract from this function the definite main orientation A_0 of the segment, we can re-form Eq. 4 into

$$a_1 = A_0 + f_1(e/L) \qquad (5)$$

and further into

$$a_1 = A_0 + qf_2(e/L) \qquad (6)$$

Here q is called the relative curvature. If we set q = 0, then a1 = A_0, and the segment becomes a straight line with orientation A_0.

In SAM, since e does not vary continuously, a_1 is calculated by sampling the segment with definite intervals, which corresponds to the segment differential. Generally, any curve can be expressed by a Fourier series, and therefore a frame function can also be expressed in the same way. Since the segment orientation does not change much during its writing process, we know through experiments that it is sufficient to represent any segment by expressing its f_2 with only the first two sine functions in the series, as shown in Fig. 6(i). Furthermore, if we give different values to q, we can generate various segments with different curvatures, as shown in Fig. 6(ii). Therefore the same segment could shape itself into different forms according to different values of q, as shown in Fig. 6(iii).

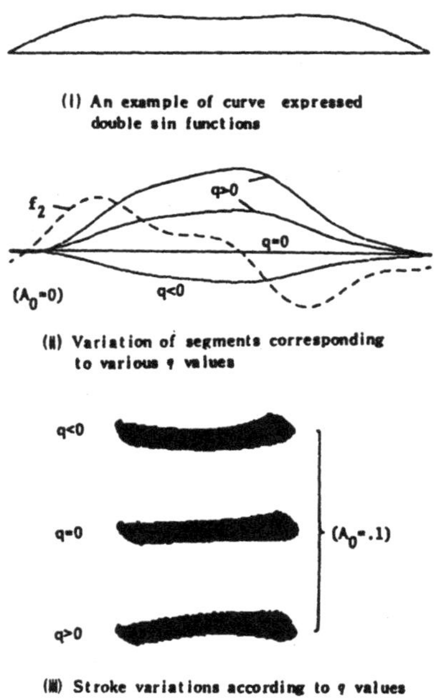

Figure 6. Frame function.

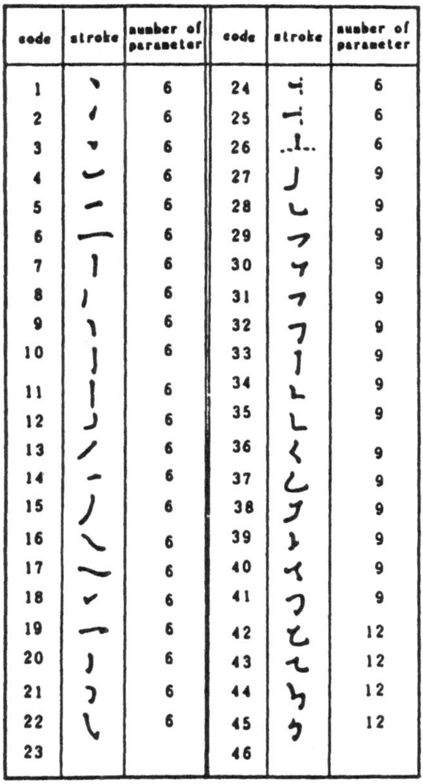

code	stroke	number of parameter	code	stroke	number of parameter
1		6	24		6
2		6	25		6
3		6	26		6
4		6	27		9
5		6	28		9
6		6	29		9
7		6	30		9
8		6	31		9
9		6	32		9
10		6	33		9
11		6	34		9
12		6	35		9
13		6	36		9
14		6	37		9
15		6	38		9
16		6	39		9
17		6	40		9
18		6	41		9
19		6	42		12
20		6	43		12
21		6	44		12
22		6	45		12
23			46		

Table 1. Classification of strokes.

The rounding-out function is used to control brush-touch size at each stroking point, and therefore to control segment thickness at each point. It is also the function of relative position of each stroking point and is defined as follows:

$$r = g(e/L) \tag{7}$$

As the case of the frame function, it can also be expressed with only two sine functions.

Various kinds of stroking parts can be represented by combining these two functions. In addition, by appropriately selecting their starting, stroking, backturning, twisting and ending parts and combining them with these two functions, the strokes necessary to generate different kinds of Chinese characters can be classified into 44 types, as shown in Table 1. These strokes are coded, and their generation subroutine libraries are produced. Here, strokes with codes from 1 to 26 are one-segment strokes which can be represented by 6 data (parameters); strokes with codes from 27 to 41 are two-segment strokes represented by 9 data; and strokes with codes from 42 to 45 are three-segment strokes represented by 12 data. These data and their geometric meanings will be discussed below.

2.3 Generation of Chinese Characters and Their Data

2.3.1 Generation of Characters

According to the shapes and postures of Chinese characters, SAM first selects appropriately from Table 1 the necessary strokes, using their codes, then characterizes them by assigning the proper data, decorates them, and finally, generates a character by composing these strokes, as shown in Fig. 7. The spirits and postures of the generated characters depend totally upon the data assigned to them.

Figure 7. A generated Chinese character.

2.3.2 Data of Chinese Characters

The data of Chinese characters should not only describe the main features of their structures, but should also include information on how to decorate those characters into beautiful postures. Furthermore, the amount of data should not be very large, but each datum should be given a clear geometric meaning.

In SAM, each stroke is expressed by a group of data, and the data for a character are a set of data groups of its corresponding strokes. The amount of data for any one stroke depends upon the number of segments in it. The data for single-segment strokes are as follows:

(i) *Stroke Code* C

(ii) *Stroke Parameters*

Feature Parameters	Coordinate pairs of Starting Point	(X,Y)
	Main Orientation	A
	Total Length	L
Decorating Parameters	Relative Curvature	Q
	Brush Pressure	D

Figure 8 shows the generation procedure of a Chinese character and explains the roles of each datum. A character consisting of only thin straight lines is formed with feature parameters as shown in Fig. 8(i). Stroke codes exhibit the careful structure of each stroke by equipping it with starting, stroking, backturning, twisting and ending parts, as shown in Fig. 8(ii). Further decorating parameters attach curvature and pressure to each stroke to make it spirited, and the whole beautiful character is finally generated as shown in Fig. 8(iii) and (iv). Figure 8(v) shows the corresponding data of the generated character. We are convinced of the small amount of data from this example.

Through the extensive experiments that were conducted using Basic and C languages on a personal computer, we have come to the conclusion that if a data base is completely arranged, all Chinese characters could be generated by our proposed model.

2.4 Summary and Remarks

Unlike other research reported in this area, our generation-model is characterized by the following results:

1. A new basic brush-touch was proposed which is controllable and is similar in shape to that of the actual basic brush-touch.

2. It was shown that a stroke could be decomposed into starting, stroking, backturning, twisting and ending parts for flexible stroke generation.
3. Two decorating functions were proposed to realize subtle changes in the stroking parts.
4. A Chinese character could be generated with a small amount of data and each datum has a clear geometric meaning.

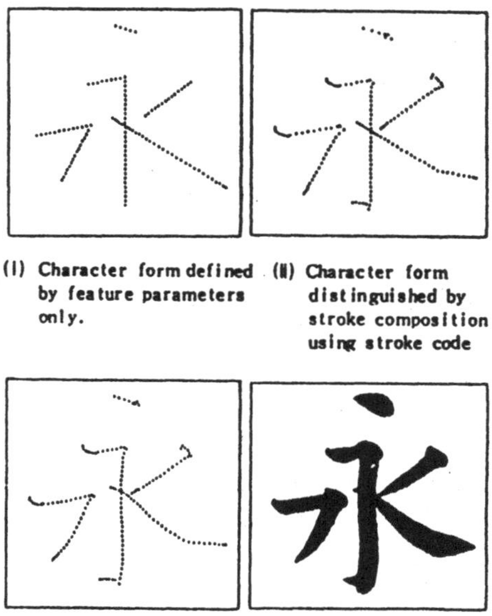

(I) Character form defined by feature parameters only.

(II) Character form distinguished by stroke composition using stroke code

(III) Frame structure refined using modifying parameters

(IV) Rounded out pattern

```
5034 DATA 1,170,25,-.3,35,1.3,0,33,130,135,.2,75,
1.0,0,-1.57,230,.06,29,40,225,.2,110,1.1,-.1,-2.05,
130,.2,14,325,125,-2.5,130,1.1,.1,16,175,195,-.55,
230,1,.4
```

(V) Character data needed to produce Chinese character "永"

Figure 8. Character data and character generation.

3. A Data Structure Suitable for Representing Calligraphic Knowledge

From the discussion in Section 2, we know that what is required for SAM to generate spirited and well-formed Chinese characters is a complete data base, which should be prepared in advance. Generally speaking, however, data setting in order to generate beautiful Chinese characters is not such an easy task, even for a calligraphist. Even if the data base is assumed to be already existent, characters generated by SAM would always be of the same shapes without being able to reflect the personalities of the users and this would make SAM only a displaying system.

A stroke parameter decision system was developed to attempt to solve this personality problem (Uchio et al., 1986). This system used a digitizer to input brush-used Chinese characters for SAM, by placing user-written characters on the paper onto the digitizer. The system was only partially successful in that it could only extract data from characters and transfer them to SAM for reproduction without any ability to modify them, regardless of whether they were poor in shape or not. Calligraphic knowledge is considered very useful in solving this remaining modification problem (Zeng et al., 1988), which we are going to discuss in the rest of this chapter. In this section, we propose a data structure which is suitable to represent calligraphic knowledge. Calligraphic knowledge will be used in SAM to evaluate user-initiated characters of their shapes.

3.1 Analysis of Calligraphic Knowledge

3.1.1 Classification of Calligraphic Knowledge

There are two kinds of calligraphic knowledge: one is textbook knowledge from calligraphic textbooks and the other is expertise that calligraphists have acquired through their practical experience (Deng, 1963). Here in this section we basically deal with the former knowledge (rules), while the latter will be discussed in Section 4.

We have collected over 90 rules from calligraphic textbooks and classified them, as shown in Table 2 (Ren et al., 1987 and Fujiwara, 1983). Nearly one third of them are rules for those characters with specific radicals and therefore their pre-conditions are clear. Among the rules whose pre-conditions

Character Type	Knowledge about General Pattern Structures					
left right	R.Z. rule Y.C. rule	Zu.Z.D. rule R.Y. rule	Y.Z.D. rule X.P. rule	Y.F. rule S.P. rule	Z.C. rule F.J. rule	Q.G. rule J.G. rule1
up-down	T.F. rule Yu. rule J.N rule	D.Z. rule1 H.C.P.D. rule X.Z.D. rule F.Y.G.Q. rule	L.D. rule G.N.rule R.H. rule	Z.S. rule S.Z.D. rule H.B. rule	Du. rule G.X. rule H.G. rule	J.G. rule2 S.X. rule P.D. rule
over-below		S.D.X.X. rule	S.G. rule			
rectangle	D.F. rule Q.J. rule G.Z. rule	P.S.J. rule H.D.Z.C. rule S.D.X.C. rule H.C.Z.D. rule	C.F. rule H.W. rule Ch. rule	Z.W. rule1 D.Z. rule2 K.L.J. rule D.F.C.S. rule	Z.B. rule J.H. rule X.Z. rule	Z.G. rule J.Z. rule R.Zh. rule
left-middle-right	S.J. rule	Z.Y.Z.D. rule		Z.Z.D. rule1		
below-over2	G.G. rule	Z.W. rule2				
up-middle-down	S.T. rule	S.X.Z.D. rule		Z.Z.D. rule2		
below-over1	C.P. rule	H.D.P.C. rule	Z.P. rule	H.C.P.D rule	H.P. rule	
	Knowledge about Special Sub-Patterns					
left-right	S.S. rule W.P. rule Li.P. rule Ji.P rule	R.P. rule E.P rule B.E. rule Be.P. rule	M.P. rule Y.P. rule Q.P. rule	S.R.P. rule G.P. rule Ri.P. rule	L.P. rule1 N.P. rule He.P. rule	J.P. rule X.P. rule Mi.P. rule
up-down	S.Y. rule S.Sh. rule	S.Yu. rule	S.H. rule	X.Y. rule	L.P. rule2	X.X. rule
over-below	Xi.Z. rule					

Table 2. Calligraphic knowledge from textbooks.

Rule Names	Contents of Rules
S.S. rule	For patterns taking three "point-strokes" as left-radicals, adjust lengths of 3rd strokes in "point-strokes" according to stroke numbers in right radicals.
L.D. rule	For patterns having equal stroke numbers and square shapes both in "up" and in "down" sub-patterns, unify the sizes of the two sub-patterns and approach them.
R.H. rule	For such patterns as examples, enlarge the strokes undertaking spirit of the patterns and reduce the others.
L.P. rule	For patterns having piling slanting strokes , write these slanting strokes both untouched and unparalleled, and lengthen the first slanting stroke and curve the second.

Table 3. Examples of calligraphic knowledge.

are described by character shape features, about two thirds have comparatively clear pre-conditions. Examples of these kinds of rules with clear and unclear pre-conditions are shown in Table 3.

3.1.2 Features of Calligraphic Knowledge

Calligraphic knowledge is a kind of know-how which has been acquired during thousands of years. This knowledge, being concerned with the structures of Chinese characters, guides the way to representing beautiful character forms. On the other hand, however, in incorporating this mass of knowledge into SAM, there are certain difficulties (Zeng et al., 1988):

(1) Calligraphic knowledge in textbooks is expressed by so much abstract terms that it is difficult to represent its pre-conditions and actions in a concrete fashion and to implement it on a computer.

(2) Some knowledge gives only several characters as examples for their pre-conditions.

(3) Some other knowledge specifies only partial pre-conditions, and leaves the rest to the judgment of the calligraphists.

(4) There is some knowledge about characters with specific radicals.

(5) Some knowledge describes relative position or size relations among specific sub-patterns.

(1) to (4) are considered as problems concerned with the representation of pre-conditions. (1) is also concerned with the representation of knowledge matching conditions and (4) and (5) are considered as problems with character generation-models. To represent calligraphic knowledge on a computer, it is especially important to select a character generation-model which should easily represent relations among sub-patterns, as described in (5). SAM is considered appropriate to serve as such a generation-model.

3.2 Data Structure

Generally, a Chinese character consists of several sub-patterns, which could also be decomposed into several strokes. We could therefore suppose that a character has several corresponding levels, and describe it hierarchically by using these levels. It may have been noted that in SAM we have defined a stroke level at which a stroke consists of several segments. These segments are expressed by groups of parameters. If higher levels of character descriptions, such as sub-patterns (or radicals) and total characters, are added to SAM, it will be much easier to represent calligraphic knowledge, especially that concerned with relations among sub-patterns or total characters. Here we propose a hierarchical data structure DS as shown in Fig. 9, which is used to organize parameters in SAM so that they can well reflect the hierarchical structures of Chinese characters. In Fig. 9, CH represents total character level, while SPi, STj and SGk represent sub-pattern, stroke and segment levels respectively. At the lowest level of DS there is a parameter set (Xk, Yk, Ak, Lk, Dk, Qk) which expresses each segment in a character. These parameters are assigned concrete values to a specific character. With this DS, it becomes possible to connect calligraphic terms with parameters in SAM and therefore to represent calligraphic knowledge on a computer.

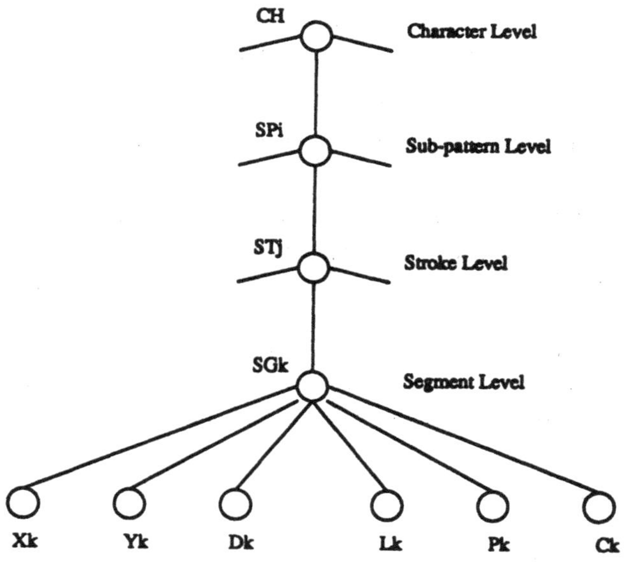

Figure 9. A data structure.

3.3 Summary

A data structure DS was proposed that is suitable for representing calligraphic knowledge for brush-used Chinese character evaluations. Evaluation experiments using DS will be discussed in Section 5.

4. A Support System for Acquiring Calligraphic Knowledge

Besides textbook knowledge, there is another kind of calligraphic knowledge that exists within the minds of calligraphists and is therefore much more difficult to be clearly described in languages (Zeng et al., 1990). This kind of knowledge is an experience-based know-how acquired through many long years of practice. Here, a strategy has been adopted to computerize this kind of knowledge by observing, recording and analyzing practical corrective actions of calligraphists. First, a correcting procedure model of calligraphists is proposed by interviewing and observing calligraphists when they are correcting students' errors. Then, a correcting support system SSACK is developed, based upon the proposed model. SSACK is designed primarily for the use of calligraphists and therefore its interface must be structured for maximum ease. SSACK internally records corrective actions and calligraphists' comments in parameters, and the recorded parameters are to be analyzed for knowledge acquisition.

4.1 Analysis and Modelling of Correcting Procedures of Calligraphists

4.1.1 Analysis of Correcting Procedures of Calligraphists

It was found that the following four items of Chinese characters would be corrected by calligraphists, after observing their actual actions and interviewing them:

(a) Balance among blanks themselves

(b) Balance between strokes and blanks

(c) Lively states of strokes in the given positions

(d) Underlying flow of strokes

By analyzing these items, we know that (a) is concerned with relative position and size of a character as a whole unit, with respect to the rectangular frame where the character is written. (b) is concerned with balance between radicals (including strokes) within a character and (c) is concerned with ways of writing each stroke. (d) concerns stroking techniques for stroke linkage. Among them, (a) to (c) are corrections relating to the shape balance of various levels of characters, while (d) is a correction relating to the stroking techniques of strokes. This means that objects for correction could be classified into the following three categorical aspects: total characters, sub-patterns and strokes, and accordingly the correcting procedures of calligraphists could also be partitioned into three corresponding parts.

A correcting example by a calligraphist is shown in Fig. 10, where the character position was adjusted as a correction of the total character, the right radical was reduced in size as a correction of the sub-pattern, and the lengths of strokes 3 and 4 were modified as corrections of strokes.

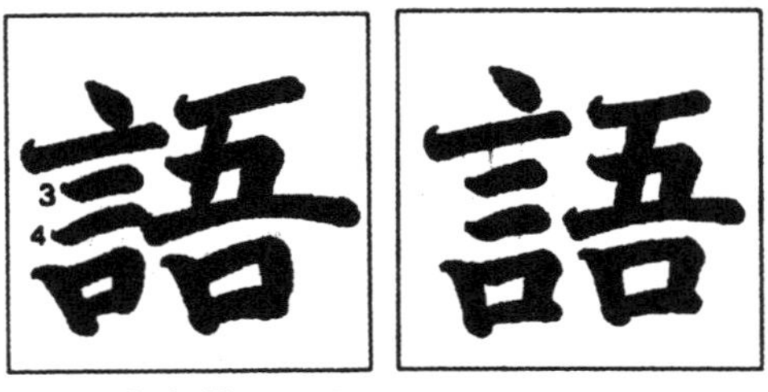

Figure 10. A correcting example by a calligraphist.

Further, calligraphists would follow some definite process in correcting as shown below, which also came from observations of their corrective actions.

Steps:

1. Evaluation of the total character
2. Assignment of an object for correction
3. Decision of the correcting action type
4. Decision of the correcting amount
5. Execution of the correction
6. Evaluation of the corrected character
7. New correction, re-correction or end of correction

This process is always kept in the order, although it is not necessary to go through all steps each time. Complex corrections could be realized by repeating this process. Comments are given after each correction. Since this process is relatively independent of particular calligraphists, it could be modelled as a general routine to support calligraphists' corrections.

4.1.2 Modelling of Correcting Procedures of Chinese Characters

In correcting a character, a calligraphist concentrates his attention on the whole character or a sub-pattern or some definite stroke. It is therefore possible to investigate calligraphists' corrective actions from these three phases of total characters, sub-patterns and strokes respectively. A correcting model of procedures of calligraphists can be then constructed by combining these phases. The model is shown in Fig. 11.

The most important component in this model is "Instruction Formulation", namely, realization methods of correcting instructions of calligraphists. Correcting instructions include those for total characters, sub-patterns and strokes. They can be classified into simple instructions that execute single-corrections each time, and complex instructions that execute multiple-corrections each time. Correcting orders, objects for correction, correcting types and the correcting amount are selected as contents of corrections.

This model is to be used by calligraphists when they correct and improve students' characters with the help of SSACK. A correcting example, by using the model, is shown in Fig. 12, where the size and position of the character on the phase of total character, the curvature of stroke No.2, the direction of stroke No. 3 and the length of stroke No. 4 on stroke phase, were selected as objects for correction and corrected in the order of (a) to (g).

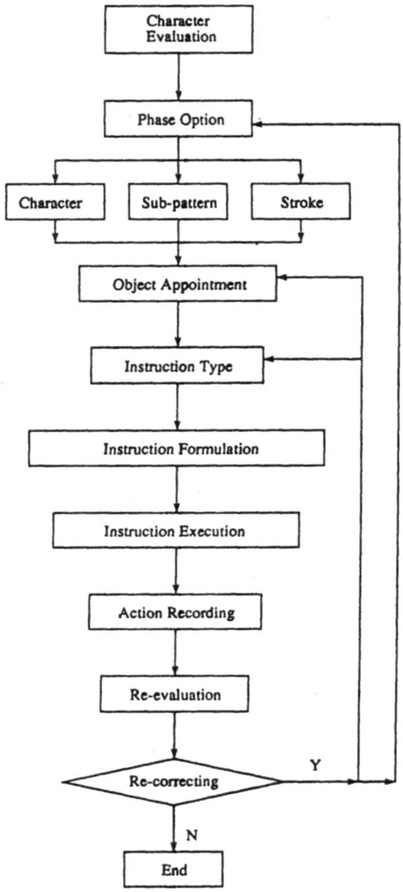

Figure 11. A model of calligraphists correcting procedures.

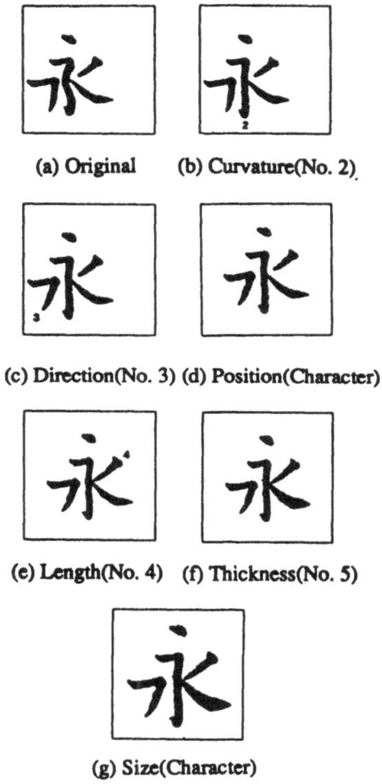

Figure 12. An example model corrected by the correcting model.

4.2 Methods to Realize Correcting Instructions of Calligraphists

4.2.1 Requirements from Calligraphists

A brush is viewed by a calligraphist as an extension to his hand when he is correcting characters. When he is given a character for correction, he first evaluates the character with his eyes, then judges its goodness with his knowledge, and organizes instructions for correction, which are passed on to his "extended hand" and finally executed. His instructions may be in various forms. He may change the length of a stroke while changing its position, or he may change the direction of some radical while changing its size. In all these cases, he can pass his instructions on to his "extended hand" in the easiest and fastest way and execute them. This means that his instructions are organized in a very effective way. Since the formulation methods of the instructions may probably be the key factor for the easy and fast passage of the instructions, what is required of the correcting model is to develop an efficient way to organize the correcting instructions of calligraphists, which will be mainly discussed next.

4.2.2 Methods of Realizing Correcting Instructions

Based upon the above analysis, the realization methods of the correcting instructions of calligraphists are described within phases of total characters, sub-patterns and strokes, respectively.

<u>1 Total characters and sub-patterns</u>

(I) Simple corrections

(a) *Sizes*. Sizes of total characters or sub-patterns could be determined by assigning rates of enlargement or reduction.

(b) *Positions*. Positions could be determined by assigning central points of total characters or sub-patterns.

(II) Complex corrections

(a) *Sizes + positions*. Sizes and positions could be determined by assigning both rates and central points simultaneously.

The above instructions are realized through calculating with the parameters of SAM. For example, the sizes of characters could be corrected by changing the coordinates, lengths and thicknesses of the corresponding strokes, according to the assigned rates. The positions of the characters could be corrected by moving the coordinates of each stroke in the characters, according to the assigned central points. Corrections within the stroke phase can be realized using the same principles.

<u>2 Strokes</u>

2.1 Balance corrections

(I) Simple corrections

(a) *Type*. Since each stroke is described by a code in SAM, the stroke type can be changed by assigning a new code.

(b) *Position.* The stroke position could be changed by assigning a new position of any point on a stroke.

(c) *Stroking order.* The stroking order can be changed by assigning the two strokes concerned.

(d) *Length.* The length can be changed by fixing the starting point or ending point and moving the other to a desired position.

(e) *Thickness.* The thickness can be changed by assigning two new waist points (two opposite points on the contour of a stroke which are near to its center).

(f) *Curvature.* The curvature can also be changed using waist points in the same way as in the case of the thickness.

(g) *Orientation.* This can also be changed by fixing one end of a stroke and moving the other in the same way as that of the length.

(II) Complex Corrections

These are corrections combined by any two simple ones among (a) to (g) except (c). They are commonly used in the actual corrective actions of calligraphists.

(a) *Type + position, type + length, type + thickness, type + curvature, type + orientation.* These corrections can be realized by assigning both new codes and one of the following corresponding components of strokes: position, length, thickness, curvature or orientation.

(b) *Position + length.* These can be corrected simultaneously by assigning both the starting and ending points on a stroke.

(c) *Position + thickness.* These can be corrected simultaneously by assigning both two new waist points and any other one point on a stroke.

(d) *Position + curvature.* These can be corrected in the same way as that of "position + thickness".

(e) *Position + orientation.* These can be corrected in the same way as that of "position + length".

(f) *Length + thickness*. These can be corrected by assigning the starting, ending and the two waist points on a stroke.

(g) *Length + curvature*. These can be corrected in the same way as that of "length + thickness".

(h) *Length + orientation*. These can be corrected in the same way as that of "position + length".

(i) *Thickness + curvature*. These can be corrected by assigning only two new waist points on a stroke.

(j) *Thickness + orientation*. These can be corrected in the same way as that of "length + thickness".

(k) *Curvature + orientation*. These can be corrected in the same way as that of "length + thickness".

2.2 Stroking techniques

Strokes in a character are not independent of each other. They are connected so that the character is constructed as a whole unit. The relations among strokes are known as the underlying flow of strokes. Several examples showing these relations are shown in Fig. 13. Although the underlying flow of strokes does not influence shape balance of a character, it plays an important role in keeping up spirits for the character. Namely, a character written without considering this flow could possibly keep its balance, but it would probably lose its spirits.

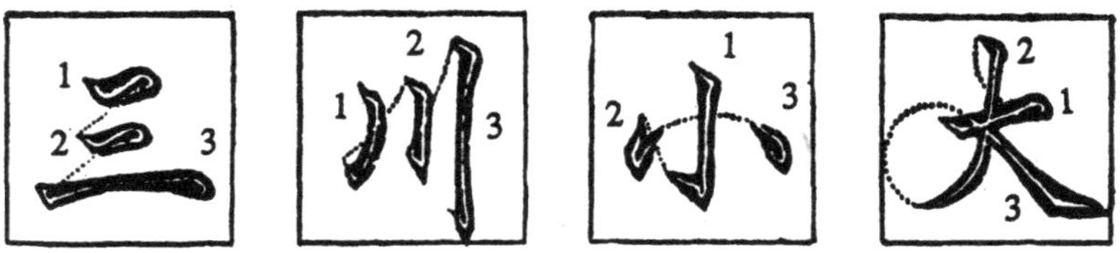

Figure 13. Examples showing underlying flow of strokes.

The underlying flow of strokes could be realized by controlling the stroking techniques which relate to the writing of the starting, turning and ending parts of the strokes. Since the starting and ending parts are the most involved in this flow, we mainly investigate here these two parts.

Typical patterns of the starting and ending parts are shown in Fig. 14, where θ and L are variables and thus various kinds of underlying flow of the strokes could be realized. Other patterns of the starting and ending parts could be obtained by rotating or mirroring these typical patterns. Here we propose to realize these typical patterns by introducing artificial segments in the starting and ending parts of the strokes as shown in Fig. 15. Since these artificial segments are short straight lines, it is easy to change their lengths and orientations. Several examples of the starting parts with different angles θ are shown in Fig. 16.

An example of a character generated by the proposed method is shown in Fig. 17 (left). The same character generated by SAM is also shown in Fig. 17 (right) for comparison. Both of these two characters keep their shape balance, but the characters are obviously different in the underlying flow of strokes 2 and 3.

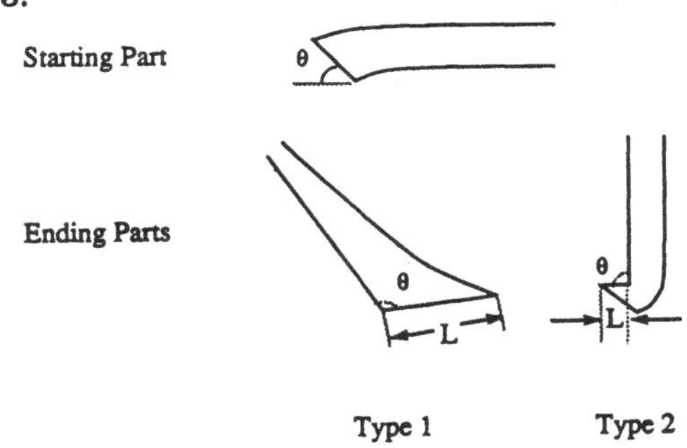

Figure 14. Typical patterns of starting and end parts.

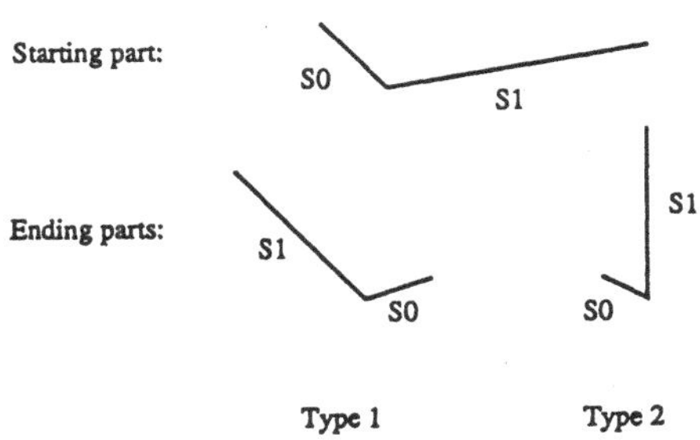

Figure 15. Realisation of typical starting and ending parts.

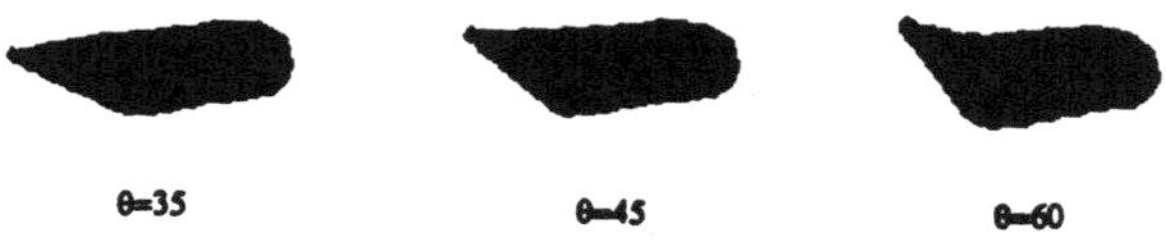

Figure 16. Examples of starting parts.

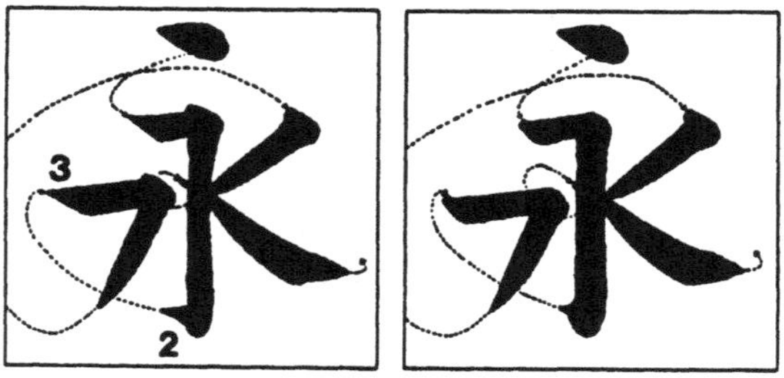

Left: Character Pattern by the Proposed Method
Right: Character Pattern by SAM

Figure 17. Comparison of underlying flow of strokes.

4.3 Configuration and Functions of the System

4.3.1 Configuration

A support system for acquiring calligraphic knowledge using the model described above was developed, and its configuration is shown in Fig. 18. It is composed of the following six modules: interface, correcting support, correcting model, DBM for recording parameters, DB and SAM.

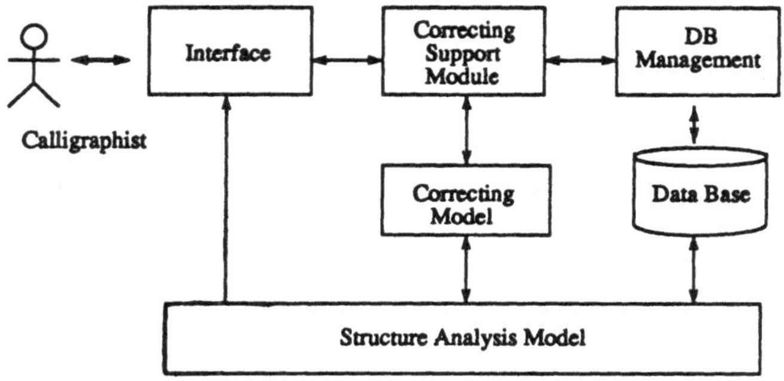

Figure 18. Configuration of the support system.

4.3.2 Correcting Functions

Correcting functions of the system include simple functions, which are movements of the total characters, sub-patterns and strokes, enlargement/reduction of the total characters and sub-patterns, stroke changes in the type, stroking order, length, thickness, curvature and orientation, and complex functions, each of which is combined from several simple functions. All of these functions are realized in the form of menu-selection with a mouse, and through interaction between the system and calligraphists.

A correcting example is shown here to explain the behaviour of the system in detail. The same character both before and after correction and the corresponding parameters are shown in Fig. 19. A calligraphist corrected the before-correcting character in the following way and gave comments with respect to each correction.

(a) Shorten (A to A') and move (B to B') stroke No. 1 (complex correction).

Comments: The starting point of stroke No. 1 should be written lower than that of stroke No. 2.

(b) Move (C to C') stroke No. 5.

Comments: Balance should be kept with the change in stroke No. 1.

(c) Move (D to D') the total character

Comments: The character should be written in the center of the frame.

(d) Change (E to E') curvature of the second segment of stroke No. 2.

Comments: Stroke No. 1 and the second segment of stroke No. 2. should be written inward or outward of each other.

(e) Move (F to F') stroke No. 4.

Comments: Touch with stroke No. 2 should be avoided.

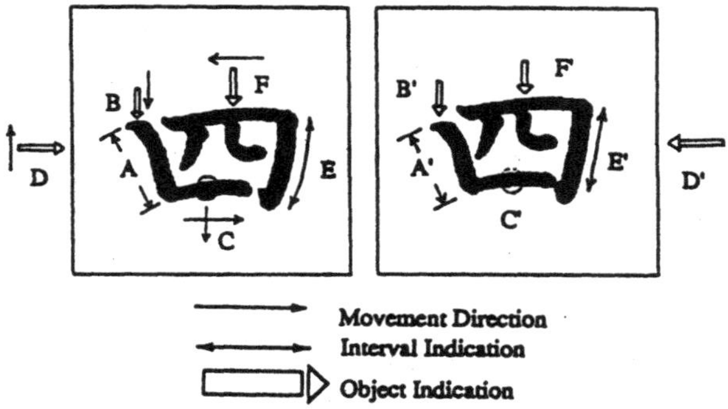

(a) Correcting Example with Explanations

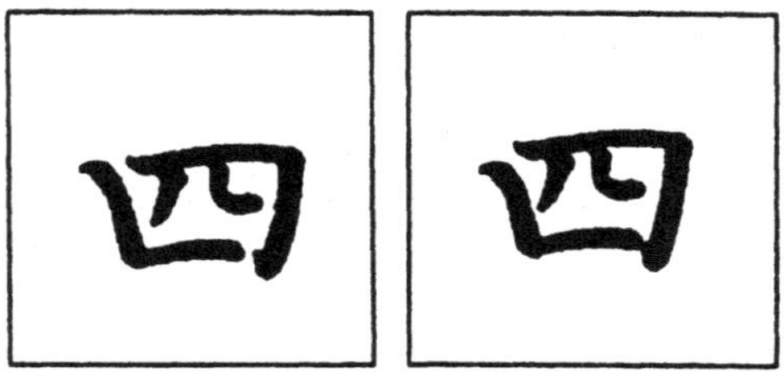

Left: ((8,84,137,-1.22,91,0.68,0.32)(32,112,141,0.07,134, 0.56,0.2,-1.75,95,0.2)(14,148,143,-1.95,47,0.58,0.2)(35, 183,133,-1.62,45,0.56,0.2,0.2,29,0)(5,112,220,0.1,87,0.51, 0.18))
Right: ((8,82,138,-1.22,79,0.68,0.23)(32,99,128,0.07,134, 0.56,0.2,-1.75,101,0.1)(14,135,130,-1.95,47,0.56,0.2)(35, 165,121,-1.62,45,0.56,0.2,0.2,29,0)(5,105,210,0.1,87,0.5, 0.18))

(b) Correcting Example with Parameters

Figure 19. An example corrected by the system.

4.4 Experiments and Evaluations of the System

The efficiency of the support system was experimentally evaluated by using the criteria of effectiveness of the model, functional richness and execution speed. As a result, it was shown that the system is feasible and efficient without overburdening the calligraphists.

This system is characterized by the correcting model of the calligraphists. Firstly, the correcting procedures of calligraphists were modelled by analyzing their corrective actions. As a result, the correcting system utilizing the model could follow the thoughts of the calligraphists and support their corrective actions logically according to their train of thought, as shown in Fig. 12. Therefore it could reasonably reduce their burden. Comparatively, if this model had not been utilized, calligraphists would have had to correct characters directly using parameters, which would have been an extremely difficult task for them. They also could only have corrected one stroke each time, without being able to treat a sub-pattern or a character as a whole unit. This implies the effectiveness of the proposed correcting model.

Furthermore, since the system has correcting functions for components at each level in the hierarchical structure of Chinese characters, it can therefore sufficiently support actual corrective actions of calligraphists from the viewpoint of correcting functions. Finally, the system behaved faster than a calligraphist could correct on paper. For example, it could display a character with ten strokes in an average time of 11 seconds. Thus the system is practical from the viewpoint of execution speed. Moreover, the system used the mouse and menu as interactive tools, which would be easy for calligraphists to use.

4.5 Summary

In this section, we first proposed a correcting model for calligraphists by abstracting their correcting procedures based upon observations of their corrective actions and interviewing with them. A support system for acquiring calligraphic knowledge was then developed, using this model. This system supports calligraphists in their thinking way and by experimentation it was shown that the system is efficient and executes the correcting instructions of calligraphists effectively without too much unfamiliarity.

We came to the following conclusions:

1. The correcting procedures of calligraphists could be partitioned into three independent phases of total characters, sub-patterns and strokes.
2. The correcting procedures of calligraphists could be modelled.
3. The corrective actions of calligraphists could be simulated using the support system.
4. The shape balance of the characters could be improved easily by using the system.
5. The corrective actions of calligraphists could be recorded in the form of parameters to be analyzed for calligraphic knowledge acquisition.

5. An Evaluation System

In this section, we show a prototype system for computer assisted calligraphic instruction (CACI), which uses calligraphic textbook knowledge to evaluate brush-used Chinese characters of primary school students, for example. This system is being developed to help promote the calligraphy education of brush-used handwritten Chinese characters.

5.1 Representation of Calligraphic Knowledge

Calligraphic knowledge in writing describes calligraphy only in terms of specific situations, in a fragmentary way, and the correlation among pieces of knowledge is very weak. The relations between elements at different levels of Chinese characters are described in the condition parts of calligraphic knowledge. In this research, production rules and the above-proposed data structure are used to represent calligraphic knowledge with such features, and the personality problem can be solved by using such calligraphic knowledge.

Here, the "S.S. rule" (SanShui rule) is used as an example to show how calligraphic knowledge can be represented. The S.S. rule is shown in Table 3 and applies to such characters that use three short strokes (which are called Sanshui) as left radicals. It states that the length of a third stroke in Sanshui is determined only by a descendent function of the number of strokes in the right radical. In this rule, calligraphic terms such as "radical", "length of a stroke" and "the number of strokes" are used, the first of which corresponds to the sub-pattern level and the second and third of which correspond to the stroke level. Thus the rule could be described by using the proposed data structure. To represent this rule, we investigated the relation between the number of strokes J and the stroke lengths U, by checking dozens of well-formed sample characters, whose numbers of strokes in right radicals were between two to sixteen. We obtained the results as shown in Table 4. Using these results, the S.S. rule can be represented in a production rule fashion as follows:

S.S. rule:

IF the number of strokes in right radical is between (J_1,J_2),

THEN restrict the length of the third stroke in Sanshui to within (U_1,U_2).

Production rules acquired in such a way could perhaps not represent the original rules completely. They may have been just prototypes, where terms were described with the aid of DS. It is thus necessary to verify and refine such prototype rules with respect to a large number of sample characters, because there exist some other conditions in the original rules, which do not appear explicitly.

(J1,J2)	*(U1,U2)*
(2, 3)	(133,136)
(4, 6)	(115,118)
(7,11)	(108,110)
(12,-)	(98, 100)

Table 4. Statistical results of the S.S. rule.

5.2 Experiments with CACI

Characters generated by SAM may be of poor shape because of the inadequate parameters. Calligraphic knowledge could be utilized to evaluate the balance of such characters through the evaluation of their parameters. While keeping their original personalities, minimal modifications could be carried out for those characters having lost their balance. That is, by taking advantage of the unique relation between the characters and parameters, calligraphic knowledge is used to evaluate and modify the shapes of the characters, through processing their corresponding parameters and therefore keeping users from the laborious processing of parameters.

An evaluation system of Chinese characters CACI using calligraphic knowledge was developed in Prolog, and some evaluation experiments of the system were carried out. The configuration of the system is shown in Fig. 20.

The inference engine searches for calligraphic rules using the parameters of an input character, and builds a list of the applicable rules. These rules are presented to the user for the identification of their application scopes and methods. When the rules are executed, characters will be generated from the evaluated parameters. An evaluation result is shown in Fig. 21. The parameters of an input character are represented as shown in Fig. 21(a), which reflects the hierarchical structure of the input character.

The input characters are first normalized into 400 x 400 sizes before their parameters are evaluated. In the above example, the output parameters evaluated by the S.S. rule are shown in Fig. 21(b).

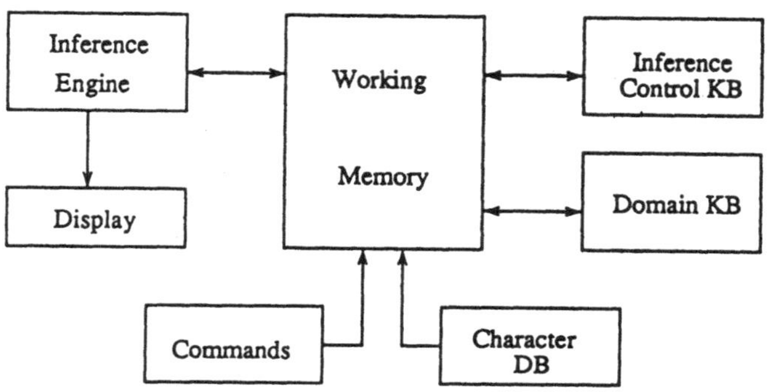

Figure 20. Configuration of the evaluation system.

```
(((100,48,66,-0.89,1,0),(70,155,49,-0.3,1,0),(95,335,145,1.16,1,0))
 ((((270,38,138,-2.09,1,0.14),(180,175,140,0.18,0.68,0)),((300,115,75,-0.87,0.92,0.3)))
  (((210,230,90,-1.37,0.85,0)),((220,240,118,0.15,0.89,0.23,80,-1.87,0.1))
   ((220,315,90,0.1,0.85,0.18)))))
```

(a) Input Character Pattern「治」and its Data.

```
(((100,48,66,-0.89,1,0),(70,155,49,-0.3,1,0),(95,335,113,1.16,1,0))
 ((((270,38,138,-2.09,1,0.14),(180,175,140,0.18,0.68,0)),((300,115,75,-0.87,0.92,0.3)))
  (((210,230,90,-1.37,0.85,0)),((220,240,118,0.15,0.89,0.23,80,-1.87,0.1))
   ((220,315,90,0.1,0.85,0.18)))))
```

(b) Evaluated Data and its Character Pattern.

Figure 21. An evaluation example.

5.3 Summary

A prototype of the computer-assisted calligraphic instruction system CACI was described in this section and the representation of calligraphic knowledge was mainly discussed. There may be some problems in the process of representing calligraphic rules and these should be dealt with further in the future.

6. Conclusions

This chapter was divided into four parts which were described in Secs. 2, 3, 4 and 5 respectively. A generation-model SAM of brush-used handwritten Chinese characters was first proposed to show that the beauty of characters could be truly expressed. A data structure DS was then proposed based upon the model. This structure organizes parameters in SAM to enable them to reflect the hierarchical structures of Chinese characters, and therefore it could be used to appropriately represent calligraphic knowledge. A support system for acquiring calligraphic knowledge SSACK was further developed by adopting a strategy that attempted to extract knowledge from calligraphists by monitoring, recording and analyzing their writing processes. Finally, a prototype system for computer-assisted calligraphic instruction CACI was shown, which could be utilized to help promote the calligraphy education of the brush-writing of Chinese characters. This system is now being developed.

We have come to the following conclusions through this research:

1. It has been shown that calligraphy can be simulated on a computer and calligraphic knowledge can be represented and utilized to generate and evaluate well-formed brush-used Chinese characters.
2. Based upon the analysis of the hierarchical structure of Chinese characters, a data structure was proposed to effectively represent calligraphic knowledge by describing its special terms at various levels.
3. A support system for acquiring calligraphic knowledge was developed which took an inductive strategy that first records the corrective actions of calligraphists in data form, and then acquires new calligraphic knowledge by analyzing the accumulated data.

Further discussion in this research would be how to acquire new calligraphic knowledge by analyzing the accumulated data (Politakis et al., 1984 and Buntine et al., 1988). Here we show a general framework which needs to be discussed in detail in the future.

(a) Classification of data

Accumulated data will be grouped into classes, and within each class the data will have similar properties, in some senses. For example, the data of characters taking Sanshui as their left radicals can be clustered into the same class. Data of one character could possibly be assigned to multiple different classes, corresponding to the different calligraphic knowledge applicable for a single character.

(b) Feature extraction of each data class

In each data class, common properties of data for each character are extracted and described as features of the data class. Hypotheses for new calligraphic knowledge are built according to these features. For example, relations between lengths of the third strokes and the numbers of strokes in the right radicals are extracted as class features of characters taking Sanshui as left radicals. Multiple features could be extracted from a single data class, corresponding to calligraphic knowledge with multiple action items.

(c) Verification of hypotheses

Hypotheses for new calligraphic knowledge are used to evaluate newly user-initiated characters. These hypotheses can be refined or verified by comparing results of the same characters evaluated by the hypotheses and by calligraphists.

*All these steps can only be smoothly realized with the cooperation of calligraphists. Calligraphists' comments on corrections are especially important for the realization of step (a). Data for analysis are to be accumulated in several primary schools where teachers can correct characters written by students with the use of SSACK, and the data are to be analyzed using the above method, in the near future.

7. References

Buntine, W. and Stirling, D., Interactive Induction, *Proc. 4th Conference on AI*, 1988, pp. 320-326.

Deng, S., *Detailed Explanation of Thirty-Six Rules on Ou-Styled Structures of Chinese Character Patterns*, People's Art Publishing, 1963.

Fijiwara, S., *A Guide to Calligraphy: Graphic Dictionary of Square-Styled Brush-Written Chinese Characters*, Seishin Book House, 1983.

Plamondon, R. and Maarse, F.J., An Evaluation of Motor Model of Handwriting, *IEEE Trans. on Syst. Man and Cybernetics*, Vol. 19, No. 5, 1989, pp. 1060-1072.

Politakis, P. and Weiss, S.M., Using Empirical Analysis to Refine Expert System Knowledge Bases, *Artificial Intelligence*, **22**, 1984, pp. 23-48.

Ren, Z. and Qian, P., *Essential Calligraphy for Square-Styled Brush-Written Chinese Characters*, Shanghai Publishing of Painting and Calligraphy, 1987.

Uchio, F., Zhang, X., Sanada, H. and Tezuka, Y., A Semi-Automatic Decision System of Stroke Parameters Used in Hierarchical Structure Analysis Method, *Trans. IEICE of Japan*, Vol. J69-D, No. 6, 1986, pp. 893-903.

Zeng, J., Inoue, T., Sanada, H. and Tezuka, Y., On the Analysis and Representation of Calligraphic Knowledge for the Design Expert System of Brush-Written Chinese Character Patterns, *Trans. IPS of Japan*, Vol. 29, No. 2, 1988, pp. 160-168.

Zeng, J., Inoue, T., Sanada, H. and Tezuka, Y., A Correcting System of Brush-Written Chinese Character Patterns Using a Model of Correcting Procedures of Calligraphists, *Trans. IEICE of Japan*, Vol. J73-D-II, No. 2, 1990, pp. 183-190.

Zeng, J., Inoue, T., Sanada, H. and Tezuka, Y., A Data Structure Suitable for Representing the Calligraphic Rules for Chinese Character Evaluation, *Proc. 9th ICPR*, 1988, pp. 181-183.

Zhang, X., Sanada, H. and Tezuka, Y., Computer Formation of Brush-Written Chinese Characters, *Technology Reports of the Osaka University*, Vol. 36, No. 1832, 1985, pp. 109-116.

AUTHOR INDEX

Abend 33
Achemlal 27, 34, 35, 40, 41, 44, 46
Ahyan 133
Amin 169
Ammar 50, 53, 54, 55, 56, 58, 62
Anderson 338
Anthony 37
Aoki 168
Apiki 90
Asada 138

Babaud 138
Baron 24
Bean 10
Beatson 3, 10, 16, 32, 35
Bechet 23, 35, 43, 46
Benson 11
Berlkamp 252
Berthod 133, 164
Bevington 95
Bills 40
Binder 9
Blesser 9, 90, 113, 169, 207, 208
Bonnefoy 12, 22, 27, 28, 34, 35, 40, 41, 46
Botnick 11
Bozinovic 168, 170, 188, 285, 298
Brady 138
Brault 10, 24, 27, 30, 31, 32, 34, 35, 40, 43, 44, 47
Brenner 29
Bridle 157
Brocklehurst 17, 18, 54, 56, 58, 59, 119, 121, 169
Brooks 30, 236, 238, 240, 245, 248, 257, 258, 267
Burr 250
Burrus 93
Bush 222
Buxton 222

Chainer 28, 31, 39, 43
Chang 338
Chen 192
Chien 27, 32, 36, 45
Chorley 16, 17, 23, 35, 40, 44, 46, 47
Chuang 55, 56, 58
Clark 222
Coleman 222
Collantier 36, 46
Conway 16
Crane 18, 24, 29, 34, 36, 40, 41, 44, 46
Crystal 277

Davis 11, 47
de Bruyne 23, 29, 32, 34, 36, 44, 51
Delaplain 236
Denier van der Gon 5
Diehl 90
Dooijes 7, 241
Doster 169, 223
Downton 23, 30, 169, 236, 238, 250
Duckworth 169

Duda 292, 355
Duvernoy 9
Dyche 36

Edelman 7
Eden 7, 9, 188, 208
Edwards 24
Eernisse 24
Ehrich 133, 274, 285, 295
Elschlager 60
Evangelisti 33
Evett 12, 287

Farag 27, 32, 36, 45, 274
Faure 339, 354
Feig 28, 31
Fejfar 4, 11, 24, 33, 36, 47
Fischler 60
Flash 7
Foodman 11
Ford 112, 170
Forney 295
Forsen 10
Francis 250, 277
Freyd 59
Frishkopf 188
Fukumura 56

Gaudaire 35
Goshtaby 274
Grossman 43
Gunderson 31, 39

Haberman 4, 11, 24, 36
Hagan 49
Hale 37, 40
Hanan 42
Hanna 261
Hanson 292, 295
Hardcastle 51, 59
Harmon 11, 188
Harrison 9, 16, 49, 59
Hart 292, 355
Hayes 294
Hearn 90
Herbst 15, 21, 24, 27, 29, 31, 32, 33, 37, 42, 43
Herkt 15
Higgins 169, 170, 274, 294, 307
Hilton 49, 59
Hin 38
Ho-Kashyap 355
Hoffman 169
Hollerbach 7
Holmes 236
Hornby 278
Howbrook 34
Hull 292, 295

Ibrahim 23, 32, 37, 44
Ikeda 250

Jakobson 208
Jarvis 222

Kadirkamanathan 134
Kamins 37, 40
Kankaanpaa 112
Kao 5
Keenan 287
Kim 23, 89
Klement 5, 54
Knuth 278, 298, 300
Koehler 133, 285
Kogure 27, 32, 33, 39, 42, 44
Komada 23
Kondo 169
Kucera 250, 277

Kuklinski 241

Lam 37, 40
Lamarche 7, 10, 28, 31
Land 29
Lankford 90
Lanshammar 95, 97
Lashley 5
Leclerc 10
Leedham 23, 24, 30, 169, 238, 244, 250, 255
Legge 9
Leng 38
Levrat 23, 32, 37, 44
Lew 24, 25, 34, 43
Licklider 222
Lin 54, 55, 56, 58, 59
Little 4
Liu 12, 15, 21, 24, 27, 29, 31, 32, 33, 34, 37, 42, 43, 44
Locard 9, 16, 49, 59
Lorette 12, 17, 21, 22, 27, 32, 33, 34, 35, 37, 40, 41, 42, 44, 51, 59, 188, 192
Lowerre 250, 296

Maamari 23, 24, 27, 33, 34, 89
Maarse 7, 23, 30, 33, 89, 90, 92, 95, 107, 171, 190, 208
MacDonald 5
Mackworth 138
Maier 4, 10, 178
Mally 55
Mandler 169
Marcozzo 51
Margolin 9
Marlowe 9
Marr 315
Martin 338
Martlew 5
Maureci 30, 38
Mendel 9
Mermelstein 188
Messmer 32
Miller 4
Mital 38, 43
Mjølnes 38, 46
Mokhtarian 138
Morasso 5, 7, 10, 133, 137, 169
Mori 188, 207
Morris 10
Mounin 208
Mourier 35
Mussa Ivaldi 10
Myers 47

Nagayama 23
Nagel 18, 51, 53, 54, 56, 58
Nagura 223
National Bureau of Standards 10, 17
Nemcek 54, 55, 56, 58, 59
Neuhoff 292, 295
Nevill 23
Newell 30, 238, 245, 248
Ney 157, 250
Nouboud 5, 55, 56, 58

Oed 169, 223
Oja 210
Oka 10, 24, 27, 33, 39, 42, 45
Olding 35
Osborn 49
Ostrem 12, 18, 24, 29, 34, 36, 40, 41, 44, 46
Oulhadj 169

Paganini 37, 40

Parizeau 11, 34, 38, 43, 46
Parks 4, 35, 93
Paterson 24
Peleg 294
Phelps 54, 55, 56
Phillips 91, 92
Pick 171
Plamondon 5, 7, 10, 11, 12, 16, 21, 22, 23, 24, 25, 27, 28, 30, 31, 32, 33, 34, 35, 38, 40, 43, 44, 46, 47, 50, 53, 54, 55, 56, 58, 59, 60, 61, 62, 89, 169, 188, 190, 208
Pobgee 35, 111
Poggio 138
Prell 10
Press 95
Price 11, 47
Pride 173

Qiao 238, 255

Raphaël 11
Raviv 274
Rayner 134
Reddy 250, 296
Reitter 15
Rengger 127
Requier 40, 44
Riganati 10
Riseman 292, 295
Robillard 30
Rosenfeld 18, 51, 54, 56, 58
Rosenthal 315

Sabourin 50, 53, 54, 55, 56, 58, 60, 61, 62
Sakai 49
Salvage 9, 36
Sampson 278
Sanada 363
Sato 27, 32, 33, 39, 42, 44
Saudek 9
Saussure 208
Sayre 274
Schappe 6
Schlosser 23
Schomaker 7, 30, 169
Schwartz 251
Shaw 10, 11, 47
Shillman 208, 211
Shinghal 207, 292, 297
Simpson 9
Skrzypek 169
Smith 6
Snelling 29
Srihari 168, 170, 188, 189, 285, 292, 295, 296, 298
Sternberg 24, 30, 39
Stevens 121
Stockolm 9
Strangio 32, 33
Stuckert 23, 33
Suen 5, 188, 207, 208
Suenaga 223
Sussenguth 280
Sutherland 222
Sykes 3
Søberg 38, 46

Tagliasco 5
Takeshi INOUE 363
Tappert 5, 23, 89, 112, 169, 188, 207
Telmosse 24, 25, 34
Teulings 7, 23, 30, 89, 90, 92, 95, 107, 169, 170, 171, 188, 189
Thomas 123
Thomassen 5, 7, 169
Thompson 10

Thompson-Titn 33
Thuring 5, 33, 188
Totty 12, 51, 59
Toussaint 292, 297
Trubetskoy 208, 211

Ullman 338

van der Gon 33, 188
Van Galen 7
Varady 27, 34, 44
Viterbi 295

Wallner 9
Wang 339, 354
Wann 134
Ward 90, 91, 92, 169
Warfel 4, 10
Watson 32, 35
Weinstein 15
Welbourn 170, 226
Wellisch 240
Wells 170, 224
Werner 34, 39
Whitrow 170, 188, 274, 294, 307
Wilson 277
Williford 31, 39
Wing 7
Wolf 9, 29
Wood 4, 10
Woodard 4, 10
Worthington 12, 27, 28, 31, 33, 39, 43, 46
Wright 169, 224, 284

Yamamoto 188
Yamaya 168
Yasahura 208
Yasuda 188
Yasuhara 5, 7, 10, 24, 27, 33, 39, 42, 45
Yergeau 5
Yoshida 56
Young 11
Yuille 138

Zeng 363
Zhang 363
Zimmerman 27, 34, 39, 44
Zweighenhaft 9

SUBJECT INDEX

2-dimensional coordinate representation of the character vector 211
26-way tree 283

Acceleration 96, 107
acceleration error 97
accuracy 90
Acquisition 244
Age 16
allographotopic maps 179
ambiguous characters 213
ascendant hierarchical classification 44
Asynchronous Sampling 94
Asynchronous-Sampling Errors 107
attributed relational tree 340
AUTHENTIC 15

ballistic movement 5
ballistic movements 21
baud-rate 99
binary n-grams 292
biological handwriting models 208
bottom-up approach 7
boundary effect 103
Boundary errors 91
brush-touch 364
brush-used handwritten Chinese characters 363

calligraphic knowledge 363
Calligraphy 363
calligraphy education 365
centroidal linkage 54
chain-coded representation 44
Classification 252
collinear arrangements 344
collinearity 340
Comparison and Decision Processes 41, 50
comparison to a reference signature 22
compounds 287
computer 99
Computer Aided Transcription 236
Computer fraud 4
computer-assisted instruction system 363
Connected word algorithm 157
connection strokes 176
context-dependent degradations 195, 196
contextual constraints 354
contextual post-processing 298
correlated 93, 105
correlated error 96
correlated errors 92, 95
cursive handwriting recognition 187, 188
cursive script 292
cursive script recognition 120

cursive writing 108
cursive-script 168

data acquisition 22, 50
Data Collection 244
data flow diagram 12
data structure 363
data-driven 344
Datapad 245
dead bands 91
decision threshold 44, 58
DEGENERATE 15
degeneration model 195
Delayed samples 101
dictionary lookup 305
dictionary search 158
differentiator 93
digital filter 93
Digitizer Characteristics. 99
digitizer performance 108
Digitizer Requirements 90
Digitizers 22, 90
digitizing tablet 341
displacement transducer 246
distance measure 44
distance metric 156
distinctive features 208
DYNAMIC 22
dynamic characteristics 21
dynamic cluster method 44
dynamic information 245
dynamic parameters 34
dynamic programming 58, 159
dynamic programming matching 42
dynamic programming template match 263
Dynamic signature verification 21

electronic ink 24
Electronic paper 111
Electronic set-ups 23
error analysis 95
Error Propagation 95
errors 93

FALSE 15
feature extraction 22, 32, 50, 54
firmware 90
Fisher linear classifier 44
forgeries 45
formal language 340
Fourier transform 40
frames 345
functional attributes 208
functions 22, 32
fuzzy classifier 256

gaussian errors 96
Gaussian function 140
gaussian random distribution 107
gaussian random processes 95
generation-model 363
genuine handwritten signatures 59
geometric parameters 34
Gesture Based Editors 222
global parameters 40
global structures 339
gradient 60
graph and subgraph isomorphism 44
Gregg shorthand 236
GTCO Digi-Pad 5A 99

Haar transform 40
Habits 16
Hadamaar transform 55
hand-drawn 112

hand-written 112
handwriting 90, 93, 95, 313, 314
handwriting generation 5
handwriting recognition 291
HANDWRITING-RECOGNITION 167
handwritten character recognition 120
handwritten input 273
handwritten mathematical expressions 338
higher level knowledge 273
histogram 50
histograms 44
human computer interfaces 112
Hysteresis errors 91

IBM PC 99
identification system 15
image-understanding system 60
IMITATED 15
Informatique et Sécurité 11
input/output device 112
instrumented pen 235
Interactive Flat-Panel Displays 22, 23
interclass or interpersonal variability 17
interface 287
intraclass or intrapersonal variability 17
irregularities 314
isotropic smoothing 139

kernel estimation 155
knowledge-driven 345
Kohonen networks 179

Laplacian or the Sobel operator 53
large vocabularies 288
large vocabulary 277, 278
Lee metric 252
letter graph 292
letters 156
lexical access device 288
lexicon 277
ligatures 156
Line thickness 245
linguistic information 287
local parameters 40
local structures 339
low-pass digital filtering 97

machine compatibility 240
machine recognition of handwritten script 273
machine recognition of handwriting 314
Machine-Readable Dictionary 283
machinography 235, 240
Markov process 274
mathematical expressions 337
maximum likelihood ratio 44
mechanisms of reading 314
memory efficiency 278
method of least squares 92, 100
methods of representing and searching large lists of words 273
Micropad 245
Missing coordinates 91
Modern Data 11
Monotonicity failure, 91
Morphological 287
motor program 9
multi-way tree 278

n-grams 274

National Bureau of Standards 10, 17
natural method of interaction 240
non-simultaneous sampling 94
non-uniform sampling period 105
nonlinearity 91
normalization 171
normalized principal component analysis 44
Normalizing handwriting patterns 170
numerical methods 95

off-line 51
off-line signature verification 49
Office Document Architecture 230
on-line 21, 22, 168
on-line cursive script recognition 134
on-line signature acquisition and verification 21
on-line system 189
orthographic information 273, 288
orthography 273, 274

Palantype 236
parameters 22, 34
path probability 250
pattern recognition 274
pattern recognizer 284
pattern recognizer for handwriting 288
pen acceleration 93
pen height 98
pen motion 97
pen pressure 248
pen velocity 93
Pen-tilt error, 91
Pencept Penpad 300 99
Performance 253
Personal identification means 10
performance evaluation 22, 50
personal identification number (PIN) 3
phonemes 208
phonetic chord keyboards 236
Physical conditions 17
Pitman's handwritten shorthand 235
Position 32
position accuracy 103
position error 103
Practical conditions 17
PREDICTION-VERIFICATION STRATEGY 187, 190, 198
preferred 278
Prejudice 173
preprocessing 22, 50, 247
preprocessing of shorthand 235
Pressure 54
primitive extraction process 60
Principal Component Analysis 210
print 313, 314
production rules 260
professional script 243
prototypical description 346
pseudo-dynamic 60, 62
Psychological or mental state 17

quantization noise 92

random error 96
random errors 94, 105
random forgeries 58
Readability 241
reading 314
reading performance 314
reading skills 314

real time 273
Recognition 244, 248
recognition system 284
recognition trials 46
reduced memory tree 283
region-growing-with-merging 54
regional correlation 43
regularity 314
relative positions 354
Residual Error 101, 105
resolution 90

sampling period 91, 93, 108
sampling rates 90
scale-space filtering 137
scale-space plot 138
scanners 51
script recognition 273, 287, 314
script variability 314
scriptor-dependent degradations 196
security devices 3
segmentation based recognition 293
segmentation into strokes 167
segmentation of cursive script 170
segmenting allographs 178
semantics 273, 287
shared memory 341
shortforms 239
signature consistency 46
signature verification processes 12
similarity measures 250
simple forgery 15
simple pattern primitives 240
Simultaneous Transcription 236
sinusoidal displacement 101
skilled forgeries 59
skilled or freehand forgery 16
smoothing functions 137
Spatial Error 100
Spatial Errors 92
specialist modules 341
Specific Input Devices 22
spectral analysis 40
speech analysis 208
speed of search 278
speedwriting systems 236
static 49, 51
statistical labelling 354
Stenotype 236
stream mode 90
stroke 133
Stroke feature extraction 170
Stroke segmentation 133
stroke structure of handprinted characters 208
Stroke thickness 239
strokes 364
Structural knowledge 250
structural recognition grammar 240
structure of reading tasks 315
structures 380
student script 243
substitution or random forgery 16
substitution set 291
support system for acquiring calligraphic knowledge 363
Symbol Recognition 226, 230
symmetric Infinite Impulse Response function 140
syntactic recognition process 241
syntax 273, 287
systematic errors 108

Tablets 23
task structure 315
template matching 235

Temporal Errors 93
test bed 93, 97, 99, 104, 107, 108
text editor 121
text insensitive 54
text sensitive 54
textured regions 54
threshold 54
Thresholding and slicing
techniques 53
Time 16
time derivatives 95
time irregularities 314
top-down approach 7
training and test subsets and users
sets 46
training step 189, 192
Transcription 257
tree 278
tree structure 287, 288
trie structure 296, 298
two-class partitioning problem 18
type I (FRR) 40, 41, 45
type I and type II errors 59
type II (FAR
False Acceptance Rate) 18
type II (FAR) 40, 41, 45
typset and impression stability of
print 314

unvoiced sounds239

Variability 241
vectors 44
velocity error 96, 97, 104, 105
velocity errors 96
verbatim reporters 244
verbatim reporting 235
verbatim written transcript 235
vidicon tube T.V. camera 50
visual perception skills 338
Viterbi algorithm 176
vocabulary 276, 277
vocalized outlines 239
voiced sounds 239

Wacom SD-421 99
Walsh power spectra 44
Walsh transform 34
weighted distances 58
well formed characters 364
word look-up 276
word recognition 314
word structure 273
Worksheet 112
writing pressure 235, 245

zero phase 142

ABOUT THE EDITORS

RÉJEAN PLAMONDON received a B.Sc. degree in Physics, and M.Sc.A. and Ph.D. degrees in Electrical Engineering from Université Laval, Québec, P.Q., Canada in 1973, 1975, and 1978 respectively.

In 1978, he joined the staff of the Ecole Polytechnique, Université de Montréal, Montréal, P.Q., Canada, where he is currently a Full Professor. From 1985 to 1986, he has been involved in several research projects while being a guest of the Computer Science Department, Concordia University, Montreal, Canada, the Motor Behavior Laboratory, University of Madison, USA, the Department of Experimental Psychology, University of Nijmegen, The Netherlands and the Laboratoire de Génie Electrique de Creteil, Université de Paris Val-de-Marne, France.

His research interests deal with the computer applications of handwriting: biomechanical models, neural and motor aspects, character recognition, signature verification, signal analysis and processing, computer-aided design via handwriting, forensic sciences, software engineering and artificial intelligence. He is the founder and director of Laboratoire Scribens at Ecole Polytechnique de Montréal, a research group dedicated exclusively to the study of these topics.

An active member of several professional societies, a senior member of IEEE ('85), Dr Plamondon is a member of the board of the International Graphonomics Society and Chairman of the Technical Committee TC-11, from IAPR (International Association for Pattern Recognition), on text processing applications. In 1989–90, he was a Fellow of The Netherlands Institute for Advanced Study, in Wassenaar. He is the author or co-author of numerous publications and technical reports.

GRAHAM LEEDHAM obtained an HND from North Staffordshire Polytechnic in 1977, a B.Sc. degree in Electrical and Electronic Engineering from Leeds University in 1979 and the degrees of M.Sc. in 1981, and Ph.D. in 1984, both from the Department of Electronics at Southampton University.

He was a student apprentice with the GEC Measurements, Stafford, Staffordshire, UK from 1975 – 1979 and a development engineer with GEC Meters, Stone, Staffordshire, UK from 1979 – 1980.

Since 1984 he has been a lecturer at the Department of Electronic Systems Engineering, University of Essex where he lectures on computer systems, human-computer interface, image processing, speech processing and signal processing. His research interests are in the human-computer interface, the automatic recognition of handwriting and computer supported co-operative working. He is a chartered engineer and a corporate member of the Institution of Electrical Engineers. In 1989 he held a short-term fellowship at British Telecom Research Laboratories.

www.ingramcontent.com/pod-product-compliance
Lightning Source LLC
LaVergne TN
LVHW071223011026
849280LV00007B/37
* 9 7 8 9 8 1 0 2 0 4 0 8 2 *